GW01607343

mdi

Assembly Line Design

MANUFACTURING ENGINEERING AND MATERIALS PROCESSING

A Series of Reference Books and Textbooks

SERIES EDITORS

Geoffrey Boothroyd

Chairman, Department of Industrial and Manufacturing Engineering
University of Rhode Island
Kingston, Rhode Island

George E. Dieter

Dean, College of Engineering
University of Maryland
College Park, Maryland

1. Computers in Manufacturing, *U. Rembold, M. Seth and J. S. Weinstein*
2. Cold Rolling of Steel, *William L. Roberts*
3. Strengthening of Ceramics: Treatments, Tests and Design Applications, *Henry P. Kirchner*
4. Metal Forming: The Application of Limit Analysis, *Betzalel Avitzur*
5. Improving Productivity by Classification, Coding, and Data Base Standardization: The Key to Maximizing CAD/CAM and Group Technology, *William F. Hyde*
6. Automatic Assembly, *Geoffrey Boothroyd, Corrado Poli, and Laurence E. Murch*
7. Manufacturing Engineering Processes, *Leo Alting*
8. Modern Ceramic Engineering: Properties, Processing, and Use in Design, *David W. Richerson*
9. Interface Technology for Computer-Controlled Manufacturing Processes, *Ulrich Rembold, Karl Armbruster, and Wolfgang Ülzmann*
10. Hot Rolling of Steel, *William L. Roberts*
11. Adhesives in Manufacturing, *edited by Gerald L. Schneberger*
12. Understanding the Manufacturing Process: Key to Successful CAD/CAM Implementation, *Joseph Harrington, Jr.*
13. Industrial Materials Science and Engineering, *edited by Lawrence E. Murr*
14. Lubricants and Lubrication in Metalworking Operations, *Elliot S. Nachtman and Serope Kalpakjian*
15. Manufacturing Engineering: An Introduction to the Basic Functions, *John P. Tanner*
16. Computer-Integrated Manufacturing Technology and Systems, *Ulrich Rembold, Christian Blume, and Ruediger Dillmann*
17. Connections in Electronic Assemblies, *Anthony J. Bilotta*

18. Automation for Press Feed Operations: Applications and Economics, *Edward Walker*
19. Nontraditional Manufacturing Processes, *Gary F. Benedict*
20. Programmable Controllers for Factory Automation, *David G. Johnson*
21. Printed Circuit Assembly Manufacturing, *Fred W. Kear*
22. Manufacturing High Technology Handbook, *edited by Donatas Tijunelis and Keith E. McKee*
23. Factory Information Systems: Design and Implementation for CIM Management and Control, *John Gaylord*
24. Flat Processing of Steel, *William L. Roberts*
25. Soldering for Electronic Assemblies, *Leo P. Lambert*
26. Flexible Manufacturing Systems in Practice: Applications, Design, and Simulation, *Joseph Talavage and Roger G. Hannam*
27. Flexible Manufacturing Systems: Benefits for the Low Inventory Factory, *John E. Lenz*
28. Fundamentals of Machining and Machine Tools, Second Edition, *Geoffrey Boothroyd and Winston A. Knight*
29. Computer-Automated Process Planning for World-Class Manufacturing, *James Nolen*
30. Steel-Rolling Technology: Theory and Practice, *Vladimir B. Ginzburg*
31. Computer Integrated Electronics Manufacturing and Testing, *Jack Arabian*
32. In-Process Measurement and Control, *Stephan D. Murphy*
33. Assembly Line Design: Methodology and Applications, *We-Min Chow*
34. Robot Technology and Applications, *edited by Ulrich Rembold*
35. Mechanical Deburring and Surface Finishing Technology, *Alfred F. Scheider*

OTHER VOLUMES IN PREPARATION

Assembly Line Design

Methodology and Applications

We-Min Chow
IBM Corporation
San Jose, California

MARCEL DEKKER, INC. New York and Basel

Library of Congress Cataloging-in-Publication

Chow, We-Min,
Assembly line design : methodology and applications / We-Min Chow.
p. cm.--(Manufacturing engineering and materials processing ; 33)
Includes bibliographical references.
ISBN 0-8247-8322-0
1. Assembly-line methods. 2. Assembly line methods--Automation.
I. Title. II. Series.
TS178.4.C56 1990
670.42'7–dc20 90-2917
CIP

This book is printed on acid-free paper.

MARCEL DEKKER, INC.
270 Madison Avenue, New York, New York 10016

Current printing (last digit):
10 9 8 7 6 5 4 3 2 1

PRINTED IN THE UNITED STATES OF AMERICA

To I-Ring Yao and Chien Tseng

Preface

Design of a manufacturing assembly line is an old problem and yet it is also a new problem. In the past several decades, an enormous amount of knowledge in design methodology and related subjects has been developed. Innovations in product and in manufacturing technology often result in new design problems. On the one hand, technological advances lead to more sophisticated manufacturing systems. On the other hand, market competition causes a short product lifetime, which forces line change or reduces the service years of a manufacturing line. Line design, therefore, plays an increasingly important role in manufacturing. Given such a complex situation, we are unlikely to reach an economical solution for line design by invoking intuition, rule-of-thumb, or trial-and-error approaches. Design optimization requires a good knowledge of methodology and a familiarity with the manufacturing environment. Design work should be guided by a systematic approach, with a careful evaluation of the design concept from both cost and line-performance viewpoints. To a practicing engineer, problems of line design may come from several different sources:

1. Evaluation techniques and design methods are often derived from advanced mathematics, and are not always easily accessible.
2. Much of the existing literature is limited to specific assumptions that may not be compatible with the designer's environment. In particular, books and articles dealing with isolated manufacturing components cannot be used directly for line design. The designer must often integrate isolated pieces of information.

3. For a new product or process, information for the design task is rarely available. A recent emphasis on manufacturability in industry demands early involvement of manufacturing in design functions. This means overlapping of product development, process definition, and line design. Consequently, the problem of information availability becomes even more acute.
4. Due to uncertainty in market demand or engineering changes, the line structure may have to be altered after installation, by adding or deleting workstations. This can be a very upsetting problem, particularly if it causes line disruption.

This book attempts to treat line design and its related subjects in a cohesive manner, with an emphasis on design applications. Mathematics in the book is at about the level of first-year college calculus. Theoretical results, when needed, are derived by heuristic approaches, rather than by rigorous mathematical manipulations. Design procedures and evaluation techniques are usually followed by examples from real-world cases. General guidelines for setting up assumptions and determining line performance parameters are discussed, based on empirical data from either literature reports or line observations. Solutions to the problem of uncertainty are also discussed.

Chapter 1, "Fundamental Concepts," gives a brief overview of assembly systems and characterizes the nature of the design problem. Individual components and their relationships are reviewed. The design and analysis work can be greatly simplified by looking at one component at a time. On the other hand, the procedure also keeps individual components subject to an integrated design concept. Cost and performance measures, which reappear in the later chapters, are introduced.

Chapter 2, "Empirical Data and Statistical Analysis," begins with a brief review of elementary probability theory and statistics, a prerequisite for data analysis and performance evaluation. Characteristics of line parameters are discussed, based on empirical data. Subjects include arrival process, operator behavior, machine reliability, product yield, and improvement (or learning) processes.

Chapter 3, "Queuing Analysis and Optimization Techniques," provides a mathematical background for design analysis and optimization. In addition to an introduction to the theory, practical queuing models, computational algorithms, and their applications are discussed.

Chapter 4, "Process and Operation Design," discusses functional design for the assembly process. The first part of the chapter reviews techniques of line balancing and the impacts of the length of assembly cycle time. To simplify material-flow and material-handling effort, an assembly process is often organized in a sequential manner. A dynamic programming procedure for sequential lines is developed for the minimal cost solution. Cost items considered in the procedure are direct labor, workstation tooling, building space, factory overhead, and work-in-process. The last part of the chapter deals with strategies for product testing and rework.

Chapter 5, "Material-Handling Systems," compares material-handling systems by looking at their cost, performance, and integration with other manufacturing components. Considered are three basic material-handling systems: manual, conveyor (continuous type), and automated storage and retrieval system (discrete type). Performance models, operating characteristics, and selection criteria for material-handling systems are discussed.

Chapter 6, "Work-in-Process Management," first deals with parts logistics problems, such as ordering, delivery, kitting, and line feeding. Two different work-in-process management policies are considered: pull and closed loop. Performance comparisons between the two policies and determination of design parameters under each policy are discussed. The last part of the chapter presents a procedure for buffer-capacity analysis and design optimization.

Chapter 7, "Capacity Planning and Human Resource Management," reviews the concept of the improvement process and its empirical models. Knowledge in this area will lead to useful results in line scale-up, capacity planning, job training, and cost estimating. The second part of the chapter discusses line capacity planning and adjustment. Finally, problems of operator selection for both training and assignment are investigated. An optimal assignment policy is obtained by using the matching algorithm, based on operator behavior, work-in-process distribution, and workstation availability.

Chapter 8, "Line Integration," discusses issues in line cost, sizing, and configuration. Since many line-design decisions are based on cost analysis, both cost estimation methods and cost items are considered. This discussion will give the reader a basic understanding about cost structure of a manufacturing line. For a fixed demand, the line cost can be regarded as a function of line size. A minimal cost solution can be found by using a dynamic programming technique. For a variable demand pattern, a design procedure, based on the concept of line flexibility, is developed. This design concept will reduce the risk and the cost resulting from uncertainty in market demand as well as engineering changes.

Chapter 9, "Line Simulation," introduces the Monte Carlo method and its applications for manufacturing lines. After design work has been laid out, performance evaluation should be the next step. At this stage, the line is usually very complex. Analytic methods may not provide detailed analysis, and therefore the designer has to rely on a simulation approach for performance evaluation. Methodology and common pitfalls of simulation models are discussed.

Finally, Chapter 10, "An Analysis of an Assembly Line Design," presents a case study, using the concepts of the previous chapters. Line assumptions and the values of line parameters are obtained from real-life assembly lines.

Much of the material in this book was developed over the past few years, when I participated in a manufacturing line design project. It is my wish that this book may encourage more theoreticians to look into practical manufacturing problems, and more practicing engineers to learn how to convert theoretical results into powerful design tools.

I am grateful for the generous support that was given to this writing project by my management in the IBM General Products Division, San Jose, California. My thanks are due particularly to F. H. Fitch and Jack O. Hildebrand for their encouragement. I am also indebted to my IBM associates who helped me in many ways in the preparation of the book. Ronald Q. Lewton and Edward A. MacNair read the manuscript and offered many suggestions. Kenneth R. Sweet did all the graphic work. Finally, I thank my wife for her understanding and patience.

We-Min Chow

Contents

Assembly Line Design

1

Fundamental Concepts

Assembly work has a very long history. Ancient people already knew how to create a useful object composed of multiple parts. However, the objective of modern assembly processes is to produce high quality and low cost products. A number of important ideas have been developed to facilitate assembly work. First of all, assembly parts are standardized. Parts of the same type must be subject to the same specifications. This also implies uniform parts quality. Parts from different sources then can be put together as a finished product.

Another innovation in the history of assembly manufacturing is the division of the assembly job. If an assembly task has a long process time or involves too many parts, the work may be broken into a number of smaller tasks. Each task builds a part of the assembly. By progressively adding parts to an assembly, a finished item is produced. Since each task has a relatively limited content, skill can be developed in a short time. Thus, assembly speed may be increased and quality improved. In many cases, nonassembly operations may be needed. For instance, parts preparation, inspection, and testing operations may be introduced to assure the product quality level, and facilitate assembly work.

Since an assembly is composed of multiple parts and parts are assembled by multiple operations, material flow becomes relatively complex. On the other hand, moving physical material is a rather simple and mechanical task, as compared with an assembly task. Operator productivity can be further improved by introducing a dedicated material-handling system and letting operators concentrate their efforts on assembly tasks where skills are needed. As a consequence, automated material-handling systems have become very popular in modern assembly lines. Manufacturing cost is reduced due to prompt material delivery, smaller space requirement, better inventory accountability, less handling damage, and less labor.

Engineers and scientists have engaged in multidisciplinary analyses, learning that proper work conditions (e.g., job content, tooling and fixtures, workstations, etc.) provide operators with safer and more productive jobs. Time and motion study, analysis of human performance, and ergonomics have been introduced to industry. At this stage, assembly job design begins to integrate human behavior into workstation design. Consequently, efficiency at the workstation level is greatly improved.

As more and more components are included, line efficiency eventually becomes a problem. Efficiency improvement at the component level does not guarantee overall performance efficiency. Line-integration concepts therefore are introduced. The line designer must take a system view and a structural approach. First, a cost objective must be defined based on both market analysis and manufacturing conditions. Then the product cost structure must be understood. Usually, this is defined by the product characteristics and the manufacturing environment. By comparing the cost objective and the cost structure, the line designer may conduct a study for production feasibility and affordability. At this stage, questions such as resource availability, production capacity, the speed of scale-up, and engineering skills must be answered. Derived from this study is a line-design concept that involves a number of interrelated subjects, e.g., tooling strategy, material-handling system, line size, line configuration, flexibility needed for future engineering changes or line-capacity adjustment, and space strategy. The mission of line design is then to convert the design concept into a physical line.

1.1 The Nature of the Line Design Problem

Although it is not difficult to develop a logical procedure for line design, problems may still come from several different sources. First, all the information needed for line design may not be available; the line designer must deal with unknowns. Then, among all the available information, some items are subject to change. Third, even if all design parameters have been firmed up, stochastic phenomena in an assembly line are inevitable. Due to lack of deterministic regularity, line problems can be very complex. Finally, problem complexity may also come from the organization of the design team. Poor communications between design groups responsible for different functions will cause schedule delay, poor design, and high cost.

None of these problems have simple solutions. However, it is not impossible to find solutions, given effective design guidance. The rest of this section will attempt to give such guidance.

1. *Unknowns* To shorten the product cycle and achieve optimum design for manufacturability, the line design and the product design functions must work in parallel and interact with each other. However, this implies that information for line design, at least at an early stage, is not totally available due to lack of a completed product definition. In reality, since a physical line can only work in a given environment, the line designer must often make assumptions for unknowns. To minimize the risk of using wrong assumptions, the designer should consider the *generality* of assumptions. That is, either the design solution is not sensitive to the assumptions, or the assumptions have a good coverage of the reality. In many cases, this can be done by conducting mathematical analysis or line simulation against different sets of line parameters.
2. *Uncertainty* Even if all the necessary information were available, it is still subject to another problem—change or uncertainty. It is too naive to believe that all the available information would remain unchanged during the line-design process. In some cases, a change is so drastic that the entire design concept must be revised.

   Two major problem areas are production demand and engineering change. The design concept of an assembly line is heavily influenced by the desired production volume. For example, automation is a common strategy only for high-volume production. Unfortunately, production demand is usually based on the market situation which is difficult to understand and well beyond the control of the manufacturer. On the other hand, engineering change is quite common when dealing with a new product. Any changes in product design, manufacturing process, tooling, fixtures, or even work methods may result in a need for line adjustment.

   Although a line can be modified by adding, deleting or rearranging workstations and altering material-handling methods, it can be a very costly and time-consuming solution without a careful plan. It is the line designer's responsibility to make sure that the line has *flexibility* to cope with any changes easily.

   Solution methods should be computerized, if possible. Whenever a change is made, the computer package can be reinvoked with a new set of parameters to find the new solution rapidly.
3. *Variation* Most line parameters that can be accurately estimated by engineers are available in terms of their average values. Examples

are the mean process time, the average yield, the mean time between failures, the mean repair time, and so on. From a capacity-planning point of view, however, variability of a parameter is as important as its average value. This thought can be clarified by a simple example as follows:

> Consider a sequential line with two workstations. Both workstations have the same process time, say 2 minutes per operation cycle. If the process times are constant, the two stations start and complete their operation cycles at the same time epochs. The line throughput is clearly 0.5 per minute. Suppose that the process times are subject to a simple distribution: with an equal probability, the process time is either 1 or 3 minutes. The mean process time remains the same. If both stations are in a 1-minute or a 3-minute cycle, they can still start and complete their cycle at the same moments. But if their process times happen to be different, then one station must wait for the other. In this case, the production cycle becomes 3 minutes. Since the probability of a 1-minute cycle is $0.5 \times 0.5 = 0.25$, the line throughput is equal to $1/(0.25 \times 1 + 0.75 \times 3) = 0.4$ per minute. If the line has more than 10 workstations, the production cycle is almost always 3 minutes and the line throughput is reduced to 0.333 per minute.

Compared to unknowns or uncertainty, the problem of variation is relatively easy to deal with. The line designer may try to reduce the variability, for instance, by introducing automation or improved work methods. In many cases, an adaptive line operation strategy can be invoked to adjust the line operation mode. One common strategy is to regulate line flow by introducing buffers for work-in-process between operations, so that operations become nearly independent of each other. Sometimes, assigning a fast operator to the operation with high variability may also help to increase line productivity. The last resort is to plan for more resources such as adding more workstations.

4. *Complexity* A line-design problem often has a complex structure due to multiple line components, e.g., tooling, operators, parts, material-handling facility, yield management, information system, and so on. For each single component, a number of design alternatives may exist. The problem can easily become unmanageable if the line designer has to consider all the possible combinations of these alternatives. Therefore, the problem of complexity must be handled

Table 1.1 Problem Solving Strategy

Problem	Solution
Unknown	Generality
Uncertainty	Flexibility
Variation	Adaptability
Complexity	Analysis, Communication

by a structural approach. For a given product and a given manufacturing environment, the design objective and constraints should be clearly defined. Then the problem may be partitioned into a number of subproblems. Since the scope of each subproblem is limited, a complete analysis becomes possible. Results of analysis should help to reduce the number of alternatives at the subproblem level and, consequently, to simplify the overall design problem.

The line design problem can also be complicated due to the requirement of multiple disciplines. For any sizable line, the design project usually involves people from different groups such as material control, quality assurance, production control, facility engineering, and information system. Consequently, communication between people is often a major obstacle. Solutions to this problem include: (i) Keep the design team small. (ii) Effectively use meeting and other communication means. (iii) Document design assumptions, and (iv) to make the direction clear to all team members.

The above discussion is summarized in Table 1.1.

1.2 Line Components

To alleviate complexity of a design job, an assembly line may be subdivided into a number of logical or physical components. The design team usually includes members from several departments; each is responsible for one or more components. The subdivision method and the definitions of components are application-dependent. One possible breakdown is given as follows:

1. Process design
2. Line balance
3. Test strategy
4. Yield management

5. Material handling
6. Maintenance policy
7. Work-in-process management
8. Parts procurement
9. Parts feeding
10. Human resource
11. Line size
12. Line layout
13. Information system

The first thing that one should consider is how a complete product can be assembled. Assembly process design is a very critical area. A good process design should lead to a well-balanced line and appropriate job content for each operation. Often assemblies must be tested or inspected for quality assurance. Through testing or inspection, assembly problems will be detected and analyzed. The results then can be used for problem corrections. Hence, a good testing strategy should reduce the amount of rework and scrap, and increase line productivity by improving product yield.

Material handling is another area that deserves intensive study. Although this function does not add value to a product, it facilitates assembly process flow. If an assembly line has two or more workstations, products must be moved from one to another. Furthermore, assembly parts also need to be delivered to workstations. Any material-handling delay or damage will degrade line performance and may become costly.

Since both workstations and material-handling systems may be subject to failure, a maintenance function becomes necessary. Compared to many other functions, a maintenance operation is relatively simple (e.g.fewer parts and fewer persons involved), but may require a high skill level.

The material-handling function is closely related to work-in-process (WIP) management. The latter is responsible for determination of in-process inventory, from the parts-staging area to product shipment. A WIP management policy also regulates line flow and parts-feeding speed.

In many assembly lines, labor is still indispensable. Problems of manual operations include the learning process, ergonomic considerations, human inconsistency, performance deviations, and job assignment. An effective training program and a good job-assignment policy are essential to efficient production.

Since market demand is based on forecasts, it is difficult to match the line capacity with the demand. To avoid the problem of overinvestment or insufficient capacity, design strategy should emphasize line flexibility, i.e., the ability to quickly adjust capacity at a minimal cost. Both line size and layout are important subjects in this area.

The rapid development of modern data processing also has a significant impact on line design. Sophisticated software programs can assist line management, including data collection, process control, failure analysis, parts tracing, and shop flow control. Information systems play a key role for both yield management and flow control.

From an analytic point of view, it is convenient to subdivide a line into a number of components, and to deal with individual components separately. On the other hand, components must be designed in an integrated manner. To achieve design integration, interrelated components must be considered under the same design function. Five functions are identified:

1. Line flow and configuration
2. Line operation
3. Material logistics
4. Product yield and quality
5. Information management

The first function deals with the physical line, where the key problems are (i) compatibility of line layout with process flow, (ii) interface between material handling-system and workstations, (iii) equipment utilization, (iv) compatibility of material handling system with WIP management policy, and (v) line flexibility.

For a given assembly line, the next problem is how the line should be operated. A common problem is that line operation policy has not been carefully analyzed during the design phase. Consequently, the actual line throughput may be considerably lower than the targeted capacity. Operation management problems, such as line scheduling, parts feeding, WIP management, job dispatching, flow control, and maintenance policy, should be regarded as a part of line design problem.

Since assembly involves multiple parts, material logistics can be a major problem. Typical logistics problems are parts ordering, delivery, parts staging, and feeding. The right kind of parts should be delivered in the right quantity to the right place at the right time in the right manner. In some cases, parts preparation serves an important function. For clean room operations, assembly parts must be cleaned before assembly. For automated stations, parts may have to be placed in a kit with the appropriate orientation.

The fourth function consists of components that may affect product yield and quality. First, a high-quality product must have high-quality parts. Next, during the assembly process, testing or inspection functions may be established to examine product quality. Test results should be used for process control so that both product quality and yield can be improved.

Table 1.2 Functional Relations Between Line Components

	Flow and configuration	Line operation	Material logistics	Yield and quality	Information management
Process design	×	×			
Line balance	×				
Test strategy	×	×		×	×
Yield management		×		×	×
Material handling	×		×		
Maintenance policy		×		×	×
WIP management	×	×	×	×	×
Parts procurement			×	×	×
Parts feeding		×	×		×
Human resource		×		×	×
Line size	×				
Line layout	×				
Information system	×	×	×	×	×

In addition, tool calibration and maintenance can be important for high-yield production. Finally, all operators must be appropriately trained and supervised to follow the process instructions.

Management of line information can be vital to the success of a modern manufacturing business. Manufacturing information systems have been widely used for many purposes. Problems in this area involve data-requirement definition, computer and network capability, information retrieval or data base system, and line management support.

The relations between line components and design functions are summarized in Table 1.2. These relations define a "map" for line integration. Each function links a number of interrelated components together, while each component may be responsible for one or more function. This map will help the line designer make sure that components are both logically and physically compatible and that design assumptions at function levels are consistent and realistic.

1.3 Measures of Effectiveness

An important problem sometimes overlooked is design optimization. For example, in high-technology industries, a company can profit by its product innovations. Even if its manufacturing efficiency is poor, the company may still have a high profit margin because it is offering a unique product. However, once losing its competitive edge, the company can quickly encounter financial problems due to high product cost. For this reason, line design should be integrated with the business plan.

A sound business plan should consider three things: (i) market, (ii) product, and (iii) production. As illustrated in Figure 1.1, among these three items the key factors are product performance, cost, and manufacturability, respectively. A product is designed not only for good performance but also for manufacturing simplicity. If a product is properly designed, an effective production line will produce it at low cost and with high quality. The relations among product performance, manufacturability, and cost should be carefully evaluated.

To assure that a line is appropriately designed, line measurements should be established to understand design effectiveness. Measurements can be classified into two categories: cost and line performance. A cost item usually measures the total amount of spending needed to produce a given amount of products. This may include product design, development, laboratory testing, and all spending to support production activity such as direct and indirect labor, equipment, facilities, and other overhead. The ratio of the total spending to the total amount of products is then the product unit cost. From an investment point of view, however, the time factor is important. To a business person, a dollar today has more value than a dollar tomorrow. Consequently, cost estimation may be based on a discount method; that is, the present value of a future cost item is estimated by multiplying the cost by a predetermined discount factor.

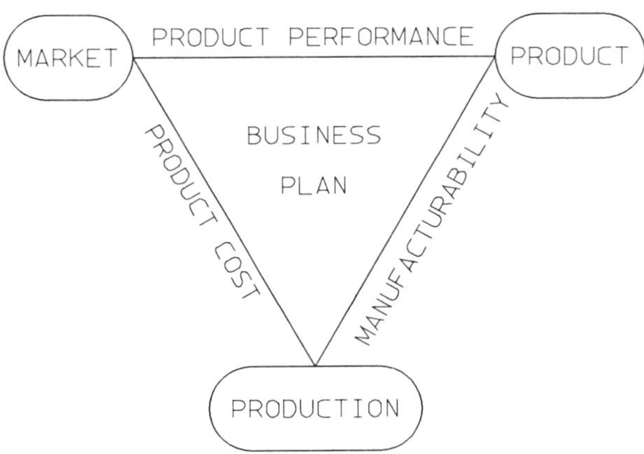

Figure 1.1 Three components of a business plan and their relationships.

Different cost estimation methods have been devised, few of them measure intangible costs such as line flexibility, job enrichment, product yield, parts quality, line balance, process time variation, and so on. Although not all these intangibles can be understood easily, their costs may be measured by indirect methods. In many cases, a cost estimation method can be derived from performance measures. For example, line flexibility would impact the capital investment plan. Yield and quality are related to line capacity and material handling cost. Process time variation may cause workstation utilization or in-process inventory problems.

Performance measures are much more sophisticated than cost measures. Often advanced mathematical tools are required. A manufacturing line can be regarded as a service system of multiple types of resources. Line performance results from resource contention due to production activities. The line designer may use *analysis* tools to characterize the line performance problems and *optimization* tools to resolve the problems. A good design should reduce contention to a minimal level and yet achieve high resource utilization. Hence, the fundamental problems are (i) to plan for an appropriate amount of resources, and (ii) to schedule and control production activities properly. These two subjects are the theme of this book and will be discussed in detail in the subsequent chapters. Basic concepts of line performance and measurement are briefly introduced in the following pages.

Line performance is determined by many parameters, such as workstation availability, operation process time, yield, process flow pattern, operator skill level, line operation management policy, and line layout. This can be explained by using the diagram shown in Figure 1.2.

First, let us consider a workstation. When a workstation breaks down, the waiting time for repair action is dependent on interfailure time, repair time, the size of the repair staff, the number of workstations served by the repair crew, and the dispatching rule of repair jobs. The sum of waiting time and repair time is then equal to workstation downtime. A workstation operating cycle is composed of an interfailure time and a downtime. The ratio of the expected interfailure time to the average operating cycle is the *reliability* of the workstation.

In addition to workstation failure, scheduled downtime may exist such as preventive maintenance, engineering changes, setup, and calibration. If it is a manual workstation, further production time may be lost because of operator's fatigue, personal delay, breaks and lunch time. The proportion of time that can be used for production is called the *availability* of a workstation.

For each operation, the process time is a function of job complexity, workstation layout, tools and fixtures, and operator skill level. The average process time may be estimated by direct measurement or by a synthetic

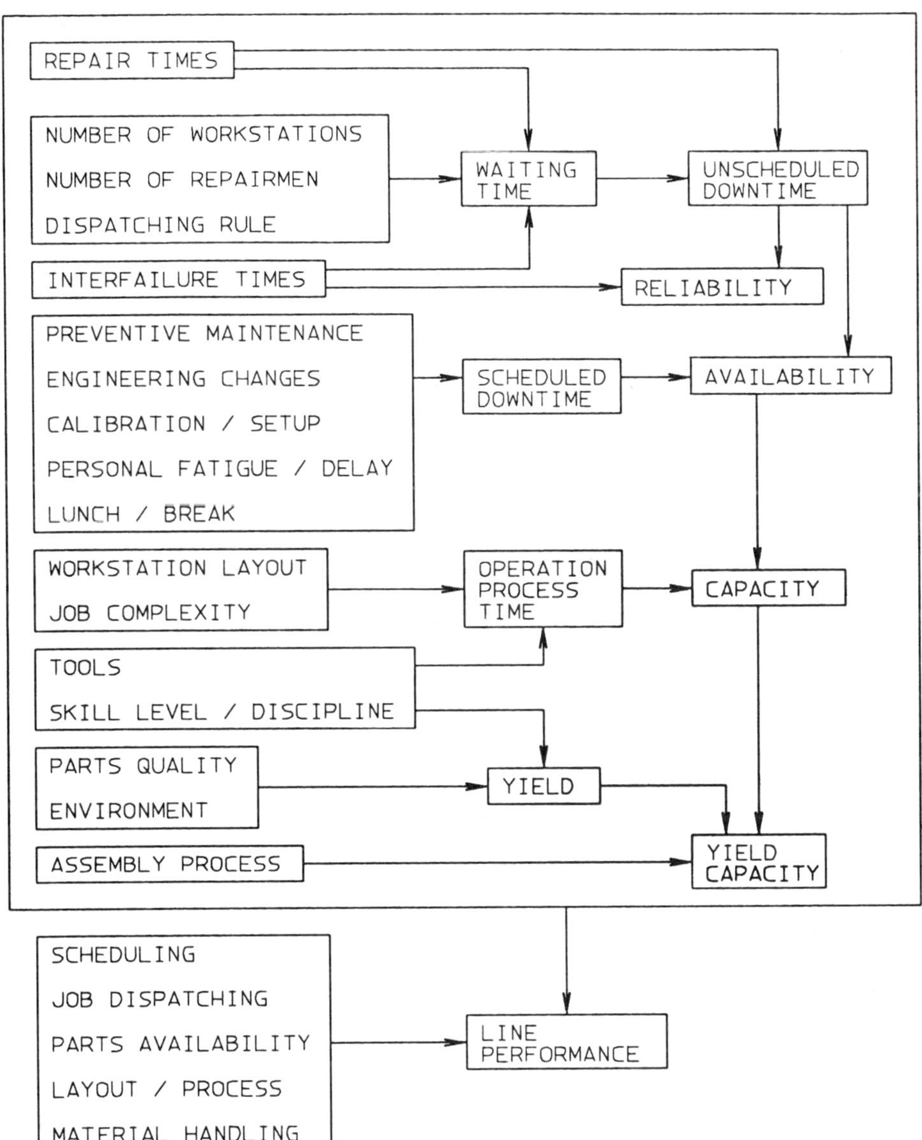

Figure 1.2 Line performance as a function of line parameters.

method based on a time and motion study. The maximum number of products that a workstation can build per unit of time (e.g., an hour or a day) is its *capacity*. The hourly capacity, for instance, is simply the ratio of availability to the average process time in hours.

Another important line parameter is the product *yield*, which is dependent on production tools, operator skill, parts quality, working environment, test strategy, and line discipline. For a given manufacturing process, operation process time and yield determine the *yield capacity*—the maximum number of good products per time unit that an operation can support.

Line measures may be derived from workstation capacity, yield, line scheduling policy, job dispatching policy, parts availability, line layout, and material-handling method. Several different measures, as follows, should be used to complete line performance evaluation.

1. *Line capacity*

   Line capacity is equal to the maximum expected number of good products that the line can produce per time unit. This number is usually close to but less than the yield capacity of the bottleneck operation, i.e., the smallest yield capacity among all the operations.
2. *Production lead time*

   This is the total amount of time that a product spends in the line, including waiting and rework. Lead time is also an important parameter for material-requirements planning.
3. *Work-in-process*

   The total number of products in the line is a measure of in-process inventory. During a sufficiently long time interval, the average amount of work-in-process is approximately equal to the product of the average lead time and line throughput. This relation is usually a good indicator of line performance. An ideal condition is high throughput, short lead time, and low work-in-process.
4. *Workstation utilization*

   This is the proportion of time that a station is busy doing useful work. For a well-balanced line, all workstations should have approximately the same utilization. If the line is running at full speed, workstation utilization will be close to 1. The average workstation utilization is defined by

$$w = \frac{1}{n}\sum_{i=1}^{n} r_i s_i \tag{1.1}$$

where n = number of workstations
s_i = average process time at workstation i
r_i = throughput at workstation i

5. *Operator utilization*
This is similar to workstation utilization, except that operators are more flexible. Sometimes, even for an unbalanced case, workload may be evenly distributed to operators. Operator utilization is a measure for the efficiency of a job assignment policy. The operator utilization is defined by

$$v = NS/mT \tag{1.2}$$

where S = expected total manual process time (including rework, if any)
N = production volume per day
m = number of operators
T = total productive time per day

6. *Line utilization*
This is the ratio of production volume to line capacity. Line utilization is given by

$$u = \frac{1}{t_o}\int_0^{t_o} \frac{r(t)}{c(t)}dt \tag{1.3}$$

where $c(t)$ = line capacity at t
$r(t)$ = line throughput at t
t_o = product life time

Since $0 \leq r(t) \leq c(t)$, then $0 \leq u \leq 1$. A line is fully utilized when $u = 1$.

7. *Line flexibility*
Line flexibility measures the difference between line capacity and production demand:

$$f = \frac{1}{t_o}\int_0^{t_o} \frac{\min\{c(t),D(t)\}}{\max\{c(t),D(t)\}}dt \tag{1.4}$$

where $D(t)$ is the demand at t.

The value of f is always between 0 and 1. If $c(t) = D(t)$ for all t, then $f = 1$. In this case, the maximum flexibility is achieved.

8. *Quality*

A conventional measure for quality is the defect rate, e.g., the average number of defective products per lot. From a line design point of view, this measure provides little information. First, product quality is a cost factor and, therefore, should be measured against cost or resource consumption rate. Next, quality should be appropriately maintained during the product-building process. Quality control must be a part of the manufacturing process. Strategies for testing, inspection, failure analysis, and rework are important design considerations. Finally, quality is closely related to product yield, and may influence productivity.

Measures for quality include:

1. Scrap rate
2. Average total rework time per product
3. Average test time per product
4. Average yield over the product life time

Clearly, all these measures are related to line capacity and resource consumption requirements.

9. *Line scale-up capability*

This measures how fast a line capacity can be increased. Suppose that line capacity reaches its peak at time y. The line scale-up capability can be measured by looking at its *scale-up index*:

$$d = \frac{1}{y} \int_0^y \frac{c(t)}{c(y)} dt \tag{1.5}$$

When $c(t) = c(y)$ for all t in $(0,y)$, d takes its maximum value at 1. This means that the line can reach its highest capacity on the first production day. Line capacity is mainly determined by process time and yield. As long as the process times are finite and yield is positive, capacity can be increased by adding more resources such as operators, workstations, etc. Therefore, this measure is meaningless without considering the affordability issue. The line designer should also examine the market demand. Even if affordability is not an issue, sometimes the demand may limit the expansion of line capacity.

The scale-up index is not only a measure for the merit of line work, but also for the entire business.

1.4 Remarks

Due to the existence of unknowns and uncertainty, line design often becomes a philosophical problem. On the other hand, line design is also a very practical problem which can greatly influence the product cost. Since a company's ultimate goal is to make profit, line design cannot be treated as an isolated engineering problem. It must be considered as a part of the entire business plan. Three things particularly important to line design are:

1. The business goal and strategy
2. Intensive cost and performance analysis
3. Line integration

The line designer must behave as a business person, an engineer, a financial analyst, a system analyst, and a line manager.

2

Empirical Data and Statistical Analysis

The analysis of line performance behavior is a key area in line design. Different design concepts and trade-offs between manufacturing components should be carefully evaluated. Unfortunately, information needed for analysis is seldom totally available during the line design process. Even when some information does become available, more often than not line designers can only deal with average values such as mean time between failures, mean repair time, average operation process time, average product yield, and average throughput. Analysis in an "average" sense, however, is by no means sufficient for line performance evaluation; more detailed information is needed for accurate and thorough analysis. This dilemma can be one of the major causes of using unverified or even improper assumptions for line studies. For instance, many line simulation studies incorrectly assume either uniform or normal assembly times. If the purpose of simulation is to investigate a detailed buffer placement policy, neither of these assumptions may honestly reflect reality.

If information is not available, line designers do not always need to "invent" information for their studies. It would be more logical to base the assumptions on experiences or observations from existing lines. Although different manufacturing lines have different performance behaviors, some commonalities can be identified and made useful to a very broad range of applications. There are two major sources for this kind of information in today's environment. One may search the literature to explore reported results, or one may collect and analyze data from existing lines. In the latter case, often the data do not directly provide useful information, either because the data volume is too large or because the data are not presented in an understandable format. Raw data must be processed by data reduction or analysis tools. Efficient tools for data analysis can be developed by using statistical techniques. The first part of this chapter gives a brief introduction to statistical concepts and techniques with which line designers should be familiar. Discussions about commonality constitute the second part of the chapter. Major parameters will be discussed, such as job or event arrival process, operation process time, product yield and equipment reliability.

2.1 Random Variable, Probability and Measures

Since the real world environment is changing constantly, it is unlikely to predict future events exactly. A convenient way to resolve this problem is to postulate some regularity that follows a statistical law. For example, process times of two operation cycles may not be identical, but the chances are identical that both take values within the same range. Outcomes of experiments (such as the process times), when measured in real numbers, are called *random variables*. There are two types of random variables: *discrete* and *continuous*. The former only takes values at isolated points on a real line.

The probability of event A is the proportion of time that A can be observed in repeated experiments. The value of a probability is between $[0,1]$. For example, A may stand for the event that a random variable X is between a and b. The probability that A occurs is equal to $P[A] = P[a \leq X \leq b]$. If X is discrete, then an event can be $\{X = a\}$ and its probability is $P[X = a]$. When dealing with assembly line problems, random variables usually have nonnegative values. If X is nonnegative, $P[X \geq 0] = 1$.

The conditional probability of A, given B, is defined by

$$P[A \mid B] = P[AB]/P[B]$$

where AB is the event that both A and B occur. As an example, let T be the time interval between machine failures, $A = \{T \leq 200 \text{ hours}\}$ and $B = \{T > 100 \text{ hours}\}$. Then $P[A \mid B]$ is the conditional probability that the machine failure would occur within 200 hours, given that the machine has been operated for 100 hours (without failure). Clearly, $P[A \mid C] = 1$ if $C = \{T < t\}$ for all $t \leq 200$ hours, and $P[A \mid D] = 0$ if $D = \{T > t\}$ for all $t \geq 200$ hours.

The *probability distribution function* of X is defined by $F(x) = P[X \leq x]$. If X is continuous, $f(x) = dF(x)/dx$ is called the *probability density function* of X. If X is discrete, it has a *probability mass function*, $p(x) = P[X = x]$. Since the distribution function is also uniquely determined by the density (or mass) function, statistical properties of a random variable can be investigated by using either its density (or mass) function or its distribution function. Both $f(x)$ and $p(x)$ are nonnegative and $F(x)$ is a nondecreasing function of x and is valued between 0 and 1.

It is not always convenient to deal with these probability functions. Some of them may have rather complex algebraic expressions. Properties of randomness sometimes can be characterized by expectations or moments. The rth *moment* of X is given by

$$\begin{aligned} E[X^r] &= \int_x x^r f(x)dx \qquad \text{if } X \text{ is continuous} \\ &= \sum_x x^r p(x) \qquad \text{if } X \text{ is discrete} \end{aligned} \tag{2.1}$$

Note that moments are averaged values with a weight function $f(x)$ or $p(x)$. When $r = 1$, $m = E[X]$ is the *mean* or the *expectation* of X. The formal definition of the expectation of X is given by (2.1) with $r = 1$. It can be shown, however, that the expectation of $g(X)$ for a function g can be obtained by using the distribution of X (instead of the distribution of $g(X)$) and is equal to

$$\begin{aligned} E[g(X)] &= \int_x g(x)f(x)dx \qquad \text{if } X \text{ is continuous} \\ &= \sum_x g(x)p(x) \qquad \text{if } X \text{ is discrete} \end{aligned}$$

The rth *central moment* of X for $r > 1$ is given by $E[(X - m)^r]$:

$$\begin{aligned} m_r &= \int_x (x-m)^r f(x)dx \qquad \text{if } X \text{ is continuous} \\ &= \sum_x (x-m)^r p(x) \qquad \text{if } X \text{ is discrete} \end{aligned} \tag{2.2}$$

The second central moment is called the *variance*, $Var[X]$. The square root of the variance, denoted by $SD[X]$, is the *standard deviation*. Both the variance and the standard deviation are measures for dispersion. The third and fourth central moments are measures for skewness and degree of peakness, respectively. A distribution is symmetric if and only if its third central moment is zero. A distribution is positive (negative) skew, if this moment is positive (negative). Normally positive (negative) skewness implies that the distribution has a right-hand (left-hand) tail. Commonly used relative measures are given by moment ratios:

$$\begin{aligned} \text{Coefficient of variation:} \quad & CV[X]=SD[X]/E[X] \\ \text{Skewness index:} \quad & SK[X]=m_3/(SD[X])^3 \\ \text{Kurtosis index:} \quad & KR[X]=m_4/(SD[X])^4 \end{aligned} \tag{2.3}$$

Statistical relation between random variables can be investigated by looking at their *joint distribution*. The joint distribution of X and Y is defined by

$$F(x,y)=P[X \le x, Y \le y]$$

Their correlation is measured by their *covariance* and *correlation coefficient*, respectively:

$$\begin{aligned} Cov[X,Y] &= E[XY]-E[X]E[Y] \\ \rho(X,Y) &= \frac{Cov[X,Y]}{SD[X]SD[Y]} \end{aligned} \tag{2.4}$$

X and Y are said to be *uncorrelated* if and only if $Cov[X,Y]=\rho(X,Y)=0$. If X and Y are independent, then $F(x,y)=P[X \le x]P[Y \le y]=F(x)F(y)$. It is easy to show that X and Y are independent if and only if $E[g(X)h(Y)]=E[g(X)]E[h(Y)]$ for almost all functions, g and h. Since independence implies that $E[XY]=E[X]E[Y]$, then $Cov[X,Y]=0$ and $\rho(X,Y)=0$. But two uncorrelated random variables are not necessarily independent of each other.

2.2 Some Theoretical Distributions

Some frequently used distributions and their statistical properties are discussed in this section. Considered are three continuous cases: (i) uniform, (ii) exponential and (iii) normal; and three discrete cases: (i) geometric, (ii) binomial and (iii) Poisson.

Uniform (or Rectangular) Distribution

A random variable is uniformly distributed over $[a,b]$, if its density is identical for all x between a and b.

$$
\begin{aligned}
f(x) &= \frac{1}{b-a}, \qquad \text{for } a \le x \le b \\
F(x) &= \frac{x-a}{b-a}, \qquad \text{for } a \le x \le b
\end{aligned}
\tag{2.5}
$$

Plots of the density and the distribution functions are given in Figure 2.1. Some moments and moment ratios are given by

$$
\begin{aligned}
E[X] &= \frac{a+b}{2} \\
m_r &= \left(\frac{b-a}{2}\right)^r \frac{1}{(r+1)} \\
CV[X] &= \frac{b-a}{b+a}\frac{1}{\sqrt{3}} \\
SK[X] &= 0 \\
KR[X] &= 1.8
\end{aligned}
$$

This distribution can be used to characterize randomness of event occurrence times. For example, the arrival epochs of a random stream are uniformly distributed over a given time interval.

Exponential Distribution

A random variable has an exponential distribution with a rate λ, if

$$
\begin{aligned}
f(x) &= \lambda e^{-\lambda x} \qquad \text{for } x > 0 \\
F(x) &= 1 - e^{-\lambda x} \qquad \text{for } x > 0
\end{aligned}
\tag{2.6}
$$

Plots of the density and the distribution functions are given in Figure 2.2. Mean and moment ratios are

$$E[X] = 1/\lambda$$

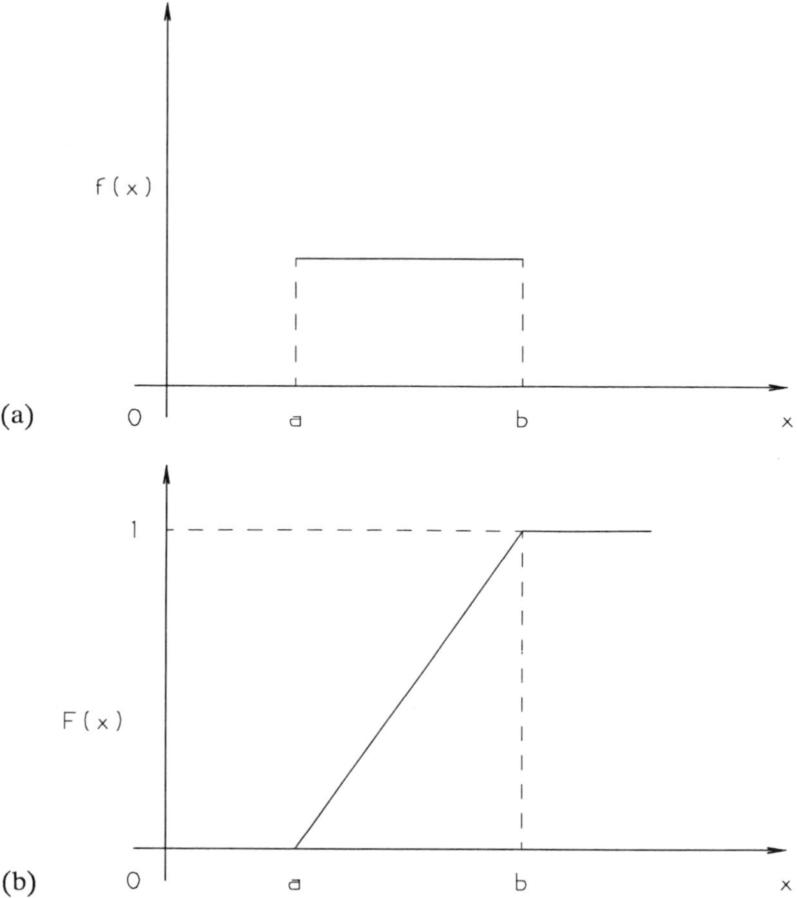

Figure 2.1 (a) Uniform density and (b) distribution functions.

$$CV[X]=1$$
$$SK[X]=2$$
$$KR[X]=9$$

Note that the value of CV is one. The density function has a right-hand tail, i.e., the distribution is positively skewed. The degree of peakness is much higher than that of uniform distribution.

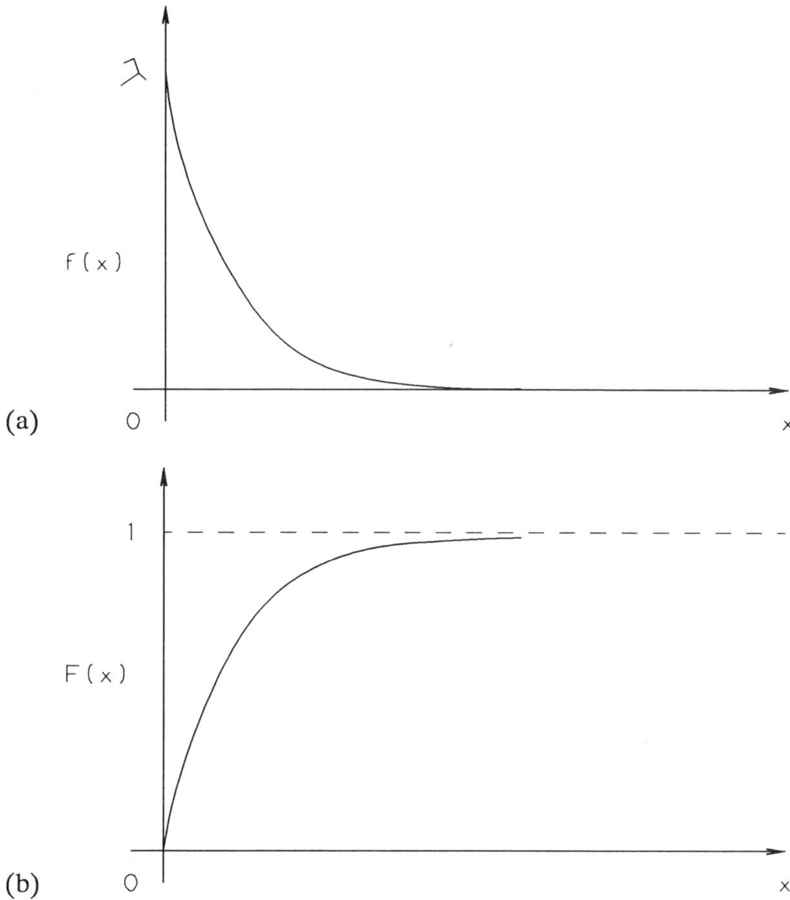

Figure 2.2 (a) Exponential density and (b) distribution functions.

An important feature of exponential distribution is its *memoryless property*. Given $\{X > s\}$, the conditional probability that $X > t + s$ is identical to the unconditional probability of $\{X > t\}$. To see this, first note that $P[X > t] = 1 - F(t) = \exp(-\lambda t)$, and $\{X > t + s\}$ implies $\{X > s\}$. Then

$$
\begin{aligned}
P[X > t + s \mid X > s] &= P[X > t + s]/P[X > s] \\
&= e^{-\lambda(t+s)}/e^{-\lambda s} \\
&= e^{-\lambda t}
\end{aligned}
$$

$$= 1 - F(t)$$
$$= P[X > t]$$

Suppose that the interfailure time of a machine is exponentially distributed and is denoted by X. If the machine has been operated for y time units, the elapsed time until the next failure is independent of y and is statistically identical to a *new* interfailure time. This property greatly simplifies our analysis work, because it asserts that what may happen in the future is independent of the past history. The distribution of remaining life of a machine is identical to an ordinary exponential lifetime. It is also possible to prove that the exponential random variable is the only continuous random variable that has the memoryless property.

If $X_1, X_2, \ldots, X_n$ are independent, identical exponential random variables, then $Y = \sum X_i$ has an *Erlang distribution* with a density function

$$f(y) = \frac{\lambda}{\Gamma(n)} e^{-\lambda y} (\lambda y)^{n-1} \qquad \text{for } y \geq 0 \tag{2.7}$$

where $\Gamma(n) = (n-1)!$ is the gamma function.

If the interarrival time is exponential, the nth arrival time is the sum of n exponential random variables and has an Erlang distribution. For noninteger n, (2.7) becomes a *gamma distribution.* In this case, $\Gamma(n)$ has a value that makes $\int f(y)dy = 1$. For both gamma and Erlang distributions, $E[Y] = n/\lambda$ and $Var[Y] = n/\lambda^2$.

Let $Z = 2\lambda Y$. The density function of Z is given by

$$f(z) = \frac{1}{2\Gamma(n)} e^{-z/2} (z/2)^{n-1} \qquad \text{for } z \geq 0 \tag{2.8}$$

In case that $2n$ is a positive integer, this form of the gamma distribution is called the *chi-square distribution* with $2n$ *degrees of freedom*. The chi-square distribution has been widely used in statistical tests of hypotheses.

Geometric Distribution

A counterpart of the exponential random variable in the discrete case is the geometric random variable. Its mass function is defined by

$$p(x) = q(1-q)^x \qquad 0 < q < 1, x = 0, 1, 2, \ldots \tag{2.9}$$

where q can be interpreted as a probability.

Suppose that the outcome of an experiment is either a success or failure. Let q be the probability of success. (The failure probability is equal to $1-q$.) Then the number of failures before the first success is geometrically distributed. Given k failures have been observed, the conditional probability of x additional failures before the first success is independent of k and is identical to the unconditional probability of x failures before the first success. The geometric distribution is the only discrete distribution that has the memoryless property. For a geometric random variable,

$$
\begin{aligned}
E[X] &= (1-q)/q \\
Var[X] &= E[X]/q \\
SK[X] &= (2-q)/\sqrt{1-q} \\
KR[X] &= 9 + (1/Var[X])
\end{aligned}
$$

The sum of n independent, identical geometric random variables is known as the *negative binomial* random variable. Let Y be such a variable. Its mass function has two parameters, q and n, and is defined by

$$p(y) = \binom{n+y-1}{n} q^n (1-q)^y \qquad 0<q<1, y=0,1,2,\ldots. \tag{2.10}$$

Thus Y is the number of failures observed before the nth success. Since the X's are independent and identical,

$$
\begin{aligned}
E[Y] &= \sum E[X_i] = n(1-q)/q \\
Var[Y] &= \sum Var[X_i] = n(1-q)/q^2
\end{aligned}
$$

Binomial Distribution

A binomial distribution has two parameters: n and q. The mass function is given by

$$p(x) = \binom{n}{x} q^x (1-q)^{n-x} \qquad 0<q<1, x=0,1,2,\ldots,n \tag{2.11}$$

If q is the probability of success, then X is the number of successes observed in n trials. Let Z be the outcome of an experiment. $Z = 1$ (or 0) if the outcome is a success (or failure). Z is usually referred to as the *Bernoulli* random variable with $P[Z = 1] = 1 - P[Z = 0] = q$. A binomial random variable is the sum of n independent, identical Z's.

Moments and moment ratios are

$$E[X] = nq$$
$$Var[X] = nq(1-q)$$
$$SK[X] = (1-2q)/SD[X]$$
$$KR[X] = 3 + [1 - 6q(1-q)]/Var[X]$$

Clearly if $q = 0.5$, the distribution is symmetric about its mean.

Poisson Distribution

A Poisson distribution is a limiting case of binomial with a fixed nq, but $n \to \infty$ and $q \to 0$. Let X be a binomial random variable. The mass function of X can be rewritten as

$$p(x) = \frac{n(n-1)\cdots(n-x+1)}{x!} q^x (1-q)^{n-x}$$

Introduce a factor n^x, let $\theta = nq$, and rearrange the terms,

$$p(x) = 1\left(1 - \frac{1}{n}\right) \cdots \left(1 - \frac{x-1}{n}\right)(\theta)^x \left(1 - \frac{\theta}{n}\right)^n (1-q)^{-x} \frac{1}{x!}$$

Letting $n \to \infty$ and $q \to 0$, and using the fact that $(1+x/n)^n \to e^x$ as $n \to \infty$, we have a Poisson distribution with a mass function:

$$p(x) = e^{-\theta} \frac{\theta^x}{x!} \qquad \theta > 0, x = 0, 1, 2, \ldots. \tag{2.12}$$

The mean and other measures are

$$E[X] = \theta$$

$$Var[X] = \theta$$

$$SK[X] = 1/\sqrt{\theta}$$

$$KR[X] = 3 + 1/\theta$$

Historically the Poisson distribution has been used to model the number of accidents, the number of particles falling in a small area, and the number of arrivals from a random stream. All these applications assumed a large number of trials and a small probability of success.

The sum of independent Poisson random variables is again a Poisson variable with a mean equal to the sum of individual means.

Normal (or Gaussian) Distribution

A normal distribution is characterized by two parameters: m and σ. Its density function is given by

$$f(x) = \frac{1}{\sigma\sqrt{2\pi}} e^{-\frac{1}{2}\left(\frac{x-m}{\sigma}\right)^2} \quad \text{for all } x \text{ on the real line.} \tag{2.13}$$

The distribution function does not have a simple algebraic form. However tabulated values for $m = 0$ and $\sigma = 1$ are available. A normal distribution is sometimes denoted by $N(m,\sigma)$. $N(0,1)$ is called the *standard form* of the normal distribution. A random variable with the standard-form normal distribution is said to be the *standard* or *unit* normal random variable. If X has $N(m,\sigma)$, then $X = \sigma U + m$, where U has N(0,1). Figure 2.3 shows the density and the distribution functions of N(0,1). Moments and moment ratios are

$$E[X] = m$$

$$SD[X] = \sigma$$

$$SK[X] = 0$$

$$KR[X] = 3$$

Normal distribution is a limiting distribution. Let Y_1, Y_2, ..., Y_n be independent, but not necessarily identical, random variables. Then $X = Y_1 + Y_2 + \cdots + Y_n$ is asymptotically a normal random variable for $n \to \infty$ with a mean $\sum E[Y_i]$ and a variance $\sum Var[Y_i]$. This is known as the *central limit theorem*. Since negative binomial, binomial and Erlang distributions can

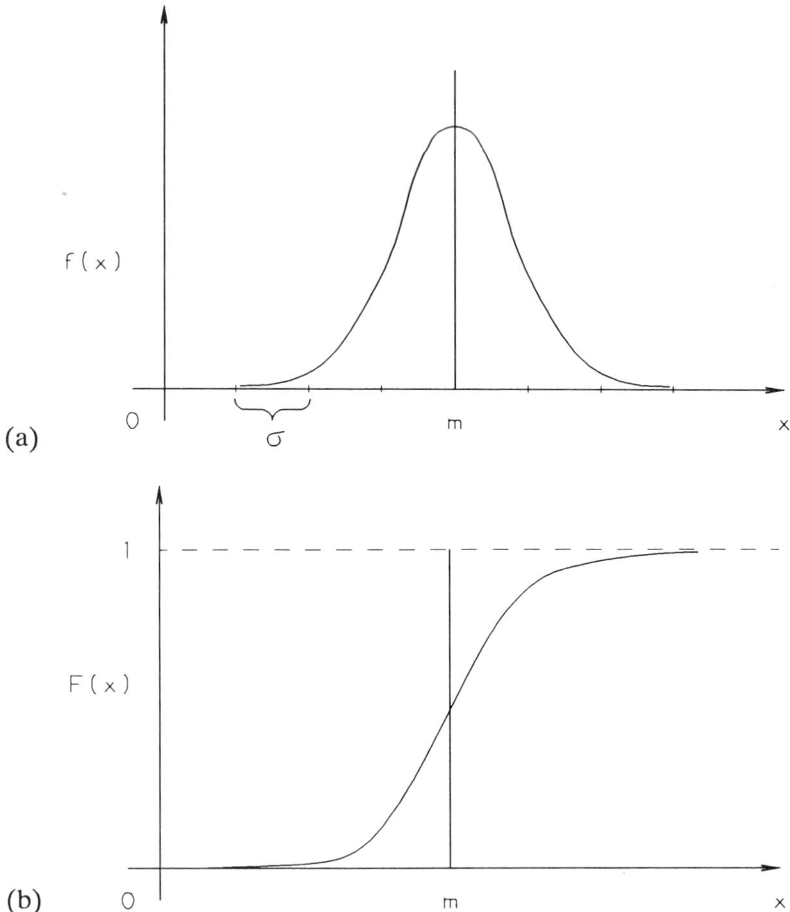

Figure 2.3 (a) Normal density and (b) distribution functions.

be viewed as the convolutions of n independent distributions, they can be approximated by normal distributions. For instance, a binomial distribution with parameters n and q is close to a normal distribution with a mean nq and a variance $nq(1-q)$, when n is large.

If X is a normal random variable, then $-X$ is also normal with the same variance but a mean having an opposite sign. The sum of independent normal variables is normal with a mean equal to the sum of the means and a variance the sum of the variances.

2.3 Sample Distribution and Statistical Inference

Distributions considered in Section 2.2 have relatively simple density functions and can facilitate analysis. When dealing with a real-world problem, no a priori knowledge may lead us to define such distributions. Inductive inference is often adopted to resolve this problem. Let $X_1, X_2, \ldots, X_n$ be the outcomes from independent observations, e.g., n assembly cycle times at the same operation. Each of X's can be regarded as a sample drawn from an unknown distribution $F(x)$. Let O_i be the number of X's of values between $I_i = (y_i, y_{i+1})$. Since $Q_i = P[y_i \leq X < y_{i+1}]$ is a relative frequency of X's in I_i, O_i should be approximately equal to $E_i = nQ_i$. A formal statistical procedure may be invoked to compare $\{O_i\}$ with $\{E_i\}$. This is done by considering

$$Z = \sum_{i=1}^{n} \frac{(E_i - O_i)^2}{E_i}$$

as a chi-square random variable with $(n-1)$ degrees of freedom. Then if Z is too large, the observed data cannot be fitted by the hypothetical distribution, F. This method is called the *goodness of fit* method.

For most practical problems, however, a simple graphic comparison is sufficient. It may take a sample of size 100 or more to characterize a distribution. A large sample is required if the sample distribution has a large variance.

Point Estimation

Moments can also be estimated by statistical inference methods. The *sample mean* and the central moments are respectively

$$\bar{X} = \frac{\sum_{i=1}^{n} X_i}{n} \tag{2.14}$$

$$\bar{X}_r = \frac{\sum_{i=1}^{n} (X_i - \bar{X})^r}{n-1} \qquad r = 2,3,\ldots \tag{2.15}$$

Note that an adjustment has been made in the last equation where a factor of $(n-1)$, not n, is used. This is necessary for unbiased estimation, that is, $E[\bar{X}_r] = m_r$ (the theoretical central moment).

Again the second central moment for the sample is called *sample variance* and is usually denoted by S^2. S is the *sample standard deviation*. Similarly,

Sample coefficient of variation: $S/\bar{X}$
Sample skewness index: $\bar{X}_3/S^3$
Sample kurtosis index: $\bar{X}_4/S^4$

Since X_1, X_2, ..., X_n are independent and identical, $Var[\sum X_i] = \sum Var[X_i] = n\,Var[X_1]$. For $\bar{X} = \sum X_i/n$, $Var[\bar{X}] = Var[X_1]/n$, i.e., the variance of the sample mean is reduced by a factor of n. Thus

$$CV[\bar{X}] = \frac{CV[X_1]}{\sqrt{n}} \qquad \longrightarrow 0 \qquad \text{if } n \longrightarrow \infty$$

The sample moments may provide useful information about distribution parameters. For instance, an exponential distribution has one parameter, λ, and its mean is equal to $1/\lambda$. If a sample is drawn from the exponential distribution, the reciprocal of the sample mean will be a good estimate for λ. This kind of statistical inference with a single-value estimate is referred to as a *point estimate*.

Confidence Interval and t-distribution

From a statistical viewpoint, a point estimate is not very meaningful without knowing the accuracy of estimation. A technique, referred to as the *confidence interval*, provides an assurance measure that the true parameter lies within a given interval. Average machining time of a part may be estimated to be 7 ± 0.5 minutes. This estimate is good if (i) the true average is unlikely to be outside the range (6.5, 7.5) minutes and (ii) the range is not too wide. If the probability that the true average lies within this interval is 0.90, then (6.5, 7.5) is said to be a *90-percent confidence interval*. A general procedure for establishment of a confidence interval is available in statistics textbooks. Since a sample mean has asymptotically a normal distribution, confidence intervals based on normal assumption are particularly useful.

Suppose that a sample $(X_1, X_2, \ldots, X_n)$ has been drawn from a normal distribution with a mean m. The sample mean and the sample variance are respectively $\bar{X}$ and S^2. The ratio

$$Y = \frac{\bar{X} - m}{S/\sqrt{n}}$$

is a random variable subject to a *t-distribution with* $(n-1)$ *degree of freedom*. The shape of this distribution is similar to that of a normal distribution, except a little flatter. If $n \to \infty$, a t-distribution is identical to a normal distribution. As with the standard-form normal distribution, tabulated values for t-distribution are also available.

Since the t-distribution is symmetric about its mean, a small probability α may be chosen so that $P[-t_{\alpha/2} < Y < t_{\alpha/2}] = 1-\alpha$. The value of $t_{\alpha/2}$ can be found from the t-distribution table. A confidence interval of m at a confidence level of $(1-\alpha)$ is $\bar{X} \pm t_{\alpha/2} \times S/\sqrt{n}$. A simple example follows.

Example 2.1 A sample of size 300 has been drawn from an exponential distribution with a unit mean. The sample mean, standard deviation, skewness index and kurtosis index are 1.0552, 0.9987, 1.5145 and 5.6314, respectively. Let X_1 be the sum of the first 30 items in the sample, X_2 be the sum of the next 30 items, and so on. Therefore $X_1, X_2, \ldots, X_{10}$ are independent identical random variables, and are approximately normally distributed due to the central limit theorem. The sample mean, standard deviation, skewness index and kurtosis index are 31.6585, 5.6935, −0.5012, and 2.1526, respectively. (Note that CV becomes smaller and SK is near zero.) Since X_i is the sum of 30 exponential random variables and each has a mean equal to one, the true mean of X_i should be 30. For a 90% confidence level and 9 degrees of freedom, $t_{0.05} = 1.83$. Hence, $\bar{X} \pm t_{0.05} S/\sqrt{10} = 31.6585 \pm 1.83 \times 5.6935/3.1623$. The 90% confidence interval is (28.3637, 34.9533). Indeed this interval does contain the true mean, and the ratio of interval width to the estimate is 6.5896/31.6585 = 20.81%. The accuracy is about ±10 with a 90% confidence level. □

Linear Regression and Least-Squares Method

Let (X_i, Y_i) be a pair of observations. The *sample correlation coefficient* of X and Y is given by

$$r(X,Y) = \frac{1}{S_X S_Y} \sum_{i=1}^{n} \frac{(X_i - \bar{X})(Y_i - \bar{Y})}{n-1} \tag{2.16}$$

where S_X and S_Y are sample standard deviations of X and Y, respectively.

Again r is a measure of relationship between two samples and always has a value between -1 and $+1$. When the magnitude of r is near 1, a functional relationship model may be established. In this section, models of linear structure will be considered.

Let $(Y_t, X_{1t}, X_{2t}, \ldots, X_{kt})$ be the outcomes at the tth observation. A linear relationship between Y and X's is defined by

$$Y = a_0 + a_1 X_1 + \cdots + a_k X_k \tag{2.17}$$

where $\{a_i\}$ are undetermined coefficients.

This relation is called a *linear regression model.* Y is said to be the *dependent variable*, and $\{X_i \mid i = 1, \ldots, k\}$ *independent variables*. The value of Y can be predicted by the model if $\{X_i\}$ can be observed. The problem is to determine $\{a_i\}$ so that the error of the regression model is minimal. A common objective is to minimize the sum of squared errors, that is, to minimize

$$z = \sum_{t=1}^{n} [Y_t - (a_0 + \sum_{i=1}^{k} a_i X_{it})]^2 \tag{2.18}$$

This method is called the *least-squares* method.

Let $Y = [Y_1, Y_2, \ldots, Y_n]$ and $a = [a_0, a_1, \ldots, a_k]$ be column vectors. The n observations on kX's is represented by a matrix:

$$X = \begin{bmatrix} 1 & 1 & \cdots & 1 \\ X_{11} & X_{12} & \cdots & X_{1n} \\ \vdots & \vdots & \ddots & \vdots \\ X_{k1} & X_{k2} & \cdots & X_{kn} \end{bmatrix}$$

It is easy to show that

$$a = (XX')^{-1} XY \tag{2.19}$$

where X' is the transpose of X.

Example 2.2 Process times of seven manual operations were observed. The sample size of each operation varies from 36 to 1691. The computed skewness and kurtosis indices are $SK = (1.41, 1.15, 1.59, 1.74, 1.87, 1.85, 2.23)$ and $KR = (5.42, 5.30, 7.55, 8.32, 7.13, 5.64, 8.71)$, respectively. The relation between SK and KR is depicted in Figure

2.4. The correlation coefficient of *SK* and *KR* is about 0.71. To construct a linear regression model, $KR = a + bSK$, let

$$X = \begin{bmatrix} 1 & 1 & 1 & 1 & 1 & 1 & 1 \\ 1.41 & 1.15 & 1.59 & 1.74 & 1.87 & 1.85 & 2.23 \end{bmatrix}$$

and Y be a column vector that is equal to *KR*. Direct calculation leads to

$$XX' = \begin{bmatrix} 7.00 & 11.84 \\ 11.84 & 20.76 \end{bmatrix}$$

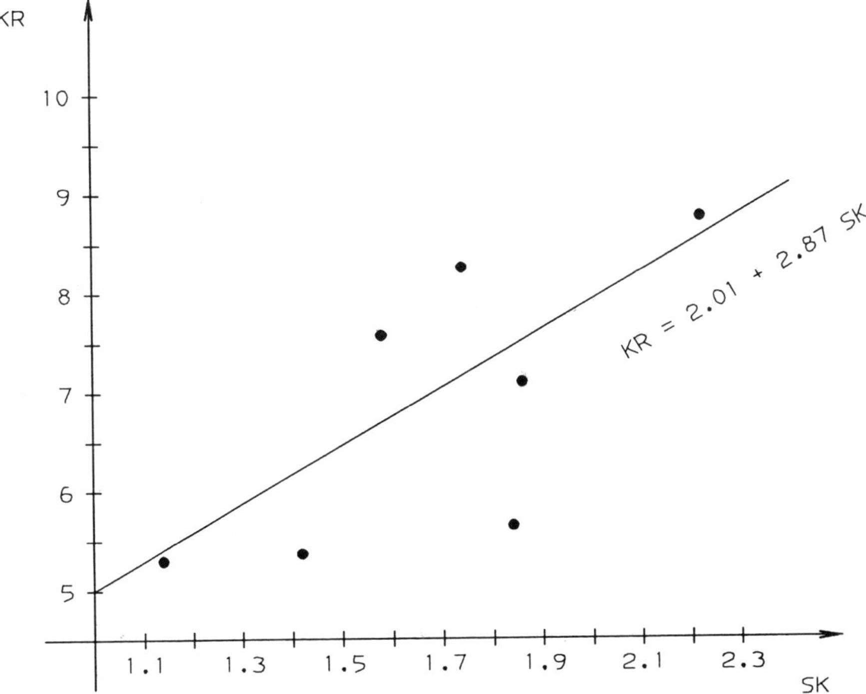

Figure 2.4 A regression analysis for manual process time.

$$XY = \begin{bmatrix} 48.07 \\ 83.41 \end{bmatrix}$$

Consequently, $(XX')^{-1}XY = [2.01, 2.87]$. The linear regression equation is given by $\hat{Y} = 2.01 + 2.87X$, where $\hat{Y}$ is the predicted kurtosis index that corresponds to an observed X. The sum of squared errors in this case is $\sum(KR_i - 2.01 - 2.87SK_i)^2 = 1.10$.

The regression model is depicted by a straight line in Figure 2.4. It can be seen that the model correctly characterizes the trend for a large correlation coefficient and $r(Y,\hat{Y}) = 0.71$. A linear regression model is accurate if the sum of squared errors is small as compared to $\bar{Y}$, and $r(Y,\hat{Y})$ is close to one, e.g., greater than 0.9. □

2.4 Poisson Process

A *stochastic process* $\{N(t), t \text{ in } T\}$ is a family of random variables. For each t in the *index set*, T, $N(t)$ is a random variable. In manufacturing problems, t usually refers to time and $N(t)$ has nonnegative values. For example, $N(t)$ can be the arrival count by time t. When $N(t)$ represents the total number of events that have occurred by time t, then $\{N(t), t \geq 0\}$ is a *counting process*. A counting process is said to be a *Poisson process* if

1. $N(0) = 0$
2. The numbers of events occurring in disjoint time intervals are independent of each other (this property is said to be the *independent increment* property)
3. The number of events occurring in $(s, s+t)$ has a Poisson distribution with a mean λt, that is,

$$P[N(t+s) - N(s) = x] = e^{-\lambda t}\frac{(\lambda t)^x}{x!} \qquad x \geq 0 \tag{2.20}$$

where λ is said to be the *rate of the process*.

Here the random variable $N(t+s) - N(s)$ is independent s. This property is sometimes recognized as *stationary increments*, that is, the number of occurrences is a function of the length of the observation interval and is independent of when the observations are made. This suggests that events occur "uniformly" along the time axis and the past history has no impact on what will be happening. The latter implies a "memoryless property." Indeed the time interval between two occurrences is exponentially distributed, and occurrence intervals are independent and identical. Suppose that at time t a person starts to monitor a work center until the next job arrival time. Let X be the monitoring interval. To say that no job arrives during $(t,t+x)$ is equivalent to saying that X is greater than x, i.e., $\{N(t+x)-N(t)=0\} = \{X>x\}$. Therefore, by (2.20) it follows that

$$P[X>x]=e^{-\lambda x}$$

and X has an exponential distribution with a rate λ.

Let S_n be the nth arrival time, i.e., the sum of n independent, identical exponential random variables. Since $\{N(t)\geq n\} = \{S_n \leq t\}$ and S_n has an Erlang distribution, it follows that

$$\sum_{i=n}^{\infty} e^{-\lambda t}\frac{(\lambda t)^i}{i!} = \int_0^t \frac{\lambda}{(n-1)!}e^{-\lambda x}(\lambda x)^{n-1}dx \qquad (2.21)$$

The equivalency of these two expressions can also be established by using integration by parts.

The stationary and independent increment assumption also suggests that events should occur in a uniform fashion along the time axis. Let X_i denote time from the $(i-1)$st and ith events. $\{X_1 \leq s\}$ means that an event occurs in $(0,s)$ and no events occur in (s,t). For $s\leq t$,

$$\begin{aligned}
&P[X_1\leq s \mid N(t)=1]\\
&\quad=\frac{P[N(s)=1, N(t)-N(s)=0]}{P[N(t)=1]}\\
&\quad=\frac{P[N(s)=1]P[N(t)-N(s)=0]}{P[N(t)=1]} && \text{(independent increment)}\\
&\quad=\frac{P[N(s)=1]P[N(t-s)=0]}{P[N(t)=1]} && \text{(stationary increment)}\\
&\quad=\frac{\lambda s e^{-\lambda s} \quad e^{-\lambda(t-s)}}{\lambda t e^{-\lambda t}}\\
&\quad=\frac{s}{t}
\end{aligned}$$

Given that a job arrived at a work center by t, the arrival epoch is uniformly distributed over $(0,t)$. This result can be generalized for an arbitrary number of occurrences in a finite interval.

For a Poisson process, the following three statements are equivalent:

1. The number of occurrences in a given interval has a Poisson distribution that is determined by the length of the interval and the rate of process, λ.
2. The intervals between successive occurrences are independent exponential random variables with a common rate λ.
3. Given that a number of events occur in an interval, their occurrences are uniformly distributed over the interval.

Therefore, a job arrival stream that follows a Poisson process is also said to have a *random arrival* pattern. For many practical problems, the event occurrence pattern can be modeled by a Poisson process. Suppose that there are n arrival sources, each with an arrival time distribution $F(x)$. The probability that an arrival occurs from a specific source by time t is $F(t)$, i.e., the probability that the arrival time is less than t. The number of arrivals from all sources by t, denoted by $N(t)$, is then subject to a binomial distribution (n independent trials, each with a probability of success $q = F(t)$):

$$P[N(t)=k] = \binom{n}{k} [F(t)]^k [1-F(t)]^{n-k}$$

As discussed earlier, for a fixed mean, $nF(t)$, this distribution becomes a Poisson distribution for $n \to \infty$. If the number of arrival sources is large, then the combined arrival source tends to be a Poisson process. This result still holds, even if individual arrival sources are not identical. When analyzing practical problems, the assumption of Poisson process is often valid, if the number of sources is O(10) or greater. Illustrated in Figure 2.5(a) is a superposed process, composed of 10 subprocesses. The interarrival time of each subprocess is generated in accordance with the following assumptions:

1. Constant interval of 5
2. Constant interval of 10
3. Uniform interval between (2, 8)
4. Uniform interval between (8, 12)

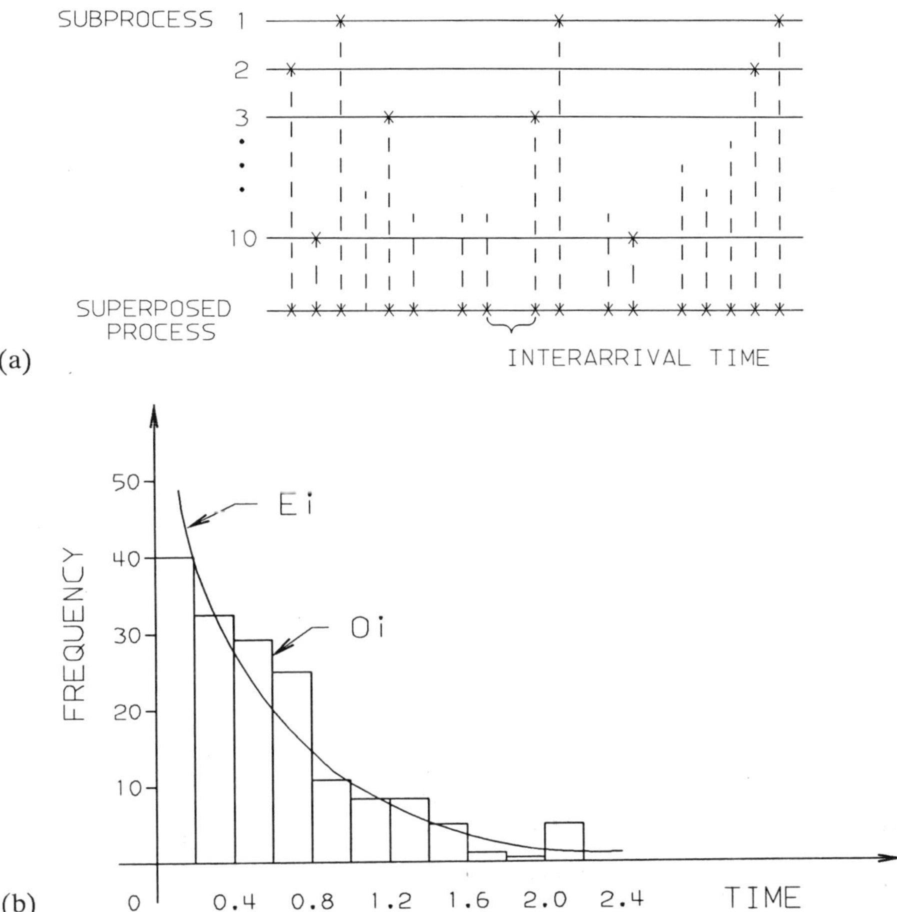

Figure 2.5 (a,b) Comparison of interarrival times from a superposed process with a theoretical exponential distribution. (a) Superposition of processes. (b) Interarrival times of the superposed process.

5. Normal interval with a mean 4 and a standard deviation 0.5
6. Normal interval with a mean 11 and a standard deviation 3
7. Erlang interval with a mean 6 and a coefficient of variation 0.5
8. Erlang interval with a mean 9 and a coefficient of variation 0.333
9. Erlang interval with a mean 3 and a coefficient of variation 0.707
10. Erlang interval with a mean 8 and a coefficient of variation 0.408

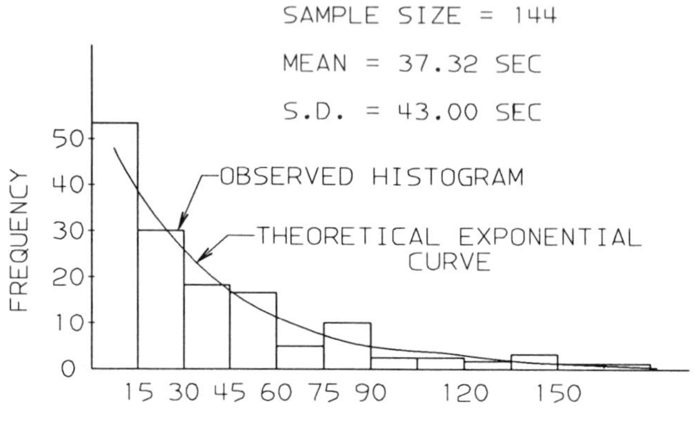

Figure 2.6 Interarrival time distribution.

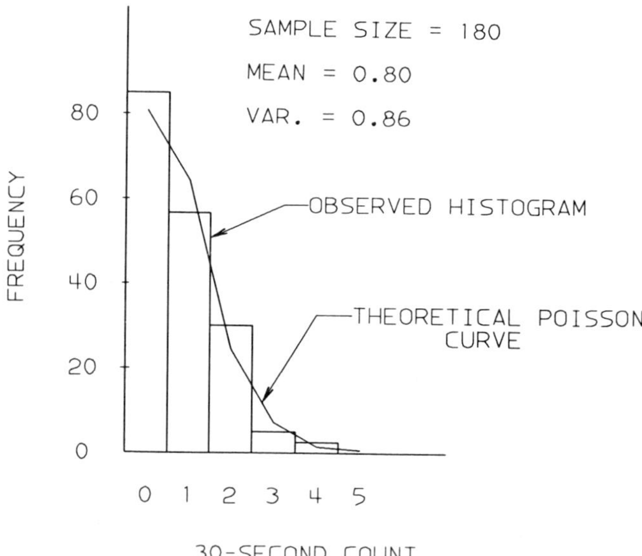

Figure 2.7 Distribution of 30-second count.

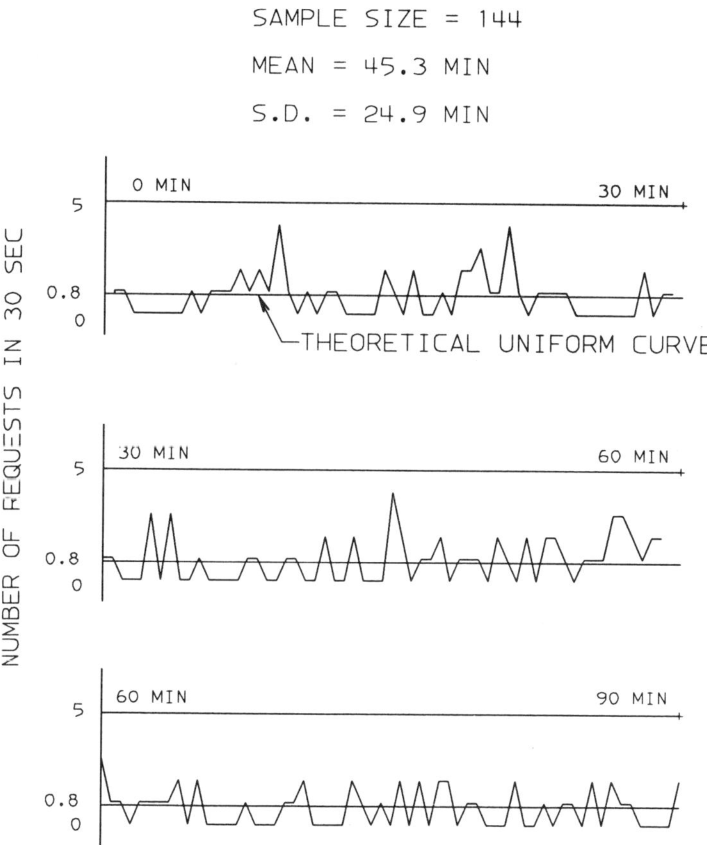

Figure 2.8 30-second counts over 90-minute period.

Note that none of the above cases is even close to exponential. Each subprocess is simulated for 100 time units. For instance, the first subprocess will have exactly 20 events, 5 time units apart. The average rate of the superposed process is 1.677 (= 1/5 + 1/10 + ··· + 1/3 + 1/8), that is, the average interarrival time is 1/1.677 = 0.5963. In Figure 2.5(b) the simulated interarrival time of the superposed process is characterized by the histogram, while the piecewise curve is the theoretical exponential density with the same mean. The average interarrival time from the simulated results is 0.5866 and the *CV* value is 0.8447. The latter is somewhat less than the *CV* of a true exponential distribution.

A good example of a real-life superposition effect can be found in a material handling system. A material handling system usually serves a number of workstations. Service requests generated from workstations, as seen by the system, would follow a Poisson process. Figures 2.6, 2.7 and 2.8 illustrate a service request pattern observed from a conveyor system installed for an existing line. The traffic on the conveyor is controlled by an IBM Series 1 computer. A material handling service request is recognized by the object ID number, the origin and the destination, and the arrival time epoch. Intervals between successive arrivals are calculated. A sample of 144 intervals was observed between 9:20 and 10:50 on a Wednesday morning. Figure 2.6 depicts the histogram of interarrival times. The distribution is nearly exponential. The 30-second arrival counts are also considered. Thus a 90-minute period consists of 180 samples. The sample mean and the sample variance are 0.80 and 0.86, respectively. The frequencies of the 30-second count have approximately a Poisson distribution, while the arrival epochs are uniformly distributed over the 90-minute period. (See Figures 2.7 and 2.8)

2.5 Operation Process Time

One of the key parameters in an assembly line is the amount of time required to accomplish an operation, or the *operation process time*. Problems related to process time behavior include line throughput, work-in-process, in-line buffer sizes, workstation utilization, manpower requirements and line configuration. For instance, a short process time may imply high throughput, less manpower, simple operation, fast learning period and better product quality.

An assembly line may consist of a number of operations, each with an independent process time. Theoretically, these process times can be treated as random variables. For an automated operation, the process time typically has a very small variance due to good repeatability. From a performance point of view, the process time can be considered a constant. It is the manual operation that is worth further discussion.

The manual assembly time, X, of an experienced operator typically has the following properties:

$$0.10 \le CV[X] \le 0.65$$

$$0.60 \le P[X \le E[X]\} \le 0.70$$

$$0.40 \le \min(X)/E[X] \le 0.80$$

$$1.50 \le \max(X)/E[X] \le 2.50$$

Based on a limited number of studies, it seems also true that

$$1.00 \leq SK[X] \leq 3.00$$

$$5.00 \leq KR[X] \leq 12.00$$

Figures 2.9, 2.10 and 2.11 present sample distributions of three different manual operations in disk-drive assembly lines. Both Figure 2.9 and Figure 2.10 show very short process times; their means are 16.05 and 8.74 seconds, respectively. The former is composed of five operation steps and has a relatively small variance. The latter involves insertion of a bearing into a flange. Sometimes this operation is jammed and the operator must use a special tool to complete the assembly task. When this happens, the work time becomes considerably longer. Consequently a large coefficient of variation is expected. Figure 2.11 reports sample distributions of two different operators at the same operation; one is nearly 25% faster than the other. This operation assembles a spindle and normally averages 4 to 5 minutes. Statistics associated with Figures 2.9 through 2.11 are given in Table 2.1.

The statistics basically fall into the ranges defined previously. It can be seen that the process time behavior varies from operation to operation and that different operators may behave differently at the same operation. This implies that the line balancing problem for manual lines must consider

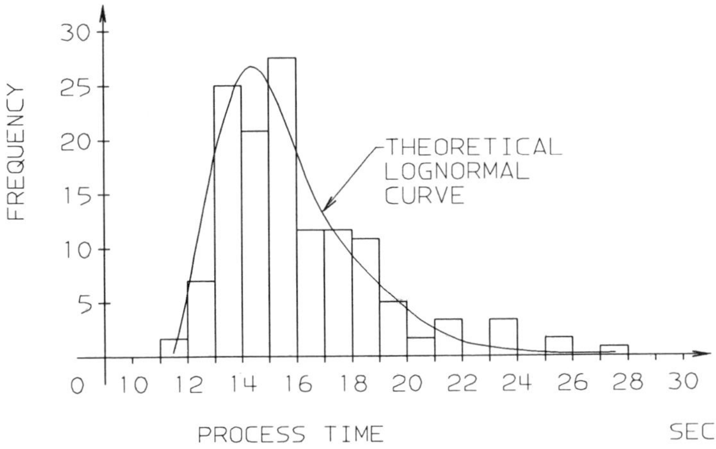

Figure 2.9 Manual process time—short cycle, low variability.

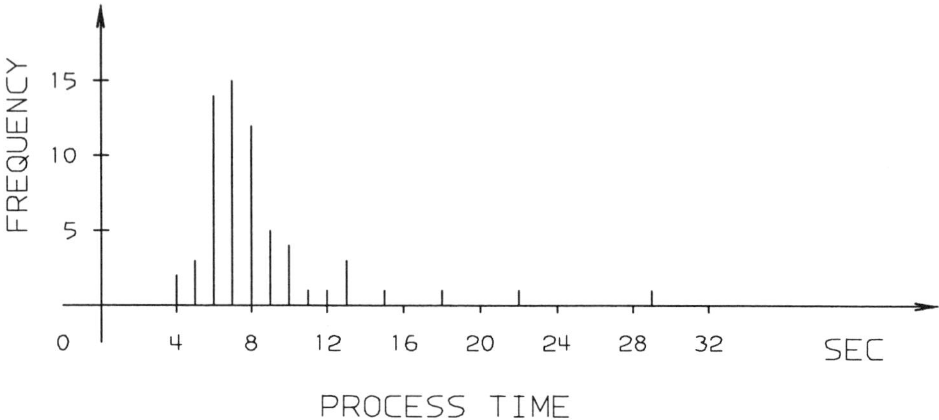

Figure 2.10 Manual process time—short cycle, high variability.

two things: (i) fair workload distribution so that each operation takes approximately the same amount of time, and (ii) good skill match between operators and jobs so that every operator works at approximately the same pace.

To facilitate manufacturing analysis, it would be very useful to construct a theoretical distribution for operation process time. Unfortunately, no single acceptable distribution can be found for all manual operations. Some possible candidates are the following:

1. *Gamma distribution*

   The gamma density function given in Section 2.2 cannot be directly used. Since the minimum value of a process time is always positive, a shifted distribution should be considered, i.e., the process time is $X = Y + s$, where s is the minimum process time and Y is a gamma random variable. The density function is given by

$$f(x) = \frac{\lambda}{\Gamma(n)} e^{-\lambda(x-s)} [\lambda(x-s)]^{n-1} \qquad \text{for } x \geq s \tag{2.22}$$

2. *Beta distribution*

   If Y and Z are independent (not necessarily identical) gamma random variables, then $Y/(Y+Z)$ has a beta distribution. Again a

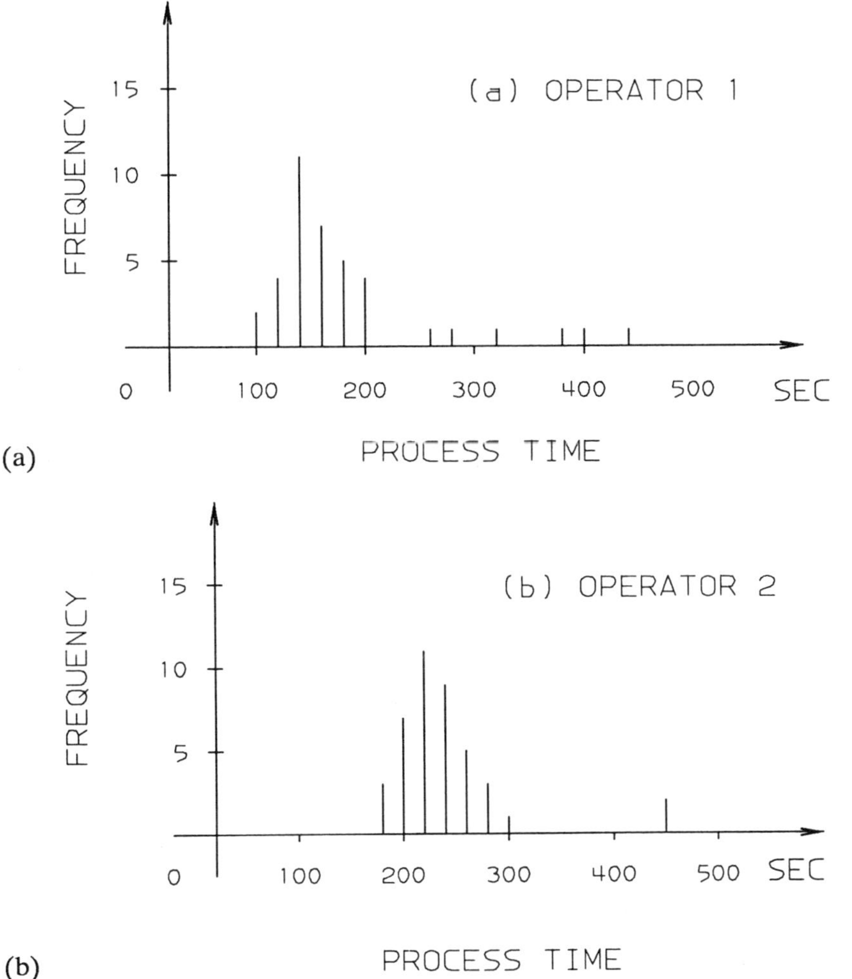

Figure 2.11 (a,b) Comparison of manual assembly times from two different operators.

shifted distribution should be considered:

$$f(x) = \frac{\Gamma(n+m+1)}{\Gamma(n)\Gamma(m)} \left(\frac{x-s}{t-s}\right)^m \left(\frac{t-x}{t-s}\right)^n \frac{1}{t-s} \qquad \text{for } s \le x \le t \tag{2.23}$$

3. *Weibull distribution*

If Y is exponentially distributed with a unit mean and $X = [(Y - \mu)/\sigma]^c$, then Y has a Weibull distribution:

$$f(x) = \frac{1}{\sigma}(\frac{x-\mu}{\sigma})^{c-1} \exp[-(\frac{x-\mu}{\sigma})^c] \qquad \text{for } x \geq \mu \tag{2.24}$$

4. *Log-normal distribution*

If $\log(X)$ is normally distributed, then X has a log-normal distribution:

$$f(x) = \frac{1}{(x-s)\sigma\sqrt{2\pi}} \exp\left[-\frac{(\log(x-s)-\mu)^2}{2\sigma^2}\right] \qquad \text{for } x \geq s \tag{2.25}$$

As a practical problem, the real process time distribution is unknown during the line design stage. If a line analysis is to be performed, engineers or line analysts should at least determine the average process time and its coefficient of variation. An estimated average time can be obtained by either simulating the physical motions or invoking a synthesis method by adding up predetermined standard times for individual motion elements. The coefficient of variation may be set to 0.15, 0.35 or 0.55, for examples, for different degrees of variation: low, medium and high, respectively. The variability of a manual assembly operation is a function of product design, assembly procedure, parts, tools and operator behavior. Grasping a tiny part from a flat surface, inserting a large object with two hands, positioning from a long distance are examples of operations with large variable process times.

Table 2.1 Statistics of Some Manual Assembly Times (Seconds)

Data Source	Mean	*CV*	*SK*	*KR*	Min	Max	$P[X \leq \text{Mean}]$
Figure 2.9	16.5	0.18	1.41	5.42	11	27	0.62
Figure 2.10	8.7	0.54	2.87	11.70	4	29	0.71
Figure 2.11(a)	179.4	0.45	1.85	5.61	87	439	0.72
Figure 2.11(b)	230.3	0.22	2.45	8.70	175	417	0.63

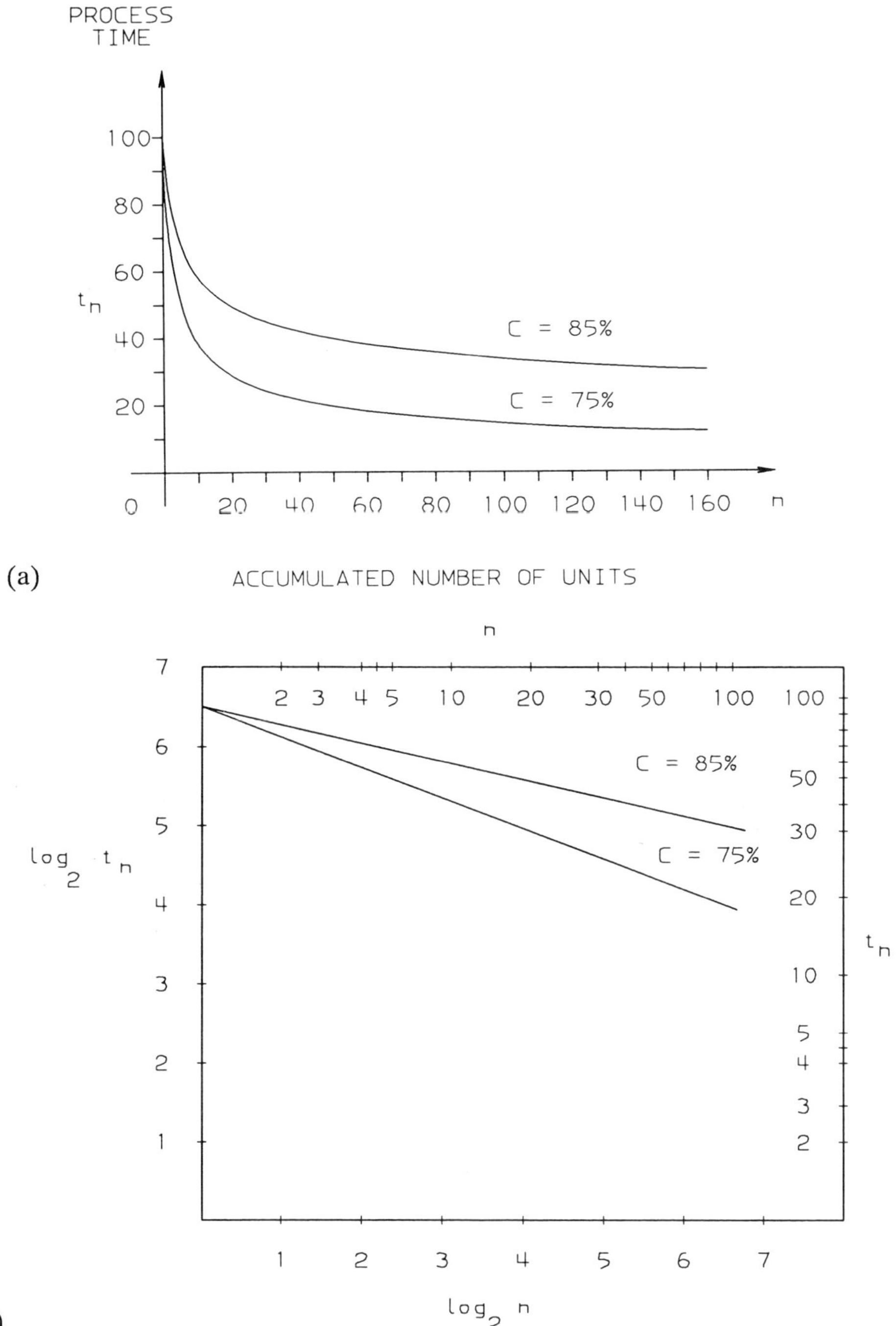

Figure 2.12 (a,b) Two learning models plotted in different scales.

Other important results in process time, as reported in literature, are summarized as follows:

Operators will perform differently on different operations, in terms of both speed and variability.
Different operators on the same job also perform differently.
There is a time period for which an operator can maintain concentration on the task at hand. After this period, process time and its variability will be increased.
Warm-up periods exist in many operations. Typically it takes an operator more than two hours to reach the maximum speed.
For a paced operation (i.e., constant feeding cycle), the quality of assembly work may deteriorate. The problem becomes more severe for shorter feed cycles.
Compared with an experienced operator, a trainee has a longer average process time with a more symmetric distribution. There is always a learning period for a new operator.

The learning speed of an operator is a function of working environment, task complexity, operator's motivation and experience. For simplicity, the learning period can be modeled by a power function:

$$t_n = t_1 n^a \tag{2.26}$$

where t_n is the process time for the nth unit.

The exponent, a, characterizes the learning capability and is a function of task complexity and operator's skill level. For convenience a can be set equal to $\log c/\log 2$, so that the ratio

$$t_{2k}/t_k = 2^{\log c/\log 2} = c \qquad \text{for all } k$$

If the number of units is doubled, the process time is reduced by a factor of c. For most manual tasks, the value of c ranges from 75 to 85%. This value can be regarded as the speed of learning. Figure 2.12 depicts two theoretical learning curves, with $c = 0.75$ and 0.85 respectively, plotted in both linear and logarithmic scales. It can be seen from Equation (2.26) that $\log t_n$ is a linear function of n. Therefore the learning model presents a straight line in logarithmic scale.

2.6 Yield Factor

Yield factor represents the proportion of produced units that meet the product specifications. A high yield often means low raw material consumption, small inventory, quick turnaround time, low labor hours, high productivity, and consequently low cost. Two major issues, particularly important in yield management, are rate of yield improvement and yield fluctuation.

When a manufacturing line begins to make a new product, the yield is relatively low due to lack of understanding of both product and manufacturing process. Poor knowledge in product performance leads to unrealistic product specifications and causes excessive rejects. On the other hand, improper process design and manufacturing procedure cannot meet product specifications. Modifications cause engineering changes and affect product yield. Like the operator's learning behavior, usually there exists a yield improvement period after the line has begun its operation. At first, yield can be improved very quickly by eliminating obvious problems. Then the remaining problems may not be trivial any more, and the rate of improvement decreases. Eventually improvement in product yield becomes very slow or even stops. At this stage, the manufacturing process reaches its ultimate yield. Figure 2.13 illustrates three yield improvement processes observed from an existing line over a period of 39 months. These processes represent cases of high, medium and low yield, respectively. Case 1 shows a good yield at the very beginning and has reached its ultimate yield after 15 months. The starting yield of case 3 is as low as 30–40%. It has been improved slowly. Case 2 is an intermediate case. All three yield processes become stable after 33 months.

The average yields for cases 1, 2 and 3 over the period of 39 months are 92.23%, 85.44% and 72.64%, respectively. This implies that the efficiencies of cases 1 and 2 are respectively about 27% (= 92.23/72.64 − 1) and 18% higher than that of case 3.

Yield improvement means cost reduction. Suppose that a piece of assembly work takes one hour. For an average yield of 80%, the average total amount of assembly time would be approximately increased by a factor of 1.25 (= 1/0.8). If the line throughput (i.e., the number of good products per day) is fixed, a low yield situation would require more work-in-process, more labor or machine time, more factory space, more material handling effort. If the defects are not reworkable, i.e., become scrap, additional raw material cost would be incurred. Consequently, a key problem in yield management is how to shorten the yield improvement period.

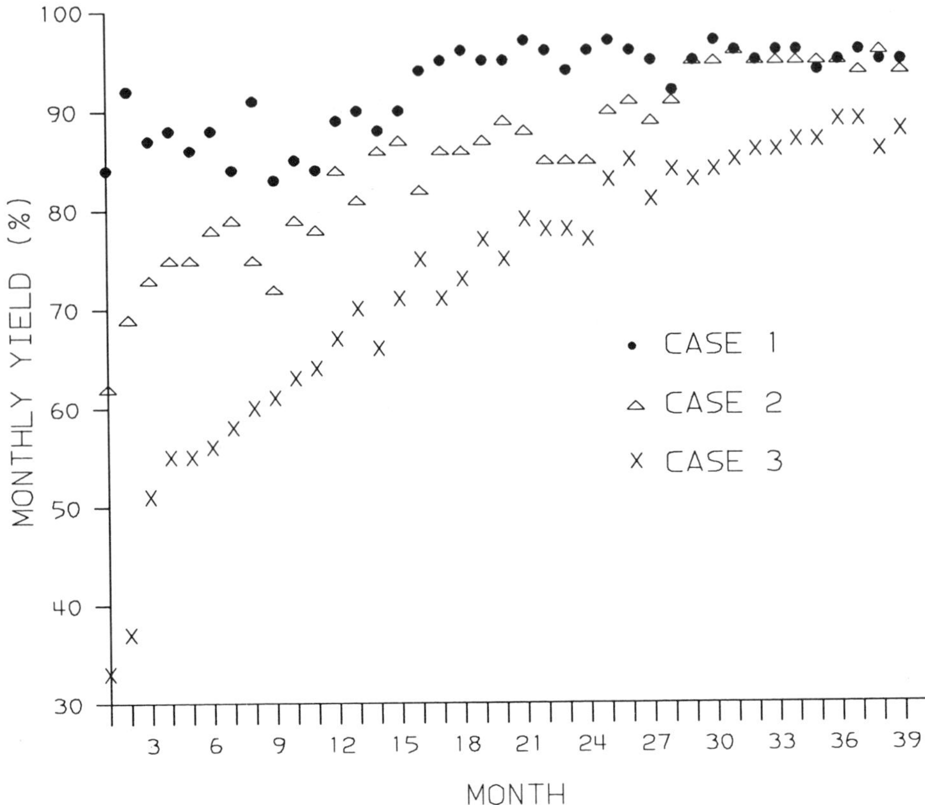

Figure 2.13 Yield improvement process.

From a resource planning point of view, on the other hand, it is important to characterize the process of yield improvement. Line capacity is a function of resource, productivity and yield. For a fixed line capacity, resource consumption is roughly inversely proportional to yield. Dynamic behavior of yield improvement may be understood by learning models similar to (2.26). Given yield data for an early production period, it is possible to construct a regression model to predict the yield improvement trend. Therefore, the resource requirement can be accurately estimated at a given production level. This subject will be treated in a subsequent chapter.

Since line throughput is directly related to yield, daily yield fluctuation may complicate line operation problems, such as line scheduling, work-in-

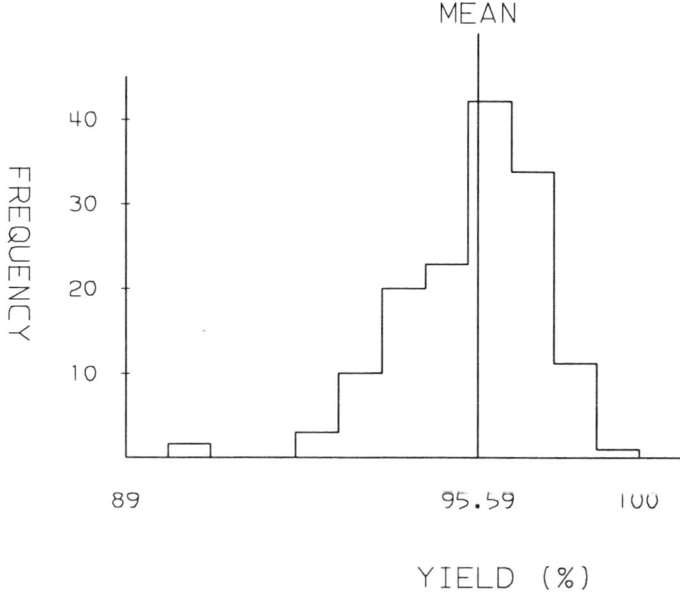

Figure 2.14 Daily Yield Distributions. (a) Case 1. (b) Case 2. (c) Case 3.

process management, parts inventory and quality control. The daily yield distributions of the previous three cases, after the manufacturing process has reached a stable condition, are shown by histograms in Figure 2.14. Computed statistics are given in Table 2.2.

All three distributions are slightly negatively skew. Case 3 distribution is flatter than the other two and has a relatively wider range. If the line is operating at its full capacity and yield is the same as shown by case 3, one may expect a span of 18% difference in daily production quantity.

The daily yield distributions may be modeled by beta distributions.

Table 2.2 Some Daily Yield Statistics

Case	Mean	*CV*	*SK*	*KR*	Min	Max	Range
1	0.9559	0.017	−1.01	3.38	0.89	0.99	0.10
2	0.9522	0.020	−0.41	4.94	0.89	1.00	0.11
3	0.8759	0.040	−0.18	2.92	0.78	0.96	0.18

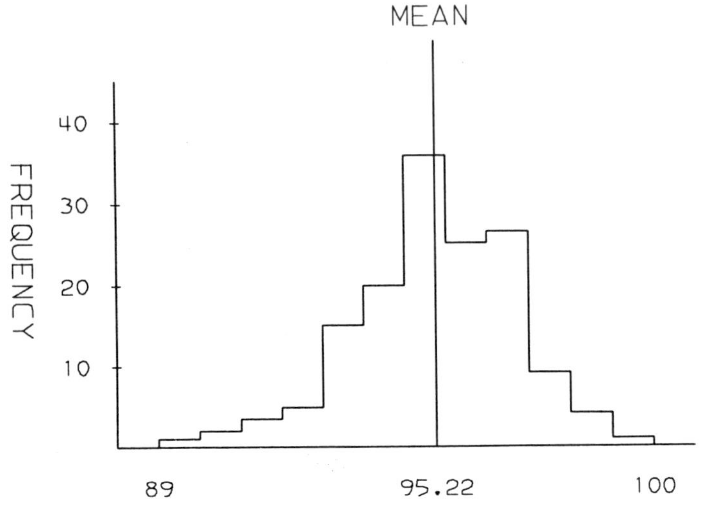

(b)

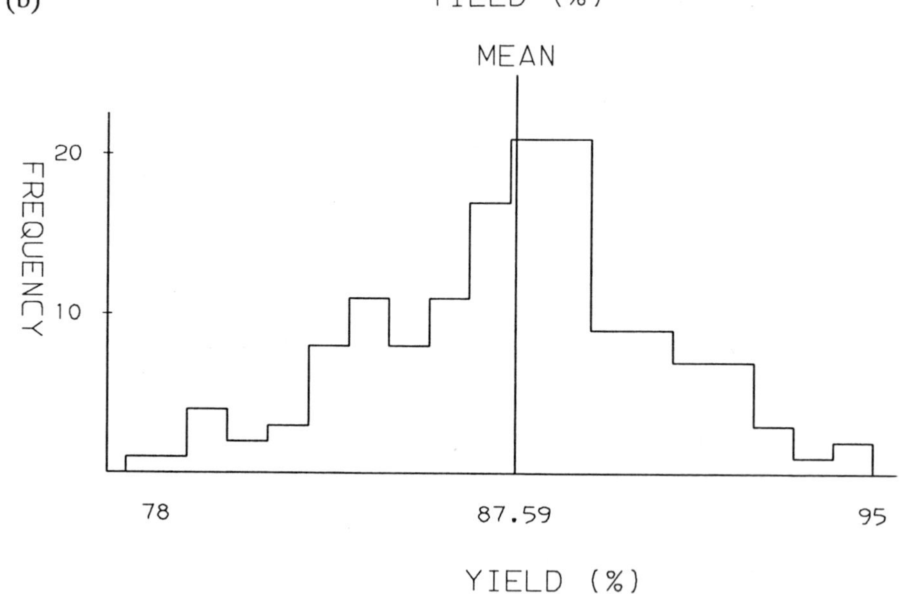

(c)

2.7 Equipment Reliability

Reliability of equipment is the proportion of time that the equipment is functioning. For a given period of time, reliability is estimated by taking the ratio of period uptime to the total period length. The status of a piece of equipment can be considered as an alternating process, composed of uptime and downtime (see Figure 2.15). A long run average reliability is equal to the mean time between failures divided by the sum of the mean time between failures and the average downtime per failure. To see this, let T be the observation period and N be the number of failures observed during T. By definition, reliability, r, is the ratio of total uptime during T to T. If N failures occur during T, then the mean time between failures and the average downtime per failure are respectively given by

$$\mathrm{MTBF} \approx \frac{\text{total uptime}}{N}$$

$$\mathrm{AVDT} \approx \frac{\text{total downtime}}{N}$$

Therefore, reliability can be characterized by

$$\begin{aligned} r &= \frac{\text{total uptime}/N}{(\text{total uptime} + \text{total downtime})/N} \\ &\approx \frac{\mathrm{MTBF}}{\mathrm{MTBF} + \mathrm{AVDT}} \end{aligned} \tag{2.27}$$

If failures occur randomly, the interfailure times should have an exponential distribution. Figure 2.16 presents histograms of interfailure times of four different types of equipment from an existing line during the period from June 1984 to May 1985. It can be seen that their shapes are similar to exponential tails. However, the sample moment ratios, i.e., CV, SK and KR, are much higher than those of the exponential case. Since higher moments

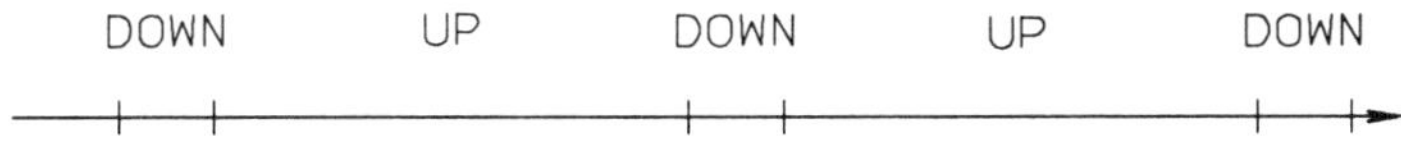

Figure 2.15 Equipment status.

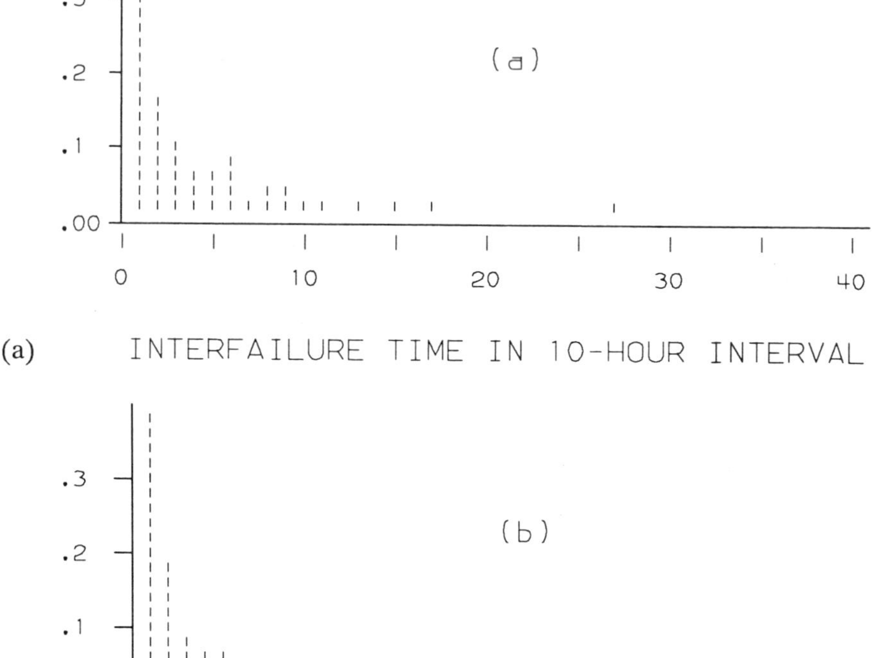

Figure 2.16 (a,b,c,d) Interfailure time distributions.

tend to be sensitive to large values in the sample, large values should be regarded as exceptional cases and eliminated from computation. Statistics shown in Table 2.3 are the computed results after having discarded values greater than a mean plus four standard deviations.

For a true exponential distribution, *CV*, *SK* and *KR* are 1, 2 and 9, respectively.

In reality, failure occurrences may deviate from a purely random pattern because of uneven equipment utilization, difference in machine quality, or an improper failure detection system. One way to derive a better model

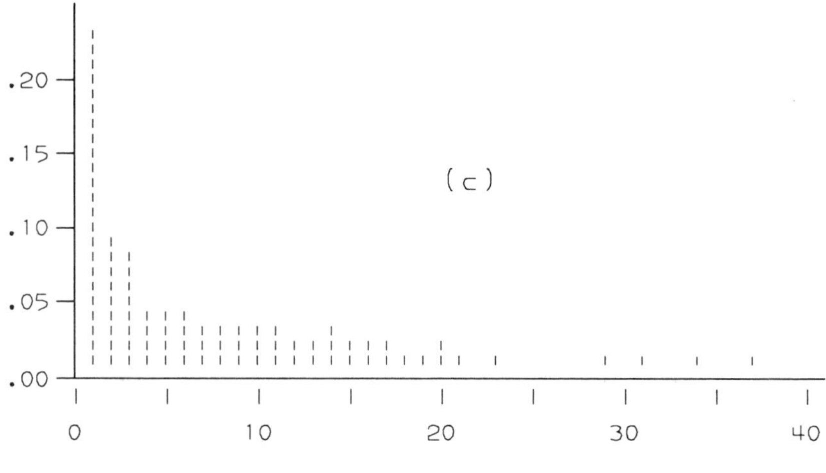

(c) INTERFAILURE TIME IN 10-HOUR INTERVAL

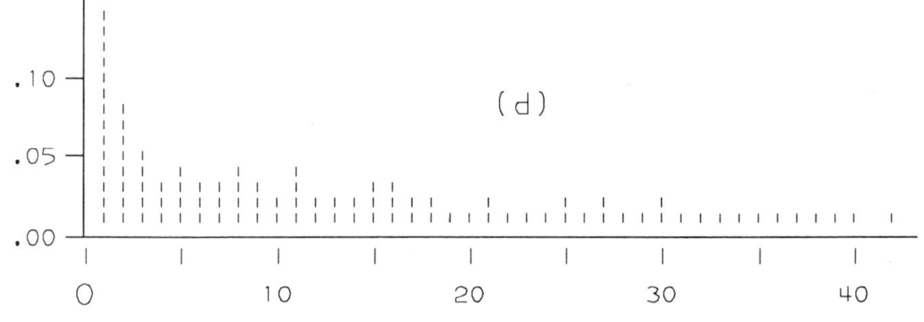

(d) INTERFAILURE TIME IN 10-HOUR INTERVAL

Table 2.3 Statistics of Some Interfailure Times (Hours)

Data Source	Mean	*CV*	*SK*	*KR*
Figure 2.16(a)	40.1	1.35	2.82	13.25
Figure 2.16(b)	35.1	1.30	2.20	8.55
Figure 2.16(c)	105.9	1.27	2.09	7.78
Figure 2.16(d)	158.7	1.12	1.78	6.48

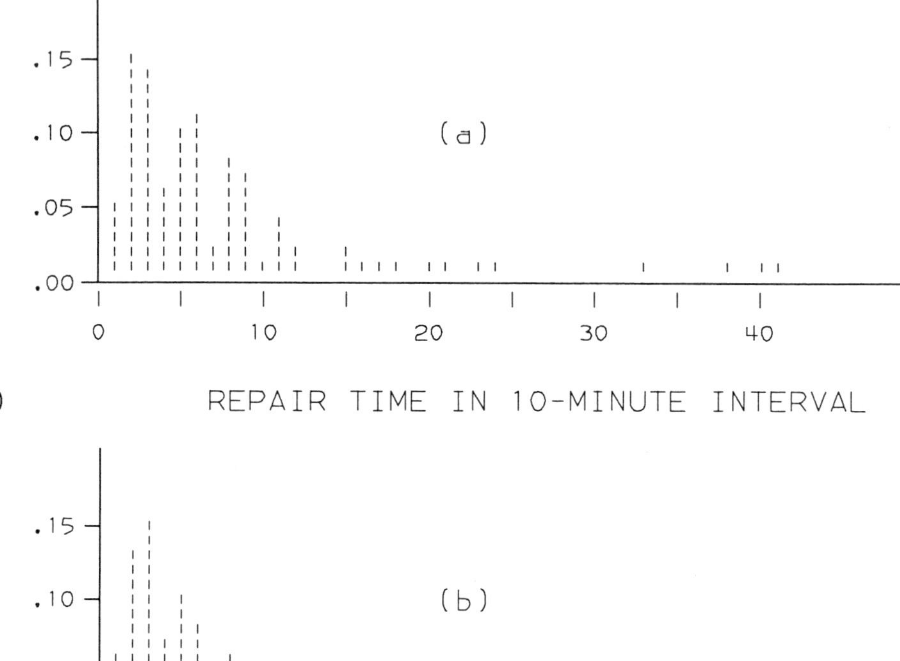

Figure 2.17 (a,b,c,d) Repair time distributions.

for interfailure time is to employ the concept of *compound distribution*. Assume that the interfailure time is exponential but the failure rate is subject to a gamma distribution. The compound distribution becomes a *Pareto distribution*. The distribution function is given by

$$F(t) = 1 - \left(\frac{\lambda}{\lambda + t}\right)^{c} \qquad \lambda, c, t > 0 \tag{2.28}$$

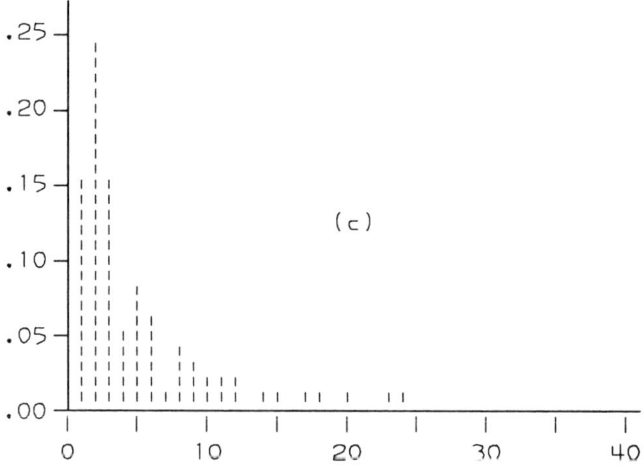

(c) REPAIR TIME IN 10-MINUTE INTERVAL

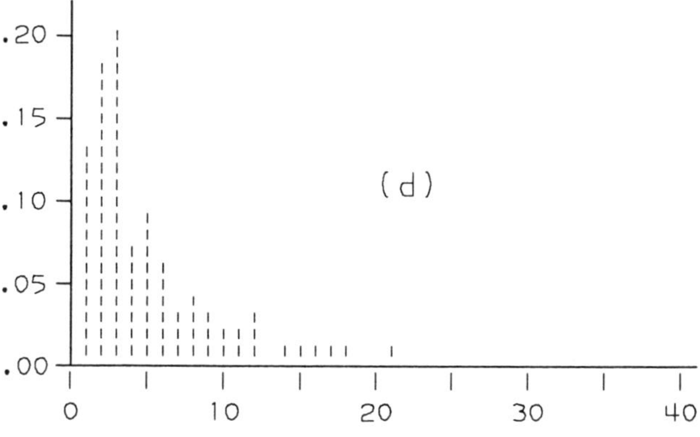

(d) REPAIR TIME IN 10-MINUTE INTERVAL

where both λ and c are parameters in the gamma distribution with a mean c/λ.

The mean and the coefficient of variation of the interfailure time are respectively

$$E[T] = \lambda/(c-1)$$

Table 2.4 Statistics of Some Repair Times (Hours)

Data Source	Mean	*CV*	*SK*	*KR*
Figure 2.17(a)	1.19	1.15	3.23	16.62
Figure 2.17(b)	1.33	1.08	2.33	9.27
Figure 2.17(c)	0.99	1.37	2.82	12.20
Figure 2.17(d)	0.91	1.29	3.60	21.57

$$CV[T] = \sqrt{c/(c-2)}$$

The value of λ and c can be estimated by sample moments. For example, the case associated with Figure 2.16(a) has $\lambda \approx 137.54$ and $c \approx 4.43$. The parameter c is a measure of dispersion and CV is decreasing in c. If $c \to \infty$, a Pareto distribution converges to an exponential. It is found that all four cases presented above can be fitted very well by Pareto distributions.

Equipment downtime consists of two parts: (i) waiting time for the repairman and (ii) the repair time itself. The waiting time is a function of the number of repairmen, the number of machines, the duration of repair time, availability of spare parts and the dispatching rule for repair jobs. It is closely related to line operation policy. From an analysis viewpoint, waiting time need not be treated as an independent variable. Instead, it may be used as a measure of maintenance efficiency. Repair time, on the other hand, is a characteristic parameter for both man and machine performance. Repair time often has large skewness and kurtosis indices. Repair time distributions that correspond to the previous cases are shown in Figure 2.17. The computed sample moments and moment ratios are given in Table 2.4.

The repair time distributions may be modeled by Weibull or log-normal distributions.

2.8 Remarks

The theoretical materials covered in this chapter will serve as a basis for later discussions, particularly in the field of line performance analysis. Further reading materials may be found in textbooks on probability theory, statistics and stochastic processes.

Many probability distributions can be derived from the six basic distributions discussed above. For relatively complete discussions about distribution types and their characteristics, the reader may consult the three volumes by Johnson and Kotz: Johnson (1969), Johnson (1970a) and Johnson (1970b).

Point estimation, confidence interval and regression analysis are classic techniques for statistical inference. Regression models will be used to predict learning and improvement processes. The concept of the confidence interval is a key subject in simulation output analysis.

When dynamic stochastic behavior is of concern, the Poisson process is often assumed because of its feasibility and simplicity. This process plays an important role in queuing analysis and simulation. Many important results from stochastic processes other than Poisson are beyond the scope of this book and have not been mentioned.

A major obstacle in line design is lack of input data for line analysis. For this reason, this chapter has attempted to provide some empirical results. Considered were three major line parameters: process time, yield and reliability. Discussions on manual process time and human behavior can be found in many published papers; see Dudley (1962), Murrell (1962), Dudley (1963), Das (1964), Sury (1964), Franks (1966), Franks (1969).

The learning model was first studied by Wright in 1936 (see Wright 1936). A number of papers have been published in this area, regarding model structures, concepts and applications. A review was given in Yelle (1979), where different models were presented and possible applications of learning models were discussed.

Since line characteristics vary from one to another, no single set of parameters can accurately reflect the most general conditions. A line designer should always try to collect his or her own information whenever possible. The reported data in this chapter should be used for reference, and may be employed if actual line information is not available.

REFERENCES

Das, B. (1964). A Statistical Investigation of Effect of Pace and Operation on Performance Rating, *International Journal of Production Research*, v. 3, pp. 65–72.

Dudley, N. A. (1962). The Effect of Pacing Worker Performance, *International Journal of Production Research*, v. 1, pp. 60–72.

Dudley, N. A. (1963). Work-time Distributions, *International Journal of Production Research*, v. 2, pp. 137–144.

Franks, I. T., G. J. Gillies, and R. J. Sury (1969). Buffer Stocks in Conveyor-based Work, *Work Study and Management Services*, v. 13, pp. 78–83.

Franks, I. T., and R. J. Sury (1966). The Performance of Operators in Conveyor-paced Work, *International Journal of Production Research*, v. 5, pp. 97–112.

Johnson, N. L., and S. Kotz (1969). *Discrete Distributions*, Houghton Mifflin, Boston.

Johnson, N. L., and S. Kotz (1970a). *Continuous Univariate Distributions—1*, Houghton Mifflin, Boston.

Johnson, N. L., and S. Kotz (1970b). *Continuous Univariate Distributions—2*, Houghton Mifflin, Boston.

Murrell, K. F. H. (1962). Operator Variability and its Industrial Consequences, *International Journal of Production Research*, v. 1, pp. 39–55.

Sury, R. J. (1964). An Industrial Study of Paced and Unpaced Operator Performance in a Single Stage Work Task, *International Journal of Production Research*, v. 3, pp. 91–102.

Wright, T. P. (1936). Factors Affecting the Cost of Airplanes, *Journal of Aeronautical Sciences*, v. 3, pp. 122–128.

Yelle, L. E. (1979). The Learning Curve: Historical Review and Comprehensive Survey, *Decision Science*, v. 10, pp. 302–327.

3

Queuing Analysis and Optimization Techniques

Although many useful techniques and principles have been developed in the past, line design is still a very complex problem. A line designer often faces a large number of design alternatives. Moreover, the relations between manufacturing components (e.g., workstation and material handling system, operation process time and learning speed, test/inspection and yield, etc.) may be different from one application to another. Past experience may not be directly transplanted into a new production line. Therefore evaluation of alternatives and trade-offs becomes inevitable. If the number of alternatives is enormous, it is too time-consuming to evaluate each individual solution. In this case, the designer's problem is to search for the optimal solution. Effective evaluation tools and searching methods may be derived from advanced mathematics, and this chapter gives a brief introduction to two of those: queuing analysis and optimization theory.

3.1 Basic Queuing Concepts

Queuing theory has many important applications in modern industry. It is a branch of mathematics dealing with systems that provide services to a well-defined population. Individuals receiving services are called *jobs* or *customers*. A service system may consist of one or more *servers*. A queuing system is defined by:

1. Job *arrival process*
2. Job *service time* (duration of service)
3. *Queuing discipline* (order of services)

Both arrival pattern and service demand can be characterized by probabilistic models. It is often assumed that an arrival stream follows a Poisson process, i.e., the interarrival times between successive arrivals are independent, identical exponential random variables. This assumption is popular because: (i) many real problems can be modeled by Poisson processes, and (ii) the memoryless property of exponential interarrival times simplifies mathematical analysis.

Although exponential service time makes our analysis even simpler, it may not always be a valid assumption. Fortunately, for single-server systems with general service time distributions, the queuing problems are still manageable. If the queuing system has a complex structure, such as multiple servers, general closed-form solutions do not exist. In such cases, approximation techniques may be used.

Two common queuing disciplines considered in manufacturing are *first-come, first-served* and *priority* (without preemption). In the former case, the order of service is identical to the order of arrival, while in the latter the job with the highest priority will be served first.

Suppose that jobs arrive at a work center at $0 = t_1 < t_2 < \cdots < t_n$ and are processed one at a time. The ith interarrival time is $T_i = t_{i+1} - t_i$. If the work center is idle at time 0 and the first job process time is S_1, then the center will be busy for S_1 time units. If $t_2 < S_1$ then the second job will experience a *delay*, $D_2 = S_1 - t_2$. If $t_2 \geq S_1$, then $D_2 = 0$. The sum of service time and delay is called *waiting time* and is denoted by $W_i = D_i + S_i$ for job i. Let t_i' be the departure time of job i. Then $t_i' = W_i + t_i$, or $W_i = t_i' - t_i$. A job's waiting time is the total amount of time the job spends in the work center. Figure 3.1 illustrates relations between arrival times, process (or service) times,

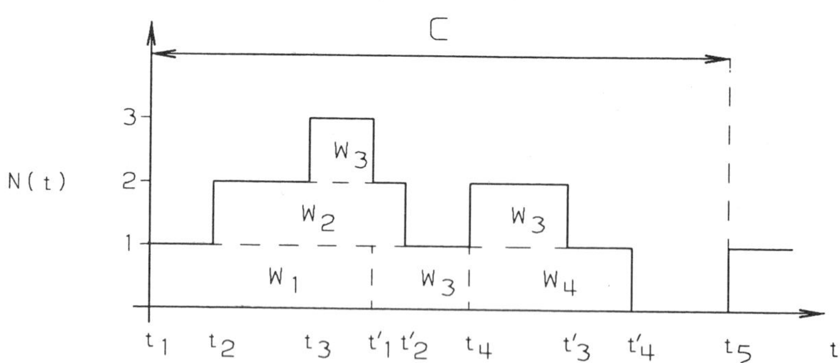

Figure 3.1 Queue length.

departure times, and waiting times. The vertical axis is the queue length or the total number of jobs in the work center, $N(t)$. $N(t)$ is a step function. Its value is increased by one at each arrival epoch and decreased by one at each departure epoch.

The system becomes busy when the first job arrives, and is idle when the last job in the queue departs. In the example illustrated in Figure 3.1, the interval between t_1 and t_4' constitutes a *busy period*. The interval between the starting times of two successive busy periods is called a *busy cycle*. It is not difficult to show that the area under the curve of $N(t)$ during a busy cycle is equal to the sum of waiting times of jobs served during the cycle. Note that job i is in the system at t if and only if t appears in the waiting time interval, W_i. Therefore $N(t)$ is simply the number of W's that contain t. From Figure 3.1, it can be seen that the area under $N(t)$ is equal to $W_1 + W_2 + W_3 + W_4$.

Let A be the area under $N(t)$, C be a busy cycle, and K be the number of jobs served during C. Since

$$A = \int_0^C N(t)dt$$

the average queue length during C is $L = A/C$. By virtue of

$$A = \sum_{i=1}^{K} W_i$$

the average waiting time becomes $W = A/K$. The ratio of $\lambda = K/C$ gives the average number of jobs served per unit of time or, equivalently, the system throughput. A relation between L, W and λ is described by the following theorem.

Theorem 3.1 Let W be the average waiting time (difference between a job's arrival and departure epochs) and λ be the throughput. If both W and λ are finite, then the average queue length (the number of jobs in the system) is also finite and $L = \lambda W$. □

This formula is sometimes called *Little's formula*.

Example 3.1 Assume that a line produces 150 products a day and the average lead time is a half-day. Then the average amount of work-in-process is $150 \times 0.5 = 75$. □

Example 3.2 A warehouse has processed 100,000 incoming items each month (λ). If the average number of items in the warehouse is estimated to be a quarter of a million (L), the inventory turnover rate is $1/W = 1/2.5$ times per month, or 4.8 times a year. □

Example 3.3 Consider a single-server system. The server is either busy (when there is a job on the server) or idle (when there is no job on the server). Then server utilization is $\rho = 1 \times P[\text{server busy}] + 0 \times P[\text{server idle}]$. Treat the server as an isolated queuing system. The queue length is the number of jobs on the server and the waiting time is the time spent for service. Therefore, $L = \rho$ and $W = E[S]$. Let λ be the throughput. Theorem 3.1 implies that $\rho = \lambda E[S]$, that is, *utilization is equal to the product of throughput and mean service time.* Note that ρ can be interpreted as utilization, or the proportion of time that the system is busy (i.e., a job is in service), or the probability that the system is busy. □

Example 3.4 In a queuing system, the job waiting time is composed of two parts: delay (waiting for service) and service time. Hence $L = \lambda W = \lambda(d + E[S])$. Using the results of Example 3.3, it follows that $L = \rho + \lambda d$. Since the utilization can be interpreted as the average number of jobs in service, $Q = \lambda d$ must be the average number of jobs waiting in the queue for their services. This relation is also a direct result from $L = \lambda W$ if the waiting queue is considered as an isolated queuing system. □

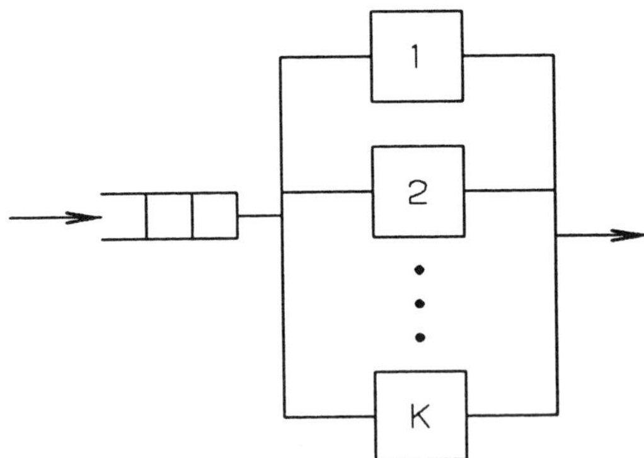

Figure 3.2 Multiple servers with a common queue.

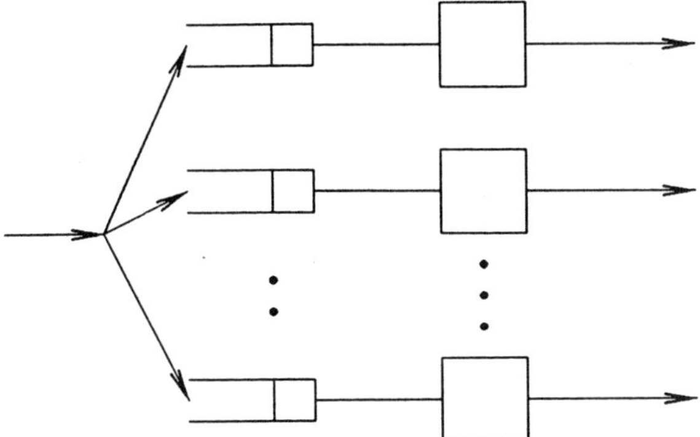

Figure 3.3 Multiple servers with dedicated queues.

Further discussion of queuing behavior under different assumptions will be given in the following sections. We shall consider systems of a single service center as well as networks of queues. For convenience, a queuing system with a single service center can be denoted by a triplet: $A/B/n$, where A is the arrival process, B the service time, and n the number of servers in the center. For example, an $M/M/k$ queue has an exponential interarrival time, an exponential service time, and k identical servers. In an $M/G/1$ queue, the service time has a general distribution. Figure 3.2 illustrates a queuing system with k parallel servers. Upon its arrival, a job has to wait in a common queue if all servers are busy. Otherwise, the job enters an idle server immediately. If each server has its own waiting line, then the system would behave as if there were k independent single-server queues, as illustrated in Figure 3.3.

3.2 $M/M/k$ Queue

Assume that X is an exponential random variable. Given that $X > t$, the conditional probability that $X > t + s$ is equal to the unconditional probability that $X > s$. This is the memoryless property introduced in Chapter 2. If a queuing system has a Poisson arrival process (i.e., exponential interarrival time), then at any given epoch the elapsed time until the next arrival is exponential, independent of the process history, and identical to an ordinary interarrival time. For this reason, problem complexity is significantly reduced by a Poisson arrival assumption. The problem is further simplified if the service time too has an exponential distribution. In this case any job in service would behave as if its service had just begun.

$M/M/1$ Systems

When a single-server queue has a Poisson arrival stream and exponential service time, the solution is rather simple. The method adopted in Example 3.3 can be used to obtain the queue length distribution. Let N be the queue length (the number of jobs in the system) and $P_i = P[N = i]$. The state in which the queue length is zero is equivalent to that in which the server is idle. Hence

$$P_0 = 1 - \rho$$

Designate each position of the queue by 0, 1, 2, ..., where the server has position 0, the first job waiting for service is located at position 1, the second at position 2, and so on. (See Figure 3.4.) An empty system means none of the positions is occupied. A single job in the system implies that the job must be in service, i.e., at position 0. In the case of two jobs, they will be located at positions 0 and 1, respectively. An arrival must go through all the positions before its departure. However, the time spent at each position is dependent on the queue length. Let N_i be the number of jobs at position i. $N_1 = 1$ if and only if there are two or more jobs in the system. Consequently,

$$\begin{aligned} E[N_1] &= 1 \times P[N \geq 2] + 0 \times P[N \leq 1] \\ &= P_2 + P_3 + \cdots \\ &= 1 - P_0 - P_1 \end{aligned}$$

The amount of time a job spends at position 1 is dependent on whether or not the server is busy. If an arrival finds an empty system, the job will enter the server immediately. Therefore its waiting time at position 1 is zero. If the arrival finds a queue length of 1, then the waiting time at position 1 is equal to the remaining service time of the job in service. Since the service time is exponential, the remaining service time is identical to an ordinary

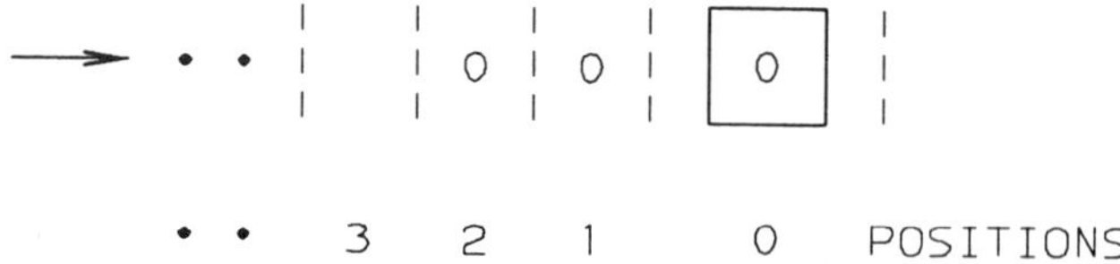

Figure 3.4 Positions in queue.

service duration. The waiting time at position 1 has a mean $E[S]$. If the queue length is greater than 1, the new job must wait for its turn to occupy the position. This will happen when the number of old jobs is reduced to 1. After having occupied position 1, the job will spend a complete service cycle there. Therefore the average waiting time at position 1 is

$$\begin{aligned} W_1 &= E[S] \times P[N > 0] + 0 \times P[N = 0] \\ &= E[S] \times \rho \end{aligned}$$

Since every job goes through position 1, the throughput at position 1 is identical to the system throughput, λ. Applying Little's formula, $E[N_1] = \lambda W_1$, it follows that

$$-P_0 - P_1 = \rho\lambda E[S]$$

Since $1 - P_0 = \rho$ and $\rho = \lambda E[S]$, the above equation can be simplified to $P_1 = \rho(1-\rho)$.

Applying the same approach to positions 2, 3, ..., in turn, the general solution for the queue length distribution is given by

$$P_i = (1-\rho)\rho^i \qquad i = 0, 1, \ldots \tag{3.1}$$

The above equation defines a geometric mass function with a mean

$$L = \sum_{i=1}^{\infty} iP_i = \frac{\rho}{1-\rho}$$

Since a Poisson arrival stream is randomly distributed over the time, P_i can be interpreted as one of the following equivalent statements:

1. Probability that an arrival finds i jobs in the system
2. Probability of i jobs in the system at a random time epoch
3. Proportion of time that the system has i jobs

Given that an arrival finds i jobs in the system, its waiting time is the sum of (i) the remaining service time of the job in service upon its arrival, (ii) total service time of $(i-1)$ jobs waiting in the queue, and (iii) its own service time. Due to memoryless property, the remaining service time is identical to an ordinary service time. Therefore the conditional waiting time is the sum of $(i + 1)$ independent, identical exponential random variables or, equivalently, an Erlang random variable (see Section 2.2). The density function of the waiting time is given by

$$\begin{aligned} f(w) &= \sum_{i=1}^{\infty}(1-\rho)\rho^i \frac{\mu}{\Gamma(i+1)}(\mu w)^{i+1}e^{-\mu w} \\ &= (\mu-\lambda)e^{-(\mu-\lambda)w} \end{aligned} \tag{3.2}$$

where μ is the reciprocal of $E[S]$ and is said to be the *service rate*.

Thus the waiting time of an $M/M/1$ queue is exponentially distributed with a mean $W = 1/(\mu-\lambda)$. Since $\rho = \lambda/\mu$, $L = \rho/(1-\rho) = \lambda/(\mu-\lambda) = \lambda W$. Once again Little's formula is verified.

Example 3.5 Consider an inventory problem at a storage area between two manufacturing departments, A and B. Products are released from Department A to the storage area and then moved to Department B. Assume that the interdeparture time from A is exponentially distributed with a mean $1/\lambda$ and that the demand interval at B also has an exponential distribution with a mean $1/\mu$. Then the inventory level at the storage area can be characterized by an $M/M/1$ queuing system. The inventory level, N, has a geometric distribution given by (3.1) with $\rho = \lambda/\mu$. If $\lambda = 8$ per hour and $\mu = 10$ per hour, then the average inventory level is $0.8/(1-0.8) = 4$. The average turnaround time is the average amount of time residing at the storage area and is given by $1/(10-8) = 0.5$ (hours). □

In the above example, if the production rate at Department A is greater than the demand rate, then the arrival flow into the storage area is larger than the departure flow. The inventory level will grow indefinitely. The turnaround time becomes infinitely long. This is equivalent to saying that an $M/M/1$ queue does not have a steady-state queue length distribution (or no steady-state condition exists) if the arrival rate exceeds the service rate.

If $\lambda = \mu$, it can be seen from (3.1) that P_i vanishes. This means that the queue length may take any nonnegative integer value and, consequently, the proportion of time that the $M/M/1$ queuing system spends in any given state is zero. Neither the average queue length nor the average waiting time exists. Since the P_0 also becomes zero, the server has a 100% utilization. In fact, it can be shown that a necessary and sufficient condition for a steady state is $\rho = \lambda/\mu < 1$.

Balance Equations

Equation (3.1) implies that for $i \geq 0$, $P_{i+1} = \rho P_i$ or equivalently

$$\lambda P_i = \mu P_{i+1} \tag{3.3}$$

Let the queue length be the state of the queuing system, i.e., state i means a queue length of i. The system state may be changed due to either an arrival or a departure. Given that the system is at state i, the rate of leaving this state is λ if $i = 0$, and $\lambda + \mu$ if $i > 0$. Since the proportion of time that the system is in state i is P_i, in a long run the system's departure rate from state i is λP_0 if $i = 0$ and $(\lambda + \mu)P_i$ if $i > 0$. By virtue of (3.3),

$$\begin{aligned} \lambda P_0 &= \mu P_1 & i &= 0 \\ (\lambda + \mu)P_i &= \lambda P_{i-1} + \mu P_{i+1} & i &> 0 \end{aligned} \tag{3.4}$$

Note that the right hand side of the equal sign is the rate of entering state i. If $i = 0$, the only possible way of entering this state is from state 1. If $i > 0$, then state i is reachable from either $(i-1)$ or $(i+1)$. Equation (3.4) is known as the *balance equation*. When the system is in an equilibrium condition, the rate of leaving a state is equal to the rate of entering the state. This relation is often used to analyze queuing systems with exponential interarrival time or exponential service time. Sometimes, relation (3.3) is referred to as the *local balance* (the rate from i to $i+1$ is equal to the rate from $i+1$ to i), and (3.4) as the *global balance*. In general, (3.3) is a sufficient condition of (3.4).

$M/M/k$ Systems

If there are k identical servers in the system, the system departure rate is a function of the number of busy servers. Suppose that the queue length is n. The number of busy servers $K = k$ if $n > k$, and $K = n$ if $n \leq k$. The memoryless property implies that the time until the next departure is the minimum of K independent identical exponential random variables. It can be shown that this minimum is also exponential. To see this, let X and Y be two independent exponential random variables with rates λ and μ

respectively.

$$
\begin{aligned}
P[\min(X,Y)>t] &= P[X>t, Y>t] \\
&= P[X>t]P[Y>t] \qquad \text{(by independence)} \\
&= e^{-\lambda t}e^{-\mu t} \\
&= e^{-(\lambda+\mu)t}
\end{aligned}
$$

Thus $\min(X,Y)$ is exponentially distributed with a rate $(\lambda + \mu)$. Thus the minimum of K independent, identical exponential variables is also exponential with a rate $K\mu$. An $M/M/k$ system would behave as if there were a single server with a queue dependent service rate, $K\mu$. According to the balance equations,

$$\lambda P_i = \min(i+1,k)\mu P_{i+1} \qquad \text{for all } i$$

It follows that the queue length distribution is given by

$$
\begin{aligned}
P_0 &= \left[\sum_{i=0}^{k-1}\frac{m^i}{i!} + \frac{m^k}{k!(1-m/k)}\right]^{-1} \\
P_i &= \frac{m^i}{i!}P_0 \qquad i \le k \\
P_i &= P_k\left(\frac{m}{k}\right)^{i-k} \qquad i > k
\end{aligned} \tag{3.5}
$$

where $m = \lambda/\mu$

M/M/∞ Systems

It is easy to show by using (3.5) that in an $M/M/\infty$ system (i.e., a system with infinitely many servers) the queue length has a Poisson distribution,

$$P_i = e^{-m}\frac{m^i}{i!} \qquad i = 0,1,\ldots \tag{3.6}$$

3.3 $M/G/k$ Queue

An $M/G/k$ queue has a Poisson arrival stream but a general service time distribution. As discussed in Section 2.4, a Poisson arrival process is often a reasonable assumption for practical problems. However, the assumption of exponential service time in many cases may not be valid at all. For this reason, an $M/G/k$ model has much broader applications than does an $M/M/k$ model. Unfortunately, the exact solution of an $M/G/k$ queue is unknown, except when $k = 1$ or ∞. A pragmatic approach is to find an approximate solution. This can be done by first solving an $M/G/1$ queue and an $M/G/\infty$ queue. Then an approximate solution of an $M/G/k$ $(1<k<\infty)$ queue can be obtained as an intermediate case. Our discussion begins with single-server queues.

$M/G/1$ Systems

The solution of an $M/G/1$ queue is dependent on the service time distribution and usually has a complex algebraic expression. A numerical solution, however, can be obtained by solving a set of linear equations similar to the balance equations. If only the first moment is of interest, the results can be rather simple. This is done by using the concept of *workload conservation*. The total workload presented at the system is the amount of time that the server must spend to complete all the existing jobs in the system. By definition, the total workload in a system is independent of the order of services. The workload at time t is also referred to as *virtual delay*. (This would be the actual delay experienced by a job, if the job arrived at the very time epoch t.)

Suppose that a system operating cost is charged at a rate equal to its workload. If the workload at time t is $w(t)$, the cost rate at t is also $w(t)$ per unit of time. Assume that a job has a service time S. When the job arrives at the system, the total workload is increased by S. Therefore the cost rate is also increased by S. The total cost contributed by this job can be partitioned into two parts: one is the cost incurred during its delay (the job waiting for its service) and the other is during its service time. Let D be its delay. Since the associated cost rate is S, its contribution during a delay period is $D \times S$. During its service, the workload imposed by the job is constantly decreasing. The associated cost rate is equal to the job's remaining service time. When the service is completed, the cost rate becomes zero. The average cost rate during its service is $S/2$, the contribution is $S^2/2$. Hence the total cost contribution by the job is

$$H = DS + \frac{1}{2}S^2 \tag{3.7}$$

Suppose that the number of arrivals by time t is $N(t)$. The average cumulative cost in $(0,t)$ is $E[N(t)]E[H]$. This quantity divided by t gives the average cost during $(0,t)$. For a large t, $E[N(t)]/t$ is arbitrarily close to the arrival rate, λ. Hence a long-run average cost is $w = E[N(t)]E[H]/t = \lambda E[H]$. By virtue of (3.7) and the fact that D and S are independent, we have

$$w = \lambda E[D]E[S] + \lambda E[S^2]/2 \tag{3.8}$$

Since the cost rate is equal to workload, w can be interpreted as the long run average workload in the system or, equivalently, the average workload observed at a random epoch. For a Poisson arrival stream, the arrival times are randomly distributed along the time axis. Therefore w is also the average workload seen by an arrival. Under the first-come, first-served discipline, this is exactly the delay experienced by an arrival, i.e., $w = E[D]$. Solving equation (3.8) with w replaced by $E[D]$, it follows that

$$E[D] = \frac{\lambda E[S^2]}{2(1-\rho)} \tag{3.9}$$

where $\rho = \lambda E[S]$ is the utilization factor.

The average waiting time is the sum of $E[S]$ and $E[D]$, and the average queue length can be obtained by Little's formula. Using the definition $CV = SD[S]/E[S]$, these two quantities can be written as

$$W = E[S] + \frac{\rho E[S]}{2(1-\rho)}(1 + CV^2) \tag{3.10}$$

$$L = \rho + \frac{\rho^2}{2(1-\rho)}(1 + CV^2) \tag{3.11}$$

It can be seen that both W and L are dependent on CV. For an exponential service time, $CV = 1$ and $L = \rho/(1-\rho)$. If $CV = 0$ (the system is denoted by $M/D/1$ and has a deterministic service time), $L = \rho + \rho^2/2(1-\rho)$.

Example 3.6 An automatic guided vehicle (AGV) serves a production line composed of 10 assembly operations. Service requests may be generated by any one of the workstations nearly independently. Consequently, the arrival stream of requests, as seen by the AGV, approximately follows a Poisson process. The line is capable of producing 30 products in each 8-hour shift. Since there are 10 operations, the number of service requests associated with a product is 9. Hence the service request rate is $(30\times 9)/(8\times 60) = 0.5625$ per minute. Based on the service pattern and the AGV configuration, the mean service time and its coefficient of variation are estimated to be 1 minute and 0.4, respectively. The average material handling delay computed by (3.10) is $W = 1 + [(0.5625)^2/2(1-0.5625)](1+0.16) = 1.42$ minutes. The average number of incomplete requests is $L = \lambda W = 0.5625 \times 1.42 = 0.80$. □

It can be seen from (3.10) and (3.11) that, for a given arrival rate, both the average waiting time and the average queue length are uniquely determined by the first two moments of the service time. For many real-life problems, the distribution of service time may not be available. However, line managers or engineers often keep good records of the mean and the standard deviation. Therefore it is desirable to characterize the queuing behavior by using the first two moments only. A simple way to obtain an approximate solution for an $M/G/1$ queue is by the combination of an $M/M/1$ queue and the $M/D/1$ queue. Let L_M and L_D be the mean queue lengths of an $M/M/1$ queue and an $M/D/1$ queue, respectively. By virtue of (3.11) it follows that the mean queue length of an $M/G/1$ queue is a linear combination of L_M and L_D,

$$L_G = (CV^2)L_M + (1-CV^2)L_D$$

Let $P_M(i)$, $P_D(i)$ and $P_G(i)$ be the probabilities of i jobs in the system under an $M/M/1$, an $M/D/1$ and an $M/G/1$ queue, respectively. Using the same approach, an approximate queue length distribution for an $M/G/1$ queue is given by

$$P_G(i) = (CV^2)P_M(i) + (1-CV^2)P_D(i) \qquad \text{for } i = 0,1,\ldots \tag{3.12}$$

The distribution for an $M/M/1$ queue is given by (3.1), that is,

$$P_M(i) = (1-\rho)\rho^i$$

It is also possible to derive a closed-form queue length distribution for an $M/D/1$ queue. The following result is given without proof:

$$P_D(i) = (1-\rho)\sum_{j=0}^{i} \frac{(-j\rho)^{i-j}}{(i-j)!}(1+\frac{i-j}{j\rho})e^{j\rho} \tag{3.13}$$

Both P_D and P_M are functions of ρ while the second moment of the service time is introduced by CV in (3.12). Thus, based on the first two moments of the service time, an approximate queue length distribution is obtained. In particular, for $CV \le 1$ (3.12) results from interpolation and usually gives a good approximation. Tables 3.1, 3.2 and 3.3 compare the exact queue length distributions with the approximation results by (3.12) for $\rho = 0.1$, 0.5 and 0.9, respectively. Service times of multiple-stage Erlang distributions are assumed (see equation 2.7). Therefore the coefficient of variation is between $(0,1)$. The value of CV is 0.7071 for the 2-stage case, 0.5 for 4-stage, and 0.3536 for 8-stage. The range of these values covers most cases found in real life. (See the empirical data discussed in Chapter 2.) The tails of queue length distributions presented in Tables 3.1, 3.2 and 3.3 are truncated after the point where the mass function is less than 0.001. It can be seen that the difference between the exact and the approximate is small.

$M/G/\infty$ Systems

If there are infinitely many servers, the queue length distribution is Poisson. To see this, let's consider an $M/D/\infty$ queue, in which the job service is deterministic, i.e., $S = s$ is a constant. Suppose that a job arrives at time t. Since there are infinitely many servers, the job will start its service instantaneously upon its arrival and will stay in the system during the interval $(t, t+s)$. Consequently, the number of jobs in the system at a

Table 3.1 $M/G/1$ Queue Length Distributions with $\rho = 0.1$

Queue length	2-stage Erlang		4-stage Erlang		8-stage Erlang	
i	Exact	Approx.	Exact	Approx.	Exact	Approx.
0	.900000	.900000	.900000	.900000	.900000	.900000
1	.092250	.091363	.093432	.092327	.094038	.093008
2	.007206	.007870	.006211	.007072	.005687	.006507

Table 3.2 $M/G/1$ Queue Length Distributions with $\rho = 0.5$

Queue length	2-stage Erlang		4-stage Erlang		8-stage Erlang	
i	Exact	Approx.	Exact	Approx.	Exact	Approx.
0	.500000	.500000	.500000	.500000	.500000	.500000
1	.281250	.287181	.300903	.305771	.312085	.315063
2	.126953	.123800	.126032	.123200	.124722	.122900
3	.053833	.050144	.046988	.043966	.042717	.040878
4	.022347	.021079	.016850	.015994	.013906	.013452
5	.009205	.009366	.005970	.006237	.004466	.004672
6	.003781	.004348	.002108	.002616	.001430	.001750

Table 3.3 $M/G/1$ Queue Length Distributions with $\rho = 0.9$

Queue length	2-stage Erlang		4-stage Erlang		8-stage Erlang	
i	Exact	Approx.	Exact	Approx.	Exact	Approx.
0	.100000	.100000	.100000	.100000	.100000	.100000
1	.110250	.117981	.125188	.131970	.134639	.138963
2	.101301	.109321	.116463	.123480	.126095	.130558
3	.089358	.093975	.100726	.104513	.107480	.109780
4	.078004	.079707	.085623	.086756	.089703	.090279
5	.067905	.067652	.072522	.071954	.074599	.074104
6	.059069	.057565	.061385	.059775	.062011	.060880
7	.051373	.048108	.051935	.049748	.051547	.050067
8	.044677	.042003	.043969	.041482	.042849	.041221
9	.038854	.036019	.037213	.034658	.035618	.033977
10	.033789	.030967	.031495	.029016	.029608	.028041
12	.025554	.023064	.022559	.020475	.020459	.019181
14	.019326	.017348	.016159	.014583	.014137	.013202
16	.014616	.013170	.011574	.010490	.009768	.009150
18	.011054	.010085	.008291	.007623	.006750	.006392
20	.008360	.007784	.005938	.005597	.004664	.004504
25	.004158	.004195	.002579	.002703	.001851	.001957
30	.002068	.002335	.001120	.001382	.000735	.000906

random epoch x is equivalent to the number of arrivals during $(x-s,x)$. For a Poisson arrival process with a rate λ, the queue length distribution is given by

$$P_i = e^{-\lambda s}\frac{(\lambda s)^i}{i!}$$

To analyze a practical problem, an arbitrary distribution can be regarded as a composite distribution of deterministic distributions: the service time may take a value of $s_1, s_2, \ldots.$ with probabilities $p_1, p_2, \ldots$, respectively. Now split the arrival process into a number of subprocesses such that arrivals from a subprocess have the same service duration, e.g., s_i. The probability that an arrival belongs to the subprocess with a service time s_i is p_i. It is possible to show that each subprocess is also a Poisson process with a *thinner* arrival rate, λp_i. (Intuitively, one can treat the subprocess as a result from random selections from a sequence of random arrivals. Consequently the arrival epochs of the subprocess are also randomly distributed over the time. This implies a Poisson arrival stream.) Since the number of servers is infinitely many, contention for servers does not exist. Therefore, each subprocess has its own queue, independent of other subprocesses; that is, for each s_i, there is an $M/D/\infty$ queue with an arrival rate λp_i. For each queue, the queue length distribution is Poisson. The queue length of the original $M/G/\infty$ queue is the sum of a number of queues generated by subprocesses. Since the sum of Poisson variables is also Poisson, the queue length distribution of an $M/G/\infty$ queue is given by

$$P_i = e^{-m}\frac{m^i}{i!} \qquad i = 0,1,\ldots \tag{3.14}$$

where $m = \lambda E[S]$.

The parameter m is the average queue length. It is also the expected number of busy servers. Equation (3.14) is identical to (3.6).

Example 3.7 A production line in an electronics company has an oven cure operation, which carries work-in-process through an oven tunnel by a conveyor device. The total cure time, or equivalently the travel time on the conveyor, is 2 minutes. Products arriving from upstream on the line form a Poisson process with a rate 3.5 per minute. Since the conveyor is constantly moving, a product enters the oven tunnel instantaneously upon its arrival. This situation can be modeled by an $M/D/\infty$ system. The total number of units in the tunnel has a Poisson distribution with a mean equal to $L = \lambda E[S] = 3.5 \times 2 = 7$. □

$M/G/k$ Systems

An approximate solution for an $M/G/k$ queue can be found by looking at two extreme cases. If the system has a long queue so that all servers are busy, the departure process can be considered as a superposed process of k independent processes, each having an interdeparture time identical to the service time. In this case, the queuing system would temporarily behave as if there were a single server with an average service time $E[S]/k$. The k jobs in service should be treated as a single equivalent job. On the other hand, if the servers are not all busy for a while, every arrival will enter its service instantaneously. The system would temporarily look like an $M/G/\infty$ queue. Let $p(i;k)$ be the probability of the queue length equal to i in an $M/G/k$ queue. Then for some a and b,

$$p(i;k) \approx \begin{cases} ap(i;\infty) & i<k \\ bp(i-k+1;1) & i \geq k \end{cases} \tag{3.15}$$

The mean service time for the queue described by $p(i;1)$ is $E[S]/k$ and by $p(i;\infty)$ is $E[S]$. The expected number of busy servers, according to Little's formula, is $m = \lambda/\mu$. By definition,

$$m = \sum_{i=0}^{k-1} ip(i;k) + \sum_{i=k}^{\infty} kp(i;k)$$

Since $\sum_{i=0}^{\infty} p(i;k) = 1$, it follows from (3.15) that

$$b\sum_{i=k}^{\infty} p(i-k+1;1) \approx 1 - a\sum_{i=0}^{k-1} p(i;\infty)$$

Hence

$$m \approx a\sum_{i=0}^{k-1} ip(i;\infty) + k\left[1 - a\sum_{i=0}^{k-1} p(i;\infty)\right]$$

Solving for a and using (3.14) for $M/G/\infty$ queues, we have

$$a=(k-m)[\sum_{i=0}^{k-1}(k-i)e^{-m}\frac{m^i}{i!}]^{-1} \tag{3.16}$$

The value of b may be obtained from the following relation:

$$\begin{aligned}\sum_{i=k}^{\infty}p(i;k) &\approx 1-a\sum_{i=0}^{k-1}p(i;\infty)\\ &\approx b\sum_{i=k}^{\infty}p(i-k+1;1)\\ &=b[1-p(0;1)]\\ &=bm/k\end{aligned}$$

Therefore

$$b=(k/m)[1-a\sum_{i=0}^{k-1}e^{-m}\frac{m^i}{i!}] \tag{3.17}$$

An approximate solution of an $M/G/k$ queue is given by equations (3.15)–(3.17). The average queue length is

$$\begin{aligned}L &\approx a\sum_{i=0}^{k-1}ie^{-m}\frac{m^i}{i!}+b\sum_{i=k}^{\infty}ip(i-k+1;1)\\ &=a\sum_{i=0}^{k-1}e^{-m}\frac{m^i}{(i-1)!}+b\sum_{j=1}^{\infty}jp(j;1)+b\sum_{j=1}^{\infty}(k-1)p(j;1)\\ &=a\sum_{i=0}^{k-1}e^{-m}\frac{m^i}{(i-1)!}+bL_1+b(k-1)m/k\end{aligned} \tag{3.18}$$

where L_1 is the expected queue length of an $M/G/1$ queue.

In the above equation, L_1 can be evaluated by (3.11) with an arrival rate λ and an average service time $E[S]/k$. Once the average queue length is available, the average waiting time follows directly from Little's formula.

There are different approximation methods for $M/G/k$ queues. Equation (3.18) does not always give the best result among all existing methods, but it is simple. Numerical results, based on a limited number of cases, showed that the error can be expected within a few percent. In particular, if $k = 1$ or $G = M$ (i.e., exponential service time), equation (3.18) presents the exact solution.

M/G/1 Priority Systems

In addition to the first-come, first-served rule, another popular queuing discipline is *priority*. Jobs belonging to different priority classes share the same queuing system. There are different kinds of priority rules. The one of interest is said to be the rule of *nonpreemptive priority* or *head of line*. Under this rule, arrival jobs will queue up at the server, if the server is busy. When the server becomes available, the first job from the highest priority class is selected. If the highest class is empty, then this selection process is repeated for the next highest class until a job is selected. Once a job begins its service, no interruption or preemption from a higher priority job is allowed. For example, if in a line two or more machine failures occur, a repairperson may first attend to the machine that causes the longer production delay. But once a repair action starts, no other repair jobs can be initiated until the current one has been completed.

The solution of an $M/G/1$ queue with nonpreemptive priority can also be derived from the concept of workload. Again for a Poisson arrival stream, the virtual delay (at a random time epoch) is identical to the workload seen by an arrival. Rewrite equation (3.8) as

$$w = \lambda E[D]E[S] + (\lambda E[S])\frac{E[S^2]}{2E[S]}$$

It is known, from Examples 3.3 and 3.4, that system utilization is $\rho = \lambda E[S]$ and the average number of jobs waiting for service is $Q = \lambda E[D]$. The above equation becomes

$$w = QE[S] + \rho\frac{E[S^2]}{2E[S]} \tag{3.19}$$

The virtual delay consists of two parts: (1) workload due to the job in service and (2) workload due to jobs waiting for service. The first term on the right-hand side is the product of the average number of jobs waiting in the queue for service and the average service time. Since all jobs have the same service time distribution, this product is the average workload contributed by the jobs in the queue (waiting for service). The second term on the right-hand side is the product of utilization and a fraction expression. Utilization can be interpreted as the probability that the system is busy. Therefore, the fraction expression must be the conditional expected remaining service time, given that the server is occupied. In fact, it is possible to show that if a position is randomly chosen from a random interval S, then the expected distance from this position to either end of the interval is equal to

$$E[S_r] = \frac{E[S^2]}{2E[S]} \tag{3.20}$$

Suppose that there are k priority classes of jobs; class 1 has the highest priority, class 2 the second highest, and so on. A job of class i is called an i-job. Class i arrivals form a Poisson process with a rate λ_i. The proportion of time that an i-job is in service is $\rho_i = \lambda_i E[S_i]$, where S_i is the service time of an i-job. Let $\lambda = \sum \lambda_i$, and $E[S]$ be the overall mean service time. Since λ_i is proportional to the relative arrival count of class i jobs, λ_i/λ can be interpreted as the probability that an arrival belongs to class i. Therefore $E[S]$ is a weighted average, i.e.,

$$E[S] = \sum_i \lambda_i E[S_i]/\lambda \tag{3.21}$$

Similarly, the rth moment is given by

$$E[S^r] = \sum_i \lambda_i E[S_i^r]/\lambda \qquad r = 2,3,\ldots \tag{3.22}$$

Equation (3.21) also implies that the overall utilization is the sum of individual utilizations:

$$\rho = \lambda E[S] = \sum_i \rho_i \tag{3.23}$$

The average delay of an 1-job, d_1, can be obtained by looking at virtual delay with respect to class 1. This delay is the sum of two parts: (i) The remaining service time of the job in service—since no preemption is allowed, the job may belong to any class. (ii) The total service demand from the existing 1-jobs that are waiting for services. Hence under the assumption of a Poisson arrival stream, the delay experienced by a 1-job is the amount of workload contributed by all 1-jobs waiting in the queue for service plus the remaining service time of the job in service, that is,

$$d_1 = w_1 = Q_1 E[S_1] + \rho \frac{E[S^2]}{2E[S]}$$

By virtue of $Q_1 = \lambda_1 d_1$, the average (actual) delay experienced by a 1-job becomes

$$d_1 = \frac{\lambda E[S^2]}{2(1-\rho_1)} \tag{3.24}$$

For a 2-job, the delay consists of workload from all 1-jobs and 2-jobs waiting for service plus the remaining service time. Hence

$$\begin{aligned} d_2 &= Q_1 E[S_1] + Q_2 E[S_2] + \lambda E[S^2]/2 \\ &= \rho_1 d_1 + \rho_2 d_2 + \lambda E[S^2]/2 \end{aligned}$$

Using the result given by equation (3.24), d_2 can be solved in a straightforward manner. Invoking this approach repetitively, d_3, d_4, ... can be obtained. As a general form, the average delay of an i-job is given by

$$d_i = \frac{\lambda E[S^2]}{2(1-\sum_{j=1}^{i-1}\rho_j)(1-\sum_{j=1}^{i}\rho_j)} \qquad \text{for } i > 1 \tag{3.25}$$

Once d_i is known, the average waiting time, W_i, and the average queue length, L_i, follow:

$$W_i = d_i + E[S_i]$$

$$L_i = \lambda_i W_i$$

Example 3.8 A repairperson serves two types of machines. Machine downtime can be converted into cost. For a type i failure, the repair time is S_i. If a type i machine is not working, a cost is incurred at a rate of c_i per time unit. Assume that the interval between two successive type i failures is exponential with a rate λ_i. Then the probability that a failure is due to type i machine is $p_i = \lambda_i/(\lambda_1 + \lambda_2)$, $i = 1$ or 2. The objective is to minimize the average delay cost, i.e., $\min z = c_1 d_1 p_1 + c_2 d_2 p_2$. The value of z will be different, contingent on the priority assignment. If type 1 failures have high priority, then

$$z_1 = \frac{\lambda E[S]}{2(\lambda_1 + \lambda_2)}[\frac{c_1\lambda_1}{1-\rho_1} + \frac{c_2\lambda_2}{(1-\rho_1)(1-\rho_1-\rho_2)}]$$

Likewise, if type 2 failures have high priority, the objective value becomes

$$z_2 = \frac{\lambda E[S]}{2(\lambda_1 + \lambda_2)}[\frac{c_2\lambda_2}{1-\rho_2} + \frac{c_1\lambda_1}{(1-\rho_2)(1-\rho_2-\rho_1)}]$$

It is straightforward to show that the difference is given by

$$\begin{aligned}\Delta = z_1 - z_2 &= K(c_2\lambda_2\rho_1 - c_1\lambda_1\rho_2)\\ &= K\lambda_1\lambda_2(c_2E[S_1] - c_1E[S_2])\end{aligned}$$

where K is a factor independent of failure types.

Thus, $\Delta > 0$ if and only if $c_2/E[S_2] > c_1/E[S_1]$. This is equivalent to saying that a higher priority should be given to the failure type with a greater value of $c_i/E[S_i]$. □

Note that the above priority assignment rule is only dependent on (i) the delay cost and (ii) the average service time or the service rate. This rule is sometimes referred to as the $c\mu$ rule, i.e., the job class with the largest $c\mu$ has the highest priority. Here μ is the service rate of a job class. It is easy to show that this rule is optimal for $M/G/1$ queues with multiple job classes.

If the objective is to minimize the average delay, c_i may be set equal to one so that $z = \sum d_i p_i$. In this case, the optimal policy will assign the highest priority to the shortest jobs. This is said to be the rule of *shortest-first*. For real life applications, if the optimal rule is not known, this rule is usually a good one to use.

3.4 Network of Queues

A network of queues is composed of a number of independent service centers. After a service completion at a center, a job may join another queue or simply leave the network. Job routing usually follows a probabilistic pattern. Material flow behavior in an assembly line sometimes can be analyzed by using a queuing network approach. Figure 3.5 shows a simplified process of four operations. The last one is an inspection operation. If a defect is found, the assembly is sent back to upstream operations for rework. Assume that the yield is 0.97 and the defect rates at the first three operations are identical. Let r_{ij} be the routing probability of a job from center i to center j. Then $r_{1,2} = r_{2,3} = r_{3,4} = 1, r_{4,0} = 0.97$, and $r_{4,1} = r_{4,2} = r_{4,3} = 0.1$, where the route $4 \to 0$ represents a job departure. Figure 3.6 presents the assembly process from a real-life assembly line, where operations 001–006 are assembly operations, 101–103 test operations and 301–304 rework operations. The numbers associated with links are the routing probabilities. Network of queues also has very important applications in computer performance analysis.

Because of its complexity, analytical results of a queuing network only exist under simplified assumptions. Therefore, our discussions in this section should be treated as a first approximation for line analysis. The key for the solution is that each service center sees a Poisson arrival stream and generates a Poisson departure process. This condition is sometimes referred to as the $M => M$ property. It can be shown that both $M/M/k$ queues (for $k \geq 1$) and $M/G/\infty$ queues have this property. For $0 < r_{ij} < 1$, the job flow from center i to center j is a split process from the departure stream at center i. If the departure process at center i is Poisson, the split process is also Poisson. On the other hand, if more than one independent Poisson arrival streams merge at center j, then the superposed arrival process is also Poisson. If each service center in a network has the $M => M$ property, one would expect that the queue length distribution should have a form similar to (3.5).

Consider a network of R service centers; each has the $M => M$ property and exogenous arrivals (i.e., arrivals from outside of the network) form a

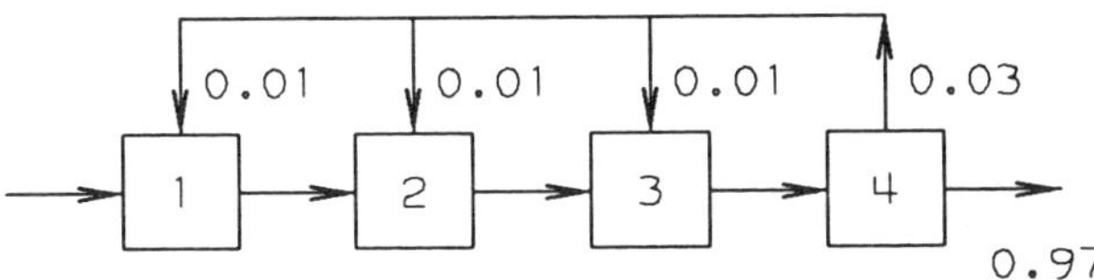

Figure 3.5 A simplified assembly process.

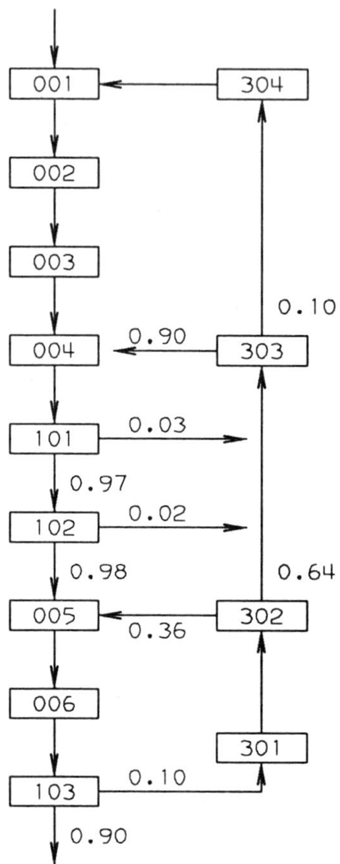

Figure 3.6 An assembly process.

Poisson process. With a probability q_i, an arrival joins center i. Because of probabilistic routing, the number of trips that a job visits a center before its departure is a random variable. Let v_j be the expected number of visits at center j. v_j is composed of (i) external visits (exogenous arrivals that enter center j) and (ii) internal visits (jobs that come from other centers). Therefore

$$v_j = q_j + \sum_{i=1}^{R} r_{ij} v_i \qquad j = 1,2,\ldots,R \tag{3.26}$$

and the value of v_j can be obtained by solving a set of linear equations. Note that v_j is the expected arrival rate at center j, given that the exogenous arrival rate is 1. v_j can also be treated as a relative throughput at center j. Let λ be the exogenous arrival rate. The (absolute) throughput or the arrival rate at center j becomes λv_j. If center j has k_j identical exponential servers of a common service rate μ_j, its queue length distribution in accordance with (3.5) is given by

$$p_j(0) = \left[\sum_{i=0}^{k_j-1} \frac{(k_j\rho_j)^i}{i!} + \frac{(k_j\rho_j)^{k_j}}{k_j!(1-\rho_j)}\right]^{-1}$$

$$p_j(n) = p_j(0)\frac{(k_j\rho_j)^n}{n!} \qquad n \le k_j$$

$$p_j(n) = p_j(0)\frac{(k_j)^{k_j}}{k_j!}\rho_j^n \qquad n > k_j \tag{3.27}$$

where $\rho_j = \lambda v_j / k_j \mu_j$

Furthermore it can be shown that the joint queue length distribution is

$$p(n_1, n_2, \ldots, n_R) = \prod_{j=1}^{R} p_j(n_j) \tag{3.28}$$

Equation (3.28) characterizes an important property of a class of queuing networks, namely, the *product form solution*. Computational effort is greatly simplified. Since the queue lengths of individual centers are independent of each other, one may evaluate one service center at a time.

As mentioned earlier in this section, an assembly flow may be analyzed by using results from a network of queues. Servers in a center may represent workstations in an operation. Queue length corresponds to work-in-process. Total waiting time in the network (amount of time that a job spends in the network) can be interpreted as the production lead time. This kind of queuing model assumes that jobs arrive from an outside source and, therefore, is called *open network*.

Another type of network is said to be *closed network*, which has a fixed number of jobs, no arrivals from outside sources and no departures from the network. A well-known model of this kind is the *machine-repairman model*. Figure 3.7 shows two sets of servers, one for the machines and the other for the repairpersons. The number of jobs is the same as the number

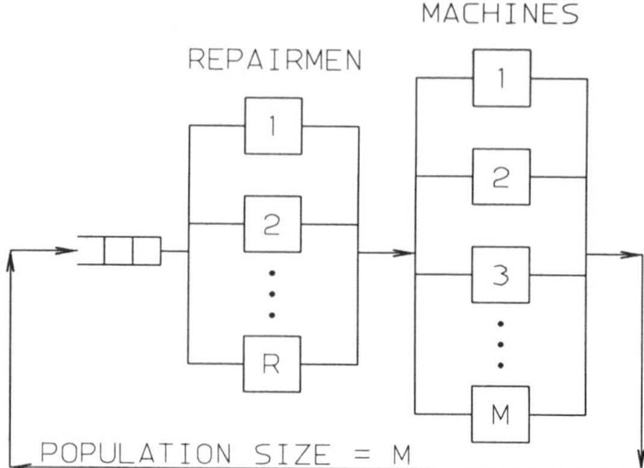

Figure 3.7 A machine-repairman model.

of machines. In this case a job can be regarded as a work order. A work order is placed on the machine as long as the machine is functioning. If a machine failure occurs, the associated work order is changed into a repair order and sent to the repair crew (i.e., the job leaves the machine queue for the repair queue). When the repair job is completed, the repair order is changed back to a work order and returned to the machine (i.e., the job enters the machine queue again). The service time in the machine queue is the time between failures while the service time in the repair queue is the repair time.

If each service center in a network has the $M => M$ property, the joint queue length distribution also has a product form. Since the total number of jobs in the network is a constant, $n_1 + n_2 + \cdots + n_R = N$ is fixed. The completed solution is a truncated distribution of (3.28), that is,

$$p(n_1, n_2, \ldots, n_R) = G(N) \prod_{j=1}^{R} p_j(n_j) \tag{3.29}$$

where $p_j(n_j)$ is given by (3.27) and $G(N)$ is a normalized factor such that $\sum p(n_1, \ldots, n_R) = 1$.

Example 3.9 Assume that a single repairperson attends 10 identical machines. The mean repair time is 2 hours and the average machine uptime interval 50 hours. Suppose that k machines are not functioning. The queue length seen by the repairperson is k and the queue length in the machine queue is $10-k$. Let $p(k)$ be the probability of this state. The computed values of $p(k)$, for $i = 0, 1, \ldots, 10$, according to (3.27)-(3.29), are respectively 0.6224, 0.2489, 0.0896, 0.0287, 0.0080, 0.0019, 0.0040, ..., 0.0000. The average queue length (i.e., the average number of machines not functioning) is $L = \sum kp(k) =$ 0.5588. The occurrence of machine failures is state-dependent. The occurrence rate, $\lambda = \sum(10-k)(1/50)p(k) = 0.1888$ per hour. Hence the average waiting time $W = L/\lambda = 2.9592$ hours. This implies that the average delay before the repairperson's response to a failure is nearly one hour ($\approx 2.9592-2$). The repairperson utilization is $1-p(0) = 0.3776$. The machine reliability is given by $50/(50+2.9592) =$ 0.9441. If an additional repairperson is hired, the computed values of $p(k)$, for $k = 0, 1, \ldots, 10$ are 0.6732, 0.2693, 0.0485, 0.0078, 0.0011, ..., 0.0000. $L = 0.3946$, $\lambda = 0.1921$, and $W = 2.0541$. The repairmen utilization is reduced to 0.1921 while machine reliability is increased to 0.9605. □

Because of the normalized factor, $G(N)$, computational complexity may cause a problem, particularly when N or R is large. Different computational methods have been proposed. A simple approximation scheme is to treat the network as if it were an open network with an external arrival rate λ. To convert a closed network into an open network, one may designate an *exit-entry* node in the closed network. Let all internal arrivals into this node leave the network and an external Poisson arrival stream enter this node. If for a specific value of λ, the expected total number of jobs in the open network is the same as the fixed population size in the closed network, then the open network solution is an approximate solution for the closed one. A stepwise procedure is as follows:

1. Pick any service center as the exit-entry node.
2. Guess a value for the Poisson arrival rate.
3. Solve the open network by equations (3.27) and (3.28).
4. If the expected total number of jobs in the open network is near the population size in the closed network, stop the procedure. An approximate solution is found.
5. Otherwise, adjust the external arrival rate and solve the open model again.

The last step requires us to change the arrival rate. Since a high arrival rate implies a large average queue length, the rate should be increased (decreased) if the expected total queue length is less (greater) than the fixed population. This method is called the *fixed population mean* method.

Example 3.10 An approximate solution of the machine-repairman model (Example 3.9) can be obtained by the fixed population mean method. Consider the case of a single repairperson. The approximated open model is illustrated in Figure 3.8, where the queuing system has two queues in tandem, one for the repairperson and the other for machine failures. Since every arrival would visit these two queues exactly once, $v_1 = v_2 = \lambda$, where λ is the arrival rate equivalent to that of the open model. These two queues look like $M/M/1$ and $M/M/10$ queues respectively. The maximal service rate of the repairperson is 0.5 per hour and the maximal expected failure count per hour is 0.2 (= 10/50). Therefore the throughput at each queue (repairperson and machine) cannot exceed the minimum of these two numbers, that is, 0.2. Let $\lambda = 0.1$. The average queue length seen by the repairperson is $\rho_1/(1-\rho_1) = (0.1/0.5)/(1-0.1/0.5) = 0.25$ and the average queue length (in this case the average number of operating machines) at the second queue is 5.0361. Since the average total queue length of 5.2861 is below the fixed population of 10 in the closed model, our second trial should increase the value of λ. A sequence of trials and the corresponding outcomes are given by $(\lambda,L) = (0.15, 8.8484)$, $(0.175, 13.4229)$, $(0.16, 10.1070)$, $(0.159, 9.9600)$, and $(0.1593, 10.0035)$. At $\lambda = 0.1593$, the queue length distribution at the first queue, $p(k)$ for $k = 0, 1, 2, \ldots$ has values 0.6814, 0.2171, 0.0692, 0.0220, 0.0070, 0.0022, 0.0007, ..., respectively. The expected number of machines for repair is 0.4676 and the average delay for a repair job is 0.9351 hours.

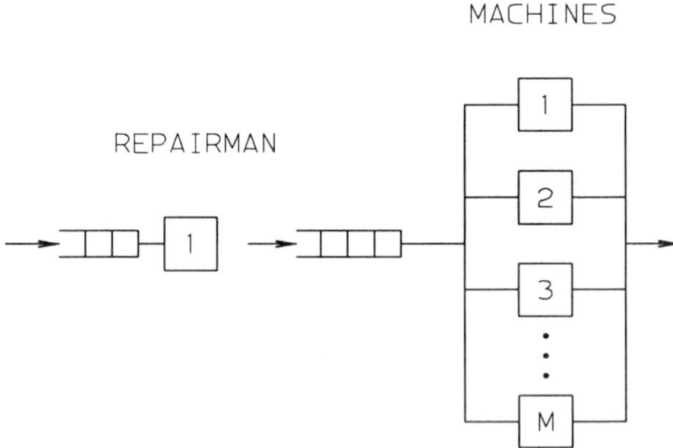

Figure 3.8 An approximated open model for a machine-repairman problem.

Machine reliability is $50/(50+2.9351) = 0.9446$. These estimates are closed to those values presented in Example 3.9. □

3.5 Optimization Problems and Basic Concepts

Although queuing analysis can be a powerful tool for manufacturing problems, many problems in design optimization must be solved by searching techniques based on optimization theory. An optimal solution is meaningful only if the objective can be well defined. To a line designer, the ultimate objective is the minimal manufacturing cost. This sometimes may be interpreted as the minimal amount of expenditure for a fixed production quantity. Sometimes, the objective means the maximal line throughput for a given amount of resources. Since a line is composed of many components, it is often necessary to break the design problem into a number of smaller subproblems. Each subproblem may have its own intermediate objective. In this section, we will briefly discuss basic concepts of optimization.

Optimization theory has developed in two stages. Many problems in science, engineering, economics and management were first solved by classic techniques of applying differential calculus. A different class of problems deals with optimal allocations of limited resources to achieve a predetermined objective. This type of problem is said to be a *mathematical programming problem*. A mathematical programming problem is usually concerned with optimizing an objective function under a given set of constraints. If both the objective function and the constraints can be expressed in linear algebraic forms, the problem is a *linear programming problem*. (Otherwise, it is nonlinear). For example, a factory may produce two different products. Let x_1 be the production quantity for product A, and x_2 for product B. Let the price for product A (product B) be p_1 (p_2). The total revenue is $p_1x_1 + p_2x_2$. Assume that a unit of product A (product B) would consume r_1 (r_2) amount of labor and the total available labor hours is R. Any feasible solution (x_1,x_2) must satisfy the inequality: $r_1x_1 + r_2x_2 \le R$. Similarly, a feasible solution should also satisfy a material constraint: $s_1x_1 + s_2x_2 \le S$. A linear programming problem can be formulated as:

$$\max p_1x_1 + p_2x_2$$

$$\text{such that } r_1x_1 + r_2x_2 \le R$$

$$s_1x_1 + s_2x_2 \le S$$

$$x_1, x_2 \ge 0$$

The above problem has two variables and two constraints. A generalized case may contains n variables (e.g., n product types) and m constraints (e.g., m different kinds of resources). Linear programming problems can be effectively solved by some well-known methods, such as the *simplex method*. Software solution packages are available for different kinds of computers, from personal computers to large host systems. Because of their general availability, solution techniques for linear programming problems will not be discussed in this book.

Effective general solution techniques for nonlinear programs are not known. Often a problem is solved by a special method that may not be used elsewhere. A powerful method is to convert a nonlinear program into an equivalent or a sequence of linear programs that can be solved by linear programming solution packages. This techniques may be particularly useful, if a programming problem has a linear structure but requires an integer solution. In some cases, it can be shown that a particular set of constraints will automatically generate integer solutions when the problem is solved without considering integer constraints. (This is said to be the *total unimodular property*.) In some case, one can simply solve the problem as if it were an ordinary linear program. If the solution does not contain integer values, an additional constraint may be added to the problem according to some criteria. The problem is then solved in a repetitive manner until an integer optimal solution is obtained. (This is called the *cutting plane method*.)

Sometimes it is possible to break a problem into a number of problem-stages. The computational procedure involves a recurrence relation that can be applied to each problem-stage and maintains an optimality condition throughout the entire procedure. This technique is known as *dynamic programming*. Take the same example of two variables, except that the objective is no longer in linear form due to price breaks and that the second constraint does not exist. The problem is to maximize $z = f_1(x_1)+f_2(x_2)$ such that $r_1x_1 + r_2x_2 \leq R$. The problem can be solved by first looking at x_1. Let y be the amount of resource allocated to product A. The first problem stage is to maximize $f(x_1)$ such that $r_1x_1 \leq y$. This should be done for all y such that $y \leq R$. Let $x_1(y)$ be the optimal solution, given that y units are allocated to product A. These results will be included at the second stage where the total resource left over is $R - y$, and the objective is to maximize $f_2(x_2)$ such that $r_2x_2 \leq R - y$. Let $x_2(R - y)$ be the optimal solution. The remaining task is to pick the value for y such that $z(y) = f_1(x_1(y)) + f_2(x_2(R - y))$ is maximum. For the problem of n products, a general procedure may be formulated. Let $f_i(x_i)$ be the revenue from x_i units of products i and $S_i(y)$ be the maximal revenue from products $1, 2, \ldots, i$ with y amount of resource. The computational procedure is composed of three parts:

1. The initial condition

$$S_1(y) = f_1(y), \qquad \text{for all } y \leq R$$

2. The recurrent relation

$$S_i(y) = \max_{x \leq y} \{S_{i-1}(y-x) + f_i(x)\}, \quad \text{for all } y \leq R \text{ and } i = 2,3,\ldots,n$$

3. The optimal value

$$\max \{S_n(x) \mid \text{for all } x \leq R\}$$

The "dynamic" structure of this procedure is realized by dealing with one product type at a time. Different resource allocation strategies have been implicitly enumerated during the process of computation. This procedure greatly simplifies total computational effort. The solution is optimal due to the fact that decisions at each stage remain optimal regardless of what would happen in later stages. This property is referred to as the *principle of optimality*. On the other hand, if a later decision would change early optimal conditions, then the principle of optimality does not hold and dynamic programming cannot lead to the optimal solution.

Another type of computational algorithm for nonlinear programming is referred to as the *gradient method*. The method starts with a feasible solution, i.e., a solution that satisfies all constraints. An iterative approach is defined by systematically moving from a given solution to a better solution. The basic problem is to find the best moving direction and the best step size. This method is based on classic optimization theory and will be discussed in the next section.

3.6 Classic Optimization and Gradient Method

Let us start with a simple case. First, note that the problem of maximizing z is equivalent to the problem of minimizing $-z$. Therefore, without losing generality only the maximization problems need to be considered.

Assume that the objective can be written as a differentiable function of a single variable. The problem is to maximize the objective function: $f(x)$ for $a \leq x \leq b$. The maximum solution, x_0, may be located at three different places: (i) $x_0 = a$, (ii) $x_0 = b$, and (iii) x_0 between a and b. If $f(x)$ takes its maximum within $[a,b]$, then for a small $h > 0$ it must be true that

$$f(x_0) - f(x_0 + h) \geq 0$$
$$\text{and } f(x_0) - f(x_0 - h) \geq 0$$

Dividing both inequalities by h and letting $h \to 0$, it follows that

$$f'(x_0) = 0 \tag{3.30}$$

Equation (3.30) is a necessary condition for the optimal solution. Thus, the optimal solution of $f(x)$ can be obtained in three steps: (i) find all roots of $f'(x) = 0$, (ii) evaluate $f(x)$ at a, b, and all roots from step i, and (iii) pick the maximal value.

Many optimization problems have multiple variables. In these cases, the solution space has multiple dimensions. The solution space, S, may be defined by a set of constraints. Since no effective general algorithms exist and mathematics for nonlinear programs is beyond our scope, our discussion is confined to unconstrained problems in the rest of this section.

Let $\vec{x} = (x_1, x_2, \ldots, x_n)$ and $f(\vec{x})$ be the objective function. A necessary condition for optimality is given by

$$\frac{\partial f(\vec{x})}{\partial x_j} = 0 \qquad j = 1, 2, \ldots, n$$

The vector, $\text{grad}[f(\vec{x})] = (\partial f(\vec{x})/\partial x_1, \partial f(\vec{x})/\partial x_2, \ldots, \partial f(\vec{x})/\partial x_n)$ is called the *gradient vector* of $f(\vec{x})$. The optimality condition can be abbreviated as

$$\text{grad}[f(\vec{x})] = \vec{0} \tag{3.31}$$

where $\vec{0}$ is the null vector of n dimensions.

Example 3.11 In Chapter 2, a linear regression model, based on the least-squares method, has been discussed. The model solution of (2.17) can be obtained by solving an optimization problem, where the objective function is

$$\min z = \sum_{t=1}^{n}\left[Y_t - \left(a_0 + \sum_{i=1}^{k} a_i X_{it}\right)\right]^2$$

Hence, the condition (3.31) leads to

$$\frac{\partial z}{\partial a_0} = -\sum_{t=1}^{n} 2\left[Y_t - \left(a_0 + \sum_{i=1}^{k} a_i X_{it}\right)\right] = 0$$

$$\frac{\partial z}{\partial a_j} = \sum_{t=1}^{n} 2\left[Y_t - \left(a_0 + \sum_{i=1}^{k} a_i X_{it}\right)\right] X_{jt} = 0 \qquad j = 1,2,\ldots,k$$

The above relations can be reduced to

$$\sum_{t=1}^{n}\left(a_0 + \sum_{i=1}^{k} a_i X_{it}\right) = \sum_{t=1}^{n} Y_t$$

$$\sum_{t=1}^{n}\left(a_0 X_{jt} + \sum_{i=1}^{k} a_i X_{it} X_{jt}\right) = \sum_{t=1}^{n} Y_t X_{jt}$$

In a matrix form, these linear equations are equivalent to

$$XX'a = XY$$

where a, X and Y are the same as the notations used in (2.19).
Multiplying both sides by $(XX')^{-1}$, (2.19) follows. □

Equations defined by (3.31) may or may not have a simple structure for solutions. Sometimes these equations are solved by numerical methods such as *Newton's method.* To simplify our notation, let $g_j(\vec{x}) = \partial f(\vec{x})/\partial x_j$. Assume that $\vec{x}_0$ is a solution to (3.31). By Taylor's theorem, there exists an α, $0 \le \alpha \le 1$, such that

$$0 = g_j(\vec{x}_0) = g_j(\vec{x}) + \text{grad}[g_j\{\alpha\vec{x}_0 + (1-\alpha)\vec{x}\}](\vec{x} - \vec{x}_0) \qquad j = 1,\ldots,n.$$

If $\vec{x}$ is near $\vec{x}_0$, a new solution may be obtained by using the following relations:

$$g_j(\vec{x}) + \text{grad}[g_j(\vec{x})](\vec{y}-\vec{x}) = 0 \qquad j = 1,\ldots,n. \tag{3.32}$$

If $\vec{x}$ is known, (3.32) defines n simultaneous linear equations with n unknowns, $\vec{y}$. Newton's method repetitively solves these equations by replacing $\vec{x}$ by $\vec{y}$ and computing new $\vec{y}$. It can be shown that if the initial point is sufficiently close to $\vec{x}_0$, this procedure would converge to $\vec{x}_0$ very quickly. Again if all the solutions to (3.31) can be obtained, the remaining task is to compare the values of $f(\vec{x})$ at these solutions.

Example 3.12 Consider two equations in two unknowns:

$$g(x,y) = x^2 + y^2 - 8 = 0$$
$$h(x,y) = x - y = 0$$

Clearly these two equations have two pairs of solutions at (2, 2) and (−2,−2). Using Newton's method, two recursive equations are obtained:

$$g(x,y) + \partial g/\partial x(\hat{x} - x) + \partial g/\partial y(\hat{y} - y) = 0$$
$$h(x,y) + \partial h/\partial x(\hat{x} - x) + \partial h/\partial y(\hat{y} - y) = 0$$

For a given (x,y), the values of $(\hat{x},\hat{y})$ can be determined by solving the above two linear equations. The iterative process with two different initial guesses is given in the following:

Initial values	(10,10)	(0,−1)
1	(5.2000, 5.2000)	(−4.5000, −4.5000)
2	(2.9846, 2.9846)	(−2.6944, −2.6944)
3	(2.1624, 2.1624)	(−2.0895, −2.0895)
4	(2.0061, 2.0061)	(−2.0019, −2.0019)
5	(2.0000, 2.0000)	(−2.0000, −2.0000)

In both cases, the process converges after 5 iterations. □

If the solution of (3.31) cannot be obtained easily, the optimization problem may be solved by a numerical procedure such as the gradient method. Let $\vec{x}_1$ be a feasible (but not necessarily optimal) solution. By Taylor's theorem, for all $\vec{x}$ in the neighborhood of $\vec{x}_1$,

$$f(\vec{x}) \approx f(\vec{x}_1) + \text{grad}[f(\vec{x}_1)](\vec{x} - \vec{x}_1) \tag{3.33}$$

The solution may be improved by moving from $\vec{x}_1$ to a new point. The question is how to determine the best moving direction to gain the maximum improvement. Since $\vec{x}_1$ is given, both $f(\vec{x}_1)$ and grad$[f(\vec{x}_1)]$ can be easily computed. The problem becomes

$$\max_{x} \{\text{grad}[f(\vec{x}_1)](\vec{x} - \vec{x}_1)\} \tag{3.34}$$

For any two n-component vectors, their inner product is equal to the product of their lengths multiplied by $\cos\theta$, where θ is the angle between these two vectors. If the vector lengths are fixed, the maximal value of an inner product is achievable at $\theta = 0$. Therefore, the optimality condition for (3.34) is that for some scale α,

$$(\vec{x} - \vec{x}_1) = \alpha\,\text{grad}[f(\vec{x}_1)]$$

The best moving direction is along the gradient. (For a minimization problem the direction is opposed to the gradient vector.) Figure 3.9 provides a geometric illustration, where the contour values of an objective function in a two-dimensional space are shown. Gradient directions at a sequence of points are displayed. The chain of arrows defines a gradient path that leads to the optimal solution.

Since condition (3.33) is valid only in the neighborhood of $\vec{x}_1$, the new (improved) solution should be kept close to $\vec{x}_1$, that is, the step size of a move should not be too large. On the other hand, if the step size is too small, the iterative process may take a long time to reach the optimal solution. Determination of the optimal step size becomes another problem:

$$\max_{\alpha} f(\vec{x}_1 + \alpha\,\text{grad}[f(\vec{x}_1)]) \tag{3.35}$$

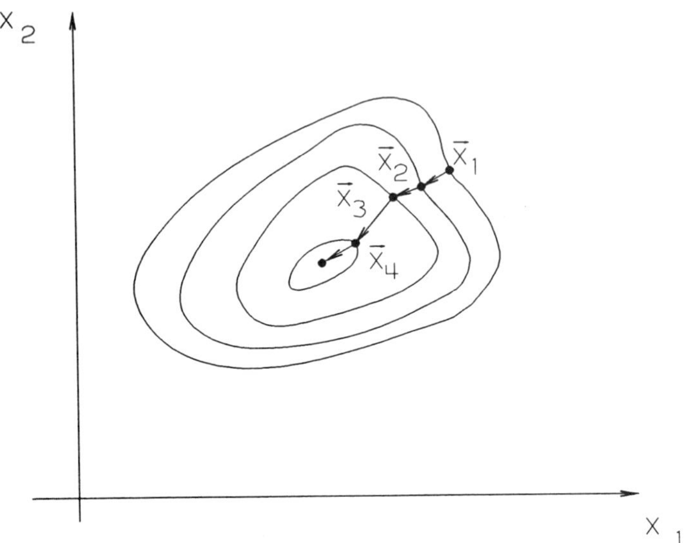

Figure 3.9 A geometric interpretation of the gradient method.

This problem has only one variable, α, and may be solved by analytic, numerical or even heuristic methods. If f has a very complex form but can be evaluated quickly, one may simply try a few possible values for α and pick the best one as the solution to (3.35).

Example 3.13 Consider an unconstrained problem: $\min f(x,y) = x^2 + y^2$. The optimal solution is $(x,y) = (0,0)$. The gradient is given by $(\partial f/\partial x, \partial f/\partial y) = (2x, 2y)$. Assume that the initial solution is (10, 100) and step size $\alpha = 0.4$ The first improved solution is located at $(x,y) - \alpha(2x,2y) = (10,100) - 0.4(20,200) = (2,20)$. A sequence of computed solutions is given by (10, 100), (2, 20), (0.4, 4), (0.08, 0.8), (0.016, 0.16), (0.0032, 0.032), (0.0006, 0.0064), (0.0001, 0.0013), (0.0000, 0.0003), (0.0000, 0.0000). □

For many real-life applications, the objective function may not be formulated in an algebraic expression. For example, one may conduct experiments on an existing line and observe line performance. In this case it is not a simple thing to construct an algebraic function. Assume that x amount of resource is consumed for production. The objective is to maximize the ratio of output to input: $r(x) = h(x)/x$, where $h(x)$ is the observed amount of output. An approximated derivative can be obtained by experimenting twice with inputs of x and $x + \epsilon$ respectively, and then computing $r'(x) \approx [r(x+\epsilon) - r(x)]/\epsilon$, for a small value ϵ. Sometimes a line experiment is too expensive and too time-consuming. One may simulate line performance by a computer program. The method of using $r'(x)$ for gradual improvement is called the *hill climbing* method. Obviously, this approach becomes very inefficient and impractical if the number of variables is large. For a problem with 10 variables, gradient vector computation would require 11 experiments. If the process converges to the optimal solution in 10 steps, the total number of experiments is 110.

3.7 Computational Algorithms

Although no general effective techniques are available for all optimization problems, many practical cases may be modeled in special structures that can be solved by efficient algorithms. This section introduces four computational algorithms that have been proven useful for line design problems. They are (i) shortest path, (ii) maximal flow, (iii) minimal cost flow, and (iv) bin packing. Other algorithms will be introduced as needed in later chapters.

Shortest Path Problem

Consider that an automatic guided vehicle, as a material handling system, serves six manufacturing departments: A, B, C, D, E and F. A layout is shown in Figure 3.10(a), where a pick/place port is indicated for each department. The vehicle route follows the boundary line between departments as illustrated in Figure 3.10(b). In addition to the six ports there are four intersections, each identified by a circle in the figure. These ports and intersections are represented by 10 nodes in a transportation network, in which the nodes are connected by links of lengths identical to the travel times between node pairs, as seen in Figure 3.10(c). A service request for the vehicle always involves two nodes: a sender (or source) and a receiver (or sink). To achieve maximal efficiency, the traveling distance between the source and the sink should be minimal. This kind of problem is known as a *shortest path problem*. An effective algorithm is as follows:

Algorithm 3.1

1. Label the source node with 0 and all other nodes with a large number (e.g.greater than the sum of all link lengths). Initially all nodes are unmarked.

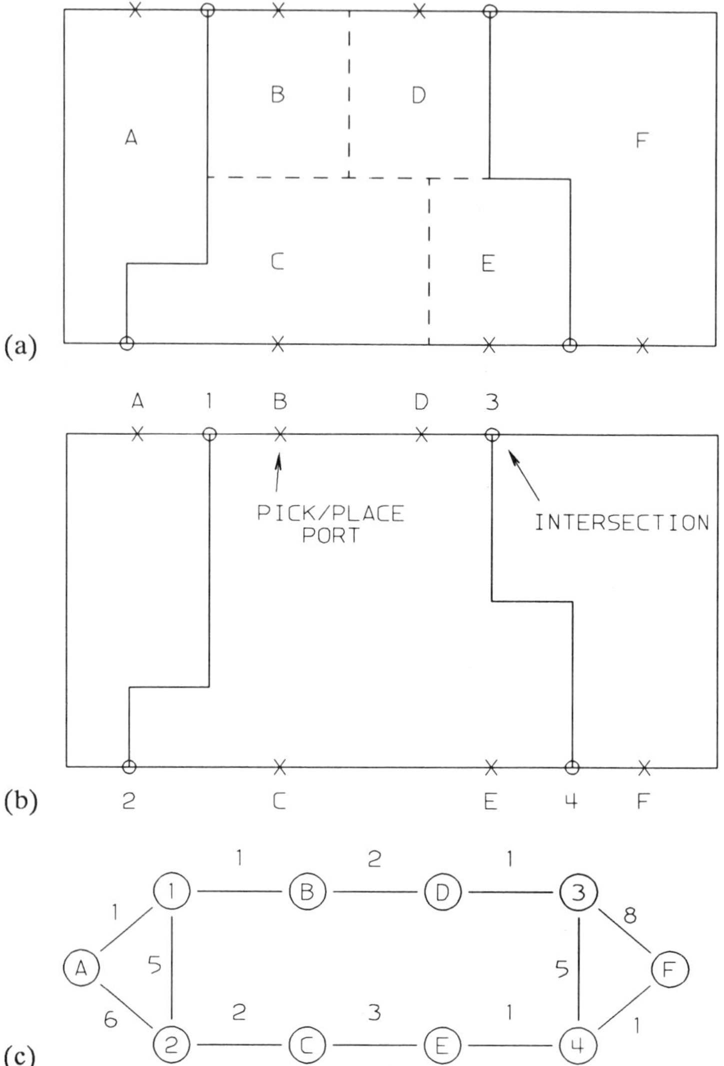

Figure 3.10 (a,b,c) AGV routing and the shortest path problem. (a) Department layout. (b) AGV routing path. (c) Network for shortest path problem.

2. If the sink node is marked, stop the procedure. The labeled value at the sink is the distance of the shortest path. Otherwise go to the next step.
3. Inspect all unmarked nodes and select the one with the minimal labeled value, say node k. Mark this node.
4. For each unmarked node, say node i, compute $u = v_k + d_{ki}$, where v_k is the labeled value of node k, and d_{ki} the length of link (k,i). If $u < v_i$, relabel node i by u. Otherwise do nothing.
5. Repeat steps (2), (3) and (4) until the sink is marked.

For a problem with n nodes, the number of computational steps is proportional to n^2.

Example 3.14 In Figure 3.10(c), the shortest path from node A to node F can be found by invoking Algorithm 3.1. The labeling process is given in the following table, where the marked node at each stage is indicated by an asterisk.

Stage	A	1	2	B	C	D	E	3	4	F
Initial	0	40	40	40	40	40	40	40	40	40
1	0*	1	6	40	40	40	40	40	40	40
2	—	1*	6	2	40	40	40	40	40	40
3	—	—	6	2*	40	4	40	40	40	40
4	—	—	6	—	40	4*	40	5	40	40
5	—	—	6	—	40	—	40	5*	10	13
6	—	—	6*	—	8	—	40	—	10	13
7	—	—	—	—	8*	—	11	—	10	13
8	—	—	—	—	—	—	11	—	10*	11
9	—	—	—	—	—	—	11	—	—	11*

The process begins with initial labels. Node A is labeled 0 and all other nodes 40. (Note that the sum of all link lengths is 37.) At the first stage, node A has the minimal label and therefore is marked by an asterisk. The adjacent unmarked nodes are node 1 and node 2. Their label values are recomputed. Since both of their new values are less, the updated labels become 1 and 6 respectively. At the next stage node 1 is marked because its labeled value is the smallest among all unmarked nodes. Node B and node 2 are adjacent nodes. The label of node B is changed into a smaller value while the label of node 2 remains the same. At the third stage, node B is marked and consequently node D is relabeled. The process continues until the sink is marked (see stage 9 in the table). The shortest distance from node A to node F is equal to 11 (which is the label value of node F).

To find the shortest path, one could backtrack the mark/label process. The sink label was last updated when node 4 was marked. Hence the link $(4,F)$ must be in the shortest path. The label of node 4 was last updated when node 3 was marked. This implies that the link $(3,4)$ should be in the shortest path. At the completion of this backtrack procedure, the shortest path, $A \to 1 \to B \to D \to 3 \to 4 \to F$, is obtained. □

Since the marked node is always labeled by the shortest distance from the source, the shortest path from the source to each marked node is also found. It should be noted however that this algorithm may not work if negative link lengths exist. Consider the graph in Figure 3.11, where the shortest path from node 1 to node 4 is $1 \to 2 \to 3 \to 4$ and the shortest distance is 1. But Algorithm 3.1 would first label nodes 2 and 3 by $v_2 = 2$ and $v_3 = 1$. Since $v_3 < v_2$, node 3 is marked first. This would cause $v_4 = 2$ and incorrectly take $1 \to 3 \to 4$ as the solution. One way to resolve the problem is to use an auxiliary tree to define the shortest path from the source. Assume that there are n nodes in the diagram; node 1 is the source and node n the sink. The algorithm is given as follows:

Algorithm 3.1(a)

1. Place a link from the source to each node. Thus a tree is formed. Label node i by $v_i = d_{1i}$. If there is no link between the source and node i, then $v_i = \infty$. Let $k = 1$.
2. $k = k + 1$.

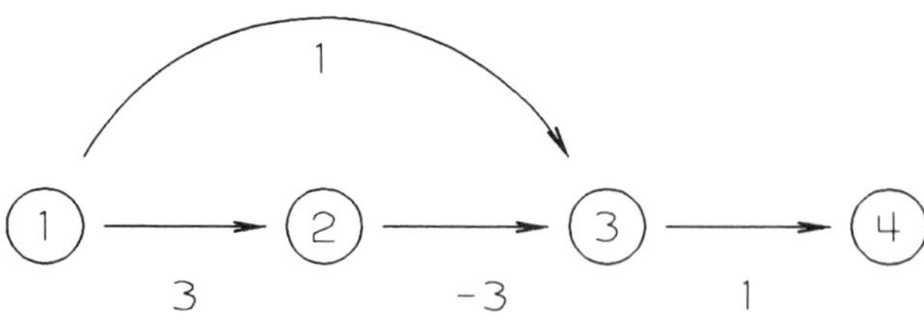

Figure 3.11 A shortest path problem with a negative link.

3. Inspect the possibility of linking node i to node k. This is done by comparing v_i to $u = v_k + d_{ki}$. If $u < v_i$, then break the link pointed into node i and put a link from node k to node i. Decrease the labeled values of node i and all its successors in the tree by an amount of $v_i - u$.
4. If every node has been inspected then stop. Otherwise go to step 2.

Applying Algorithm 3.1(a) to the problem described in Figure 3.11, the solution steps can be illustrated by Figure 3.12. The initial tree looks like the one in Figure 3.12(a). The labeled values of nodes 2, 3 and 4 are respectively 3, 1 and ∞. After having inspected node 2, it is found that node 3 should be reconnected to node 2 and node 4 should remain at the same position (see Figure 3.12(b)). Inspection of node 3 causes a reconnection of node 4, and the tree structure becomes a single path. The label at node 4 is the shortest distance.

Maximal Flow Problem

Consider that a network is composed of a number of nodes connected by links. Each link may carry an amount of flow, up to its capacity level. Different links may have different capacities. A *maximal flow problem* seeks to find the maximal amount of flow that may go from a source node to a sink node. This type of problem has many applications in line analysis and optimization. It can be used to analyze a network of conveyor systems, to determine an optimal job assignment pattern, and to understand the impact of a workstation capacity change. Let us take a simple example as follows.

Suppose that a line of three manual operations is running for three shifts a day. Each operation needs exactly one operator. The work paces of operators during the first shift are (4, 6, 5), that is, the first operator can produce 4 units, the second 6 units, and the third 5 units. The work paces for shifts 2 and 3 are respectively (5, 4, 6) and (6, 5, 4). The line throughput per shift appears to be 4 units. If, however, work-in-process can be accumulated through the next shifts, line throughput can be increased to 5 units per shift. The maximal throughput and the level of work-in-process can be analyzed by solving a maximal flow problem. In the following algorithm, nodes are numbered so that the source is node 1 and the sink is node n. Let c_{ij} be the capacity of the link connecting node i and node j and f_{ij} be the flow from node i to node j.

Algorithm 3.2

1. Let $f_{ij} = 0$ for all i and j.
2. Define a modified network as follows:

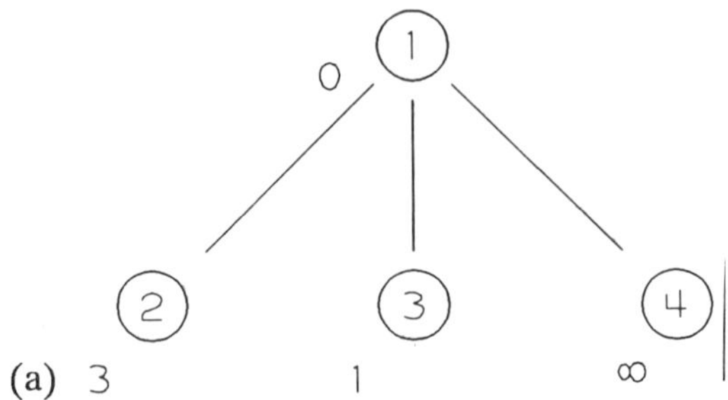

Figure 3.12 (a,b,c) Shortest path solution with auxiliary trees.

a. Remove all links from the network but preserve all nodes.
b. If $f_{ij} < c_{ij}$, place a direct link from node i to node j with a capacity of $c_{ij} - f_{ij}$ (a forward link).
c. If $f_{ij} > 0$, place a link from node j to node i with a capacity of f_{ij} (a backward link).

3. Each node in the modified network is in one of the three states: unlabeled, labeled but not scanned, and labeled and scanned. A label of node j is defined by either $[i^+, x(j)]$ or $[i^-, x(j)]$ where the former (the latter) indicates a flow of value $x(j)$ comes from node i through a forward (backward) link. First, label node 1 by $[0, \infty]$.
4. Scan the labeled but unscanned node that has the smallest index, say node j. Label all its adjacent unlabeled nodes. If node k is accessible from node j along a forward link, then the label of node k is $[j^+, g]$, where g is the minimum of the modified capacity of the forward link and the second label-element of node j. Similarly, if a backward link is used, the label will be $[j^-, g]$.
5. In step 4, if no labeled nodes remain unscanned and the sink is not labeled, there does not exist a path from the source to the sink in the modified network. The maximal flow has already been found. Stop.
6. If node n (the sink) is labeled, then a path from the source to the sink has been found. Send a flow of value $x(n)$ (the second label-element of the sink) along the path, that is, increase f_{ij} by $x(n)$ if link (i,j) is on the path. Go to step 2.
7. If the sink is not labeled, go to step 4.

Example 3.15 The three-shift production problem is formulated as a maximal flow problem. Figure 3.13(a) presents a network of eight nodes. Nodes 1 and 8 are the source and the sink respectively. There

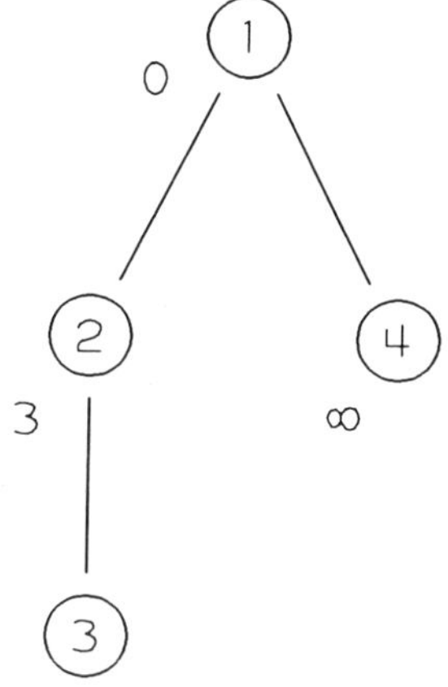

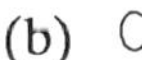

(b) 0

(c) 1

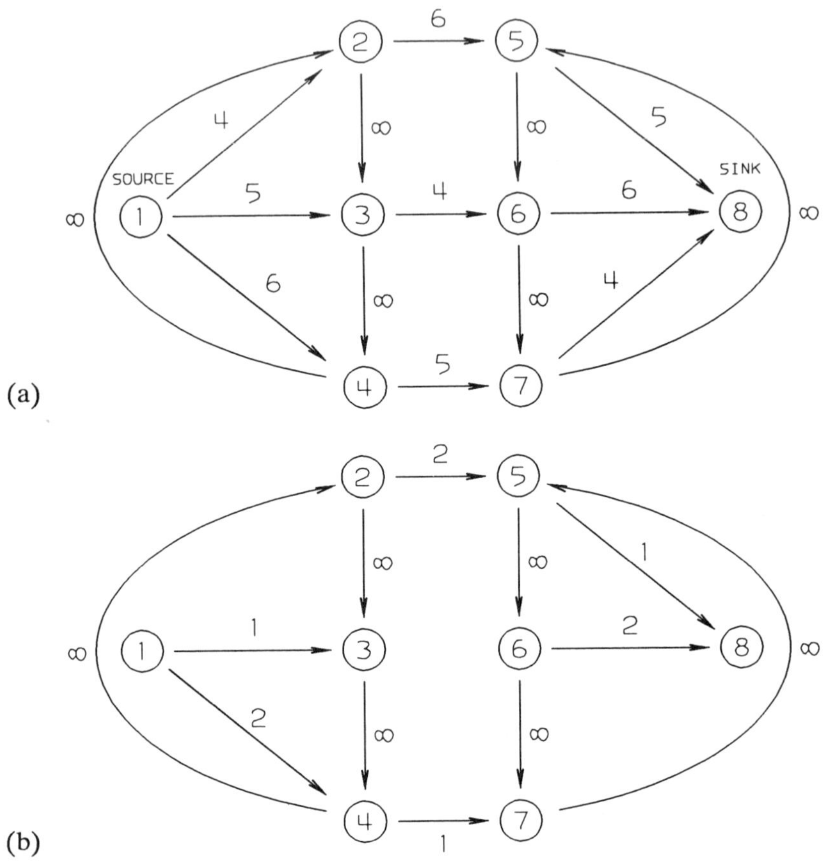

Figure 3.13 (a,b) A maximal flow model for production schedule analysis. (a) A maximal flow model. (b) A reduced maximal-flow model.

are three paths from node 1 to node 8: (i) $1 \to 2 \to 5 \to 8$, (ii) $1 \to 3 \to 6 \to 8$, and (iii) $1 \to 4 \to 7 \to 8$. Each path is equivalent to a shift, while the capacity of a link represents a work pace. For example, the three operators in the first shift can produce 4, 6, and 5 units respectively. Therefore $c_{1,2}$, $c_{2,5}$, and $c_{5,8}$ are 4, 6, and 5 respectively. Clearly each of these paths can carry a flow of value 4. If an upstream operation produces more than 4 units, the extra units must be carried over to the next shift as work-in-process. Consequently, two additional loops are needed: (i) $2 \to 3 \to 4 \to 2$ and (ii) $5 \to 6 \to 7 \to 5$. If the first operation in the third shift, represented by link $(1,4)$, produces 6 units, the extra units may flow to the next shift, that is shift 1. The links on each loop have infinite capacity.

To simplify the problem, assume that each of the three paths has already carried 4 units. Thus each link capacity can be decreased by 4, so that the maximal flow problem is reduced to the one shown by Figure 3.13(b), where the number associated with each link is the new capacity.

Node 1 is labeled and scanned. Then nodes 3 and 4 are labeled by (1,1) and (1,2) respectively. The algorithm scans node 3 first by the smallest index rule. The only accessible node from node 3 is node 4. Since node 4 has been labeled already, the next step is to scan node 4. This causes inclusion of nodes 2 and 7. The entire labeling and scanning process is illustrated in Figure 3.14(a), where the single numbers associated with links are link capacities and the pairs are node labels. After node 5 has been scanned, node 8 is labeled. Therefore a flow of one unit can be sent over the path: $1 \rightarrow 4 \rightarrow 2 \rightarrow 5 \rightarrow 8$. A modified network is constructed accordingly, as shown in Figure 3.14(b). After having sent three units from node 1 to node 8, the final modified network is as shown in Figure 3.14(c). It can be seen at this stage that no path between nodes 1 and 8 can be found and the flow is maximal. The flow diagram is given by Figure 3.14(d), where the numbers associated with links are flow values carried by the links. □

Minimal Cost Flow Problem

The problem of *minimal cost flow* involves flow costs. In addition to link capacity constraints, for each unit flowing through a link there is an incurred cost. For a fixed flow requirement (from the source to the sink), the problem is to find the flow pattern that minimizes the flow cost. In this section let us consider a robotic material handling system. A number of workstations are placed along a linear track on which a robot is installed to move work units between workstations. Figure 3.15(a) shows a material handling pattern that moves five work units between ten workstations. The problem is to maximize the number of services in one trip. The robot may pick up work units at stations 1, 2, 3, 5, and 7 and deliver them at stations 4, 6, 8, 9, and 10 respectively, that is, there are 5 service requests: (1,8), (2,10), (3,4), (5,6), and (7,9). The robot may serve any of these requests, but it cannot hold more than two units at the same time. Units carried by the robot are placed in on-board storage with a capacity of two units.

This problem can be formulated as a minimal cost flow problem. Each workstation can be considered as a node and each move is represented by a

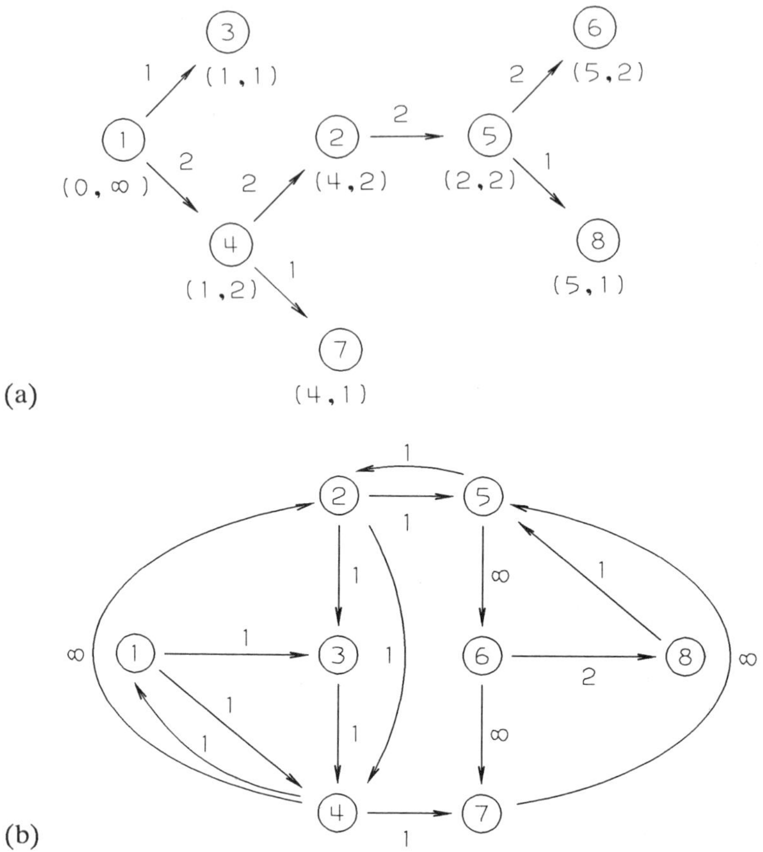

Figure 3.14 (a,b,c,d) Solution for a production schedule problem. (a) Labeling and scanning process for flow increment. (b) A modified network after the first flow increment.

link with a capacity 1 and a cost −1. The ten nodes are further connected by a directed path $1 \to 2 \to \cdots \to 9 \to 10$. Each link on this path has an infinite capacity and a zero cost. The same storage unit can be repetitively used for multiple service requests if and only if their corresponding links are not overlapping. If two links cross each other, their requests must be placed in separate storage units. Since the objective is to minimize flow cost, the links with negative costs will be chosen first. The problem is to find the minimal cost flow pattern with a flow value of 2. A network flow model is shown in

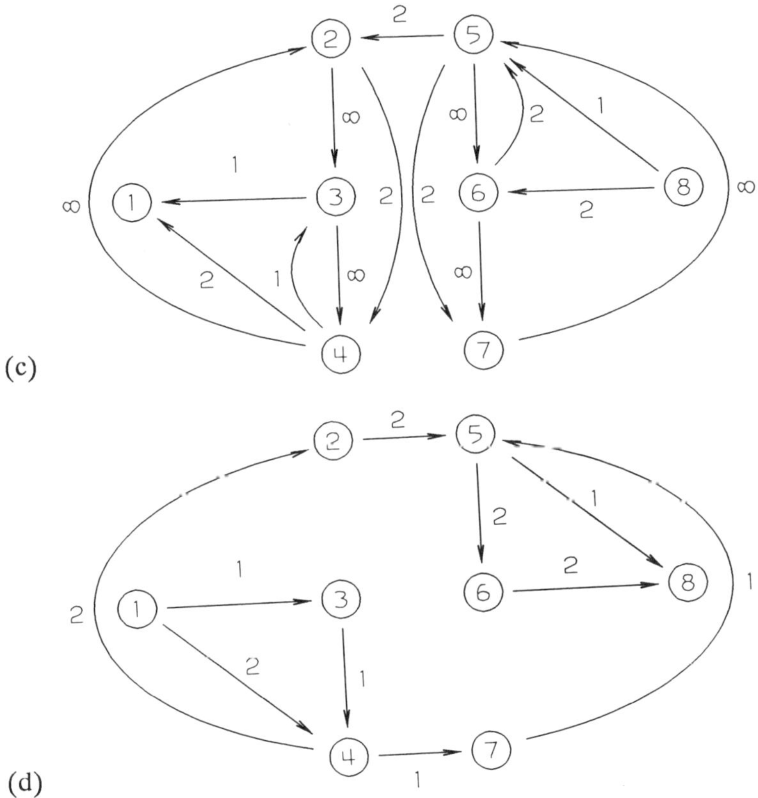

(c) The modified network after having reached the maximal flow. (d) The maximal flow pattern.

Figure 3.15(b), where the two numbers associated with each link are its cost and capacity, respectively.

There are two different approaches for the minimal-cost solution. One way is to find a sequence of shortest paths and let each path carry as much flow as possible until the flow requirement is satisfied. The other is to define a flow pattern that satisfies the flow requirement (this may be done by a maximum flow algorithm), then to improve the cost by changing the flow pattern. If the required amount of flow is small, the former is a better approach. A stepwise procedure is given by Algorithm 3.3.

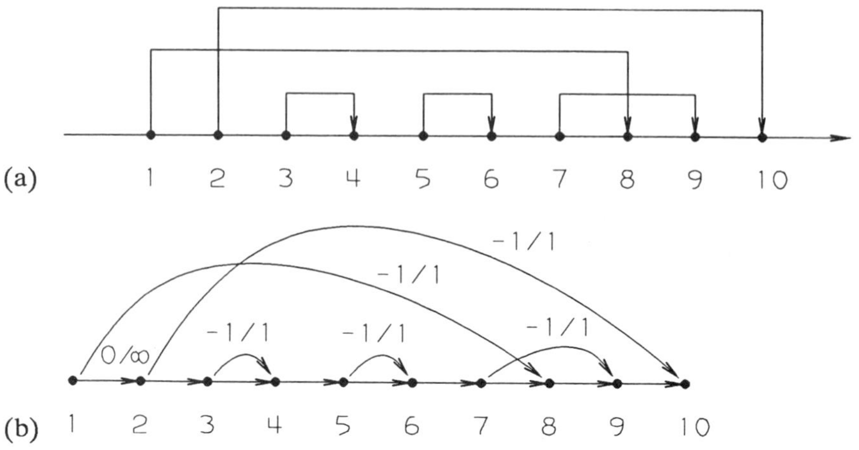

Figure 3.15 The maximal number of services and the minimal cost flow. (a) A service request pattern. (b) A minimal cost flow model.

Algorithm 3.3

1. Solve the shortest path problem by employing Algorithm 3.1 or 3.1(a) with the link lengths equal to the link costs.
2. If the smallest link capacity on the shortest path is greater than or equal to the flow requirement, then send the required amount of flow along the path. Stop further computation (the problem is solved). Otherwise go to step 3.
3. Send as much flow as possible along the shortest path. (This flow value is equal to the smallest link capacity.) Reduce the flow requirement by the value of the flow.
4. Modify the network. If a link (i,j) carries a flow of value f_{ij}, then place a reverse link from j to i with a capacity equal to f_{ij}. However, the cost of this reverse link is the same as that of link (i,j). If $f_{ij} < c_{ij}$, then let $c_{ij} = c_{ij} - f_{ij}$ and the link cost remains unchanged. Go to step 1.

Example 3.16 The problem of a robotic material handling system defined in Figure 3.15 is now solved by Algorithm 3.3. Referring to Figure 3.15(b), the first shortest path is obtained and given in Figure 3.16(a). The number associated with a link is its length, and with a node is the identification of the node. Since the smallest link capacity

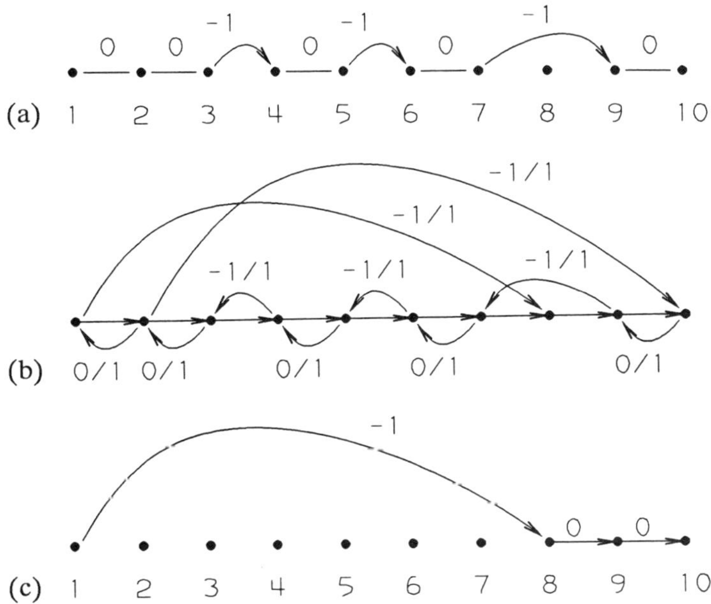

Figure 3.16 (a,b,c) A minimal cost flow solution process. (a) The first shortest path. (b) The first modified network. (c) The next shortest path.

on the path is 1, only one unit of flow can be carried by the path. Then a modified network follows, as given in Figure 3.16(b). From the modified network, the second shortest path is found and shown in Figure 3.16(c). One additional flow unit is now established. Since the total flow requirement is two, the problem is completely solved. The robot is going to serve the requests (1,8), (3,4), (5,6) and (7,9). This is the maximal number of services without violating the storage constraint. □

Bin Packing Problem

Packing problems play an important role in line capacity planning. Production activities always involve resource consumption. Suppose that the total amount of available resources can be quantified by discrete units and that the resource consumption rate of each activity is known. Each resource unit may support a number of production activities. The minimal cost policy is to "pack" production activities into a minimal number of resource units. Take labor planning as a example. In order to accomplish a given set of tasks, the line management has to determine the minimum number of workers during the regular working hours. Assume that it takes a worker t_i hours to finish task i. Then the problem is equivalent to packing the task times into a number of 8-hour slots and the total number of slots must be the minimum. This problem is said to be the *bin packing problem.* It can be stated as follows: A number of given objects of sizes $\{t_i\}$ should be packed into the smallest number of bins of uniform size, T. Clearly, the condition that $t_i < T$ must be imposed. Otherwise no solutions exist.

Unfortunately, no efficient algorithm is known. In the following a simple heuristic algorithm is introduced. This algorithm always gives a very good solution, if not optimal.

Algorithm 3.4

1. Put $\{t_i\}$ in descending order (i.e.the largest first) into a list S. Let $k = 0$.
2. Let $C = T$ (the bin size) and $k = k + 1$. (The kth bin is under consideration.)
3. Find the first element in S that is less than or equal to C. If such an element is found, then place it into the kth bin and delete it from S. (This is called the *first-fit rule.*) Reduce C by an amount equal to the element size. If S becomes empty after the deletion, a solution is found. stop.
4. If no elements are less than or equal to C, go to step 2 and start with another bin.

Example 3.17 Consider that 16 independent tasks are assigned to three operators. Each operator spends exactly eight hours; the task times are $\{1.1, 2.3, 1.5, 3.1\}$, $\{2.1, 1.9, 1.7, 1.0, 1.3\}$, and $\{0.9, 1.4, 1.2, 1.8, 0.7, 1.6, 0.4\}$. Using Algorithm 3.4 with $T = 8$ and $S = (3.1, 2.3, 2.1, 1.9, 1.8, 1.7, 1.6, 1.5, 1.4, 1.3, 1.2, 1.1, 1.0, 0.9, 0.7, 0.4)$, the solution shows that four workers are needed. Their total task times are respectively 7.9 ($= 3.1 + 2.3 + 2.1 + 0.4$), 8.0 ($= 1.9 + 1.8 + 1.7 + 1.6 + 1.0$), 7.4 ($= 1.5 + 1.4 + 1.3 + 1.2 + 1.1 + 0.9$), and 0.7. Note that the objective is to minimize the total number of workers. In this case, the ratio of the number of bins found by the algorithm to the optimal number is 4/3. From a practical viewpoint, the tasks of the last two workers may be combined so that three workers are needed and their working hours are 7.9, 8.0 and 8.1. □

3.8 Remarks

Further reading about queuing theory can be found in textbooks, e.g., Cox (1961), Newell (1971) and Kleinrock (1975). The first vigorous proof of $L = \lambda W$ is given by Little (1961) and a proof under a rather general condition (stated in Theorem 3.1) by Stidham (1974). For the concept of virtual delay the reader may consult Kleinrock (1965) and Wolff (1970). Approximations for $M/G/k$ queues are discussed in Nozaki (1978) and Boxma (1979). The exact queue length distributions in Tables 3.1, 3.2, and 3.3 are given in Hillier (1981). Queuing networks with product form solutions are discussed in Jackson (1957), Gordon (1967), Baskett (1975) and Chandy (1983). The concept of approximation of a closed queuing network by an open network is due to Whitt (1984).

In addition to the queuing approach, at least two other techniques have been used for manufacturing analysis: *simulation* and *Petri net*. Simulation can handle problems at a very detailed level, but is very time-consuming. Therefore this technique is recommended for final refinement study. Aspects of simulation technique will be introduced in Chapter 9.

A Petri net has two types of components, *states* of the system (e.g., the number of jobs at each workstation) and *transitions* that cause state changes. It is possible to construct relations between states and transitions by matrix forms. Analysis of system behavior can be conducted by using matrix state equations. Both the number of states and the number of transitions may increase very rapidly if the number of workstations or the number of jobs becomes large, say to exceed five. Applications of the Petri net have been limited to the cases where the number of stations is small (see Martinez 1986). For this reason, this approach is not considered in this book.

Classic optimization techniques are discussed in textbooks on calculus, advanced calculus or calculus of variation. The most popular solution technique in linear programming is the *simplex method* due to Dantzig. The Simplex method and its ramifications are given in Dantzig (1962). General solution techniques for integer programming can be found in Hu (1969). The principle of dynamic programming is developed by Bellman, and many applications can be found in the book by Bellman and Dreyfus (1962). For network flow problems, the reader may consult Ford (1962) and Hu (1969). The packing problem is well known in the area of *combinatorial optimization*. Christofides (1979) may be used for reference.

REFERENCES

Baskett, F., K. M. Chandy, R. R. Muntz, and F. G. Palacios (1975). Open, Closed, and Mixed Networks of Queues with Different Classes of Customers, *Journal of Association for Computing Machinery* v. 16, pp. 527-531.

Bellman, R. E., and S. E. Dreyfus (1962). *Applied Dynamic Programming* Princeton University Press, Princeton, N.J.

Boxma, O. J., J. W. Cohen, and N. Huffels (1979). Approximations of the Mean Waiting Time in an $M/G/s$ Queueing System, *Operations Research*, v.27, pp. 1115-1127.

Chandy, K. M., and A. J. Martin (1983). A Characterization of Product Form Queueing Networks, *Journal of Association for Computing Machinery*, v. 30, pp. 286-299.

Christofides, N., A. Mingozzi, P. Toth, and C. Sandi (1979). *Combinatorial Optimization*, John Wiley, New York.

Cox, D. R., and W. L. Smith (1961). *Queues*, Methuen & Co., London.

Dantzig, G. B. (1962). *Linear Programming and Extensions*, Princeton University Press, Princeton, N.J.

Ford, L. R., and D. R. Fulkerson (1962). *Flows in Networks*, Princeton University Press, Princeton, N.J.

Gordon, W. J., and G. F. Newell (1967). Closed Queueing Systems with Exponential Servers, *Operations Research*, v. 15, pp. 254-265.

Jackson, J. R. (1957). Networks of Waiting Lines, *Operations Research*, v. 5, pp. 518-521.

Hillier, F. S. and O. S. Yu (1981). *Queueing Tables and Graphs*, North Holland, New York.

Hu, T. C. (1969). *Integer Programming and Network Flows*, Addison-Wesley, Reading, Mass.

Kleinrock, L. (1965). A Conservation Law for a Wide Class of Queueing Discipline, *Naval Research Logistics Quarterly*, v. 12, pp. 181-192.

Kleinrock, L. (1975). *Queueing Systems, Volume I: Theory*, John Wiley, New York.

Little, J. D. C. (1961). A Proof of the Queueing Formula $L = \lambda W$, *Operations Research*, v. 9, pp. 383-387.

Martinez, J., H. Alla, and M. Silva (1986). Petri Nets for the Specification of FMSs, *Modelling and Design of Flexible Manufacturing Systems*, edited by A. Kusiak, Elsevier Science Publishing, Amsterdam.

Newell, G. F. (1971). *Applications of Queueing Theory*, Chapman and Hall, London.

Nozaki, S. A., and S. M. Ross (1978). Approximations in Finite-Capacity Multi-Server Queues with Poisson Arrivals, *Journal of Applied Probability*, v. 15, pp. 826-834

Stidham, S. (1974). A Last Word on $L = \lambda W$, *Operations Research*, v. 22, pp. 417-421.

Whitt, W. (1984). Open and Closed Models for Networks of Queues, *AT & T Bell Laboratories Technical Journal*, v. 63, pp. 1911-1979.

Wolff, R. W. (1970). Work-Conserving Priorities, *Journal of Applied Probability*, v. 7, pp. 327-337.

4

Process and Operation Design

An assembly line may consist of one or more operations. The total assembly time equals to the sum of the separate operation process times. The number of operations and operation contents are basically determined by the structure of the assembly and the complexity of assembly work. An assembly is usually composed of a number of components or subassemblies. If the total assembly work is too complex, engineers tend to break the entire assembly work into a number of operations, making each operation responsible for one or more subassemblies. Depending on its complexity, a subassembly may further be broken into components, and then operations may be defined for one or more components. The question is how to determine the level of complexity at each operation. Subjects to be discussed in this chapter are line balance, minimal cost process design, test/inspection strategies, and rework strategies.

4.1 Line Balance

A simple process design criterion is to balance the assembly line so that each operation takes approximately the same amount of time. A balanced line often means better resource utilization and consequently lower production cost.

Line balancing is a classic problem in manufacturing business. First the line designer or the manufacturing engineer should construct an assembly work diagram showing all work elements and their *precedence relations*. Then a *cycle time* is selected. The problem is to group work elements into a number of operations such that (i) the precedence relation is preserved, i.e., no work can be done unless all its preceding work elements have been completed, and (ii) the process time of each operation does not exceed the cycle time. Figure 4.1 illustrates a simplified precedence diagram for an assembly job. There are nine work elements, represented by nine nodes, respectively. The numbers associated with nodes are their corresponding process times. The precedence relation is shown by the directed links. For example, work element 5 has a process time of 6 units and cannot be started unless elements 1, 2, and 3 have been completed. The total assembly job takes 30 time units.

Many line-balancing algorithms are based on heuristic methods. No efficient computational methods for the exact solution are known. Comparisons between different algorithms are usually made by looking at the computational time and the *balancing efficiency*. If n workstations have been derived from an algorithm with a cycle time of T and a total assembly time of S, then the balancing efficiency is S/nT. This value can also be interpreted as the expected utilization.

Two principles have been widely used:

1. *Packing rule* The problem of grouping elements into operations with a predetermined cycle time is similar to a bin-packing problem. (See Section 3.7.) If no precedence relations exist, the two problems are identical. In order to keep a small number of operations, the largest element (i.e., the longest process time) is often considered first.

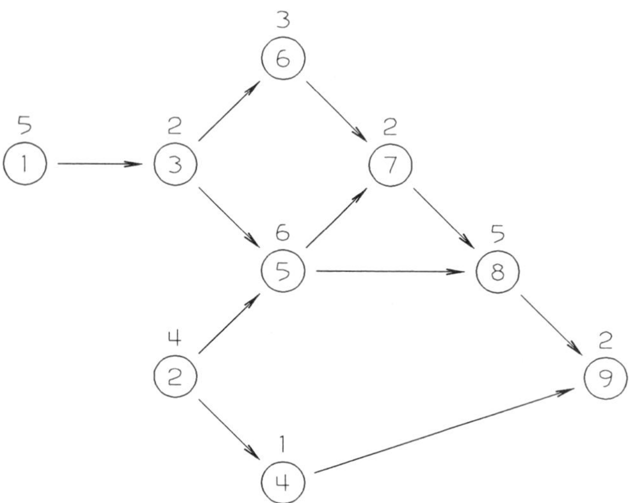

Figure 4.1 Precedence diagram for an assembly job.

2. *Proximity rule* Elements that belong to the same neighborhood in the precedence diagram are likely to be associated with the same subassembly. From a line operation point of view, related work should be placed under the same operator. Therefore the associated elements should be placed in the same operation.

Presented below are two heuristic algorithms: (i) *ranked positional weight* (RPW), and (ii) *largest set rule* (LSR). Both of them require the computation of a weight for each element. Let

t_i = process time of element i
w_i = weight of element i
p_i = set of all elements preceding element i.

The value of the weight is given by

$$w_i = \sum_{j \text{ in } p_i} t_j + t_i$$

In Figure 4.1, for example, w_7 equals $t_1 + t_2 + t_3 + t_5 + t_6 + t_7 = 5 + 4 + 2 + 6 + 3 + 2 = 22$. The RPW method is defined first by the following algorithm.

Algorithm 4.1 (Ranked Positional Weight)

1. Select a cycle time, namely T. Compute the weight factor for each element.
2. List all elements in descending order by their weights. Let S be such a list. Set $k = 0$.
3. $k = k + 1$ (A new operation will be assigned.)
   $T_k = T$
4. From S select the first-fit element, say element i, such that $t_i \leq T_k$ and no succeeding elements of i appear in S. If no such element can be found, go to step 6.
5. Assign element i to operation k. Delete element i from S. Decrease T_k by t_i. If S is empty, a completed solution is found. Stop. Otherwise, go to step 4.
6. If $T_k = T$, the selected cycle time is too small. No feasible solution exists. Otherwise, a new operation should be considered. Go to step 3.

Example 4.1 The computed weights and the ranks of the nine elements in Figure 4.1 are given by

Element i	9	8	7	5	6	3	4	1	2
t_i	2	5	2	6	3	2	1	5	4
w_i	30	27	22	17	10	7	5	5	4

The ranked list $S = (9,8,7,5,6,3,4,1,2)$ and $T = 10$. The algorithm places element 9 into operation 1. Then delete 9 from S and $T_1 = 10 - t_9 = 8$. The first element in S is 8. Since element 8 does not violate the restrictions (no successors in S and $t_8 \leq T_1$), it should be assigned to operation 1. So $S = (7,5,6,3,4,1,2)$ and $T_1 = 8 - t_8 = 3$. Element 7 can be assigned to the same operation for the same reason. Now $S = (5,6,3,4,1,2)$ and $T_1 = 1$. The next candidate is element 5. But t_5 exceeds T_1, and so does element 6. The next is element 3. Since neither element 5 nor element 6 has been assigned, element 3 cannot be selected. Element 4 does not violate the restriction and is assigned to operation 1. Now $T_1 = 0$. No more elements can be assigned to this operation. The algorithm will start assignment to operation 2 with $S = (5,6,3,1,2)$.

The completed solution consists of four operations. The elements assigned to each operation, in sequence, are (9, 8, 7, 4), (5, 6), (3, 1), and (2). Their corresponding process times are 10, 9, 7, and 4, respectively. This solution is illustrated in Figure 4.2, where each number in parentheses is the RPW associated with the element. Since the longest operation time is 10 units, the assembly line throughput is 0.1 per time unit. A material flow pattern is illustrated in Figure 4.2(b). If a line consists of exactly four workstations (one for each operation), we could expect a balancing efficiency at $(10 + 9 + 7 + 4)/40 = 0.75$. □

Since the weight of an element consists of all the process times of its predecessors, the algorithm tends to treat the elements near the end of precedence diagram first. Therefore, the problem is solved backwards. Furthermore, this method does not always place associated elements in the same operation. Consider a different precedence diagram with six elements, each with a process time of 5 units. The precedence relations are given by $1 \to 3 \to 5$, $2 \to 4 \to 5$, and $5 \to 6$. (See Figure 4.3.) For a cycle time of 10, three operations are defined as the results from Algorithm 4.1. This solution is illustrated in Figure 4.3. The elements included in the operations are respectively (1, 2), (3, 4), and (5, 6). Although the solution provides a perfect line-balance condition, it may not be appropriate from a material-handling viewpoint. Elements 1 and 2 respectively belong to two

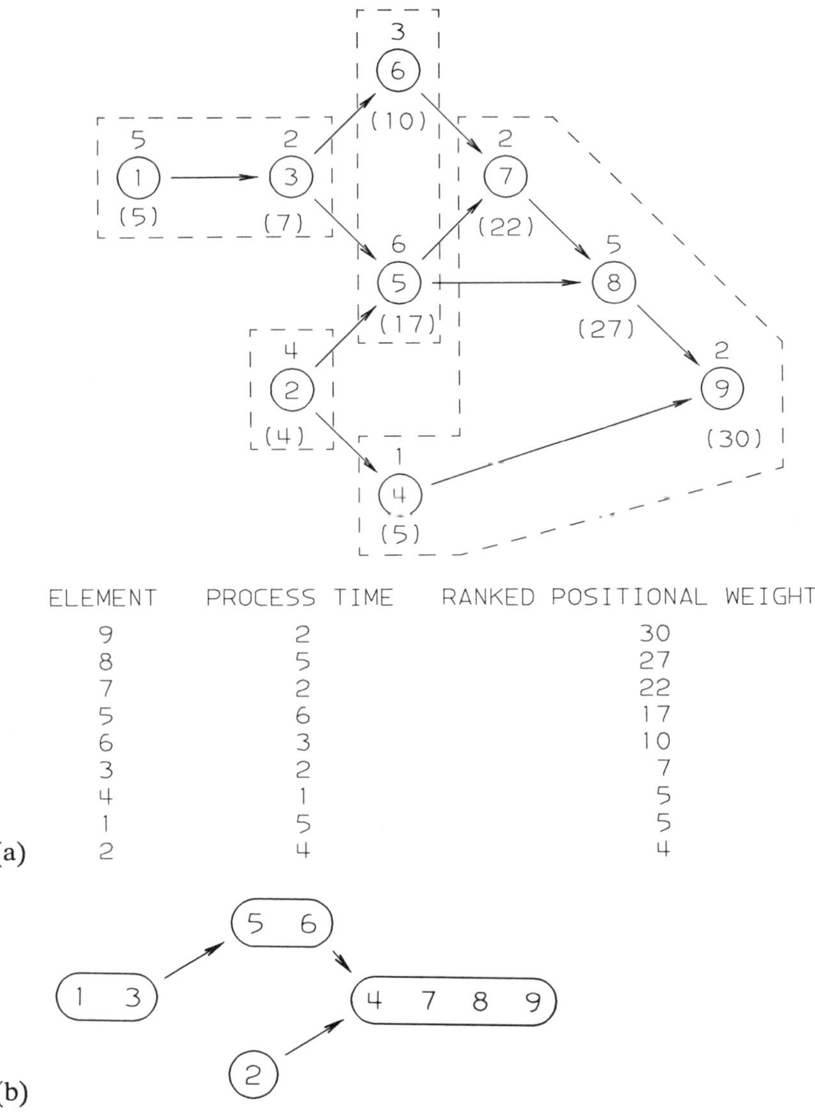

ELEMENT	PROCESS TIME	RANKED POSITIONAL WEIGHT
9	2	30
8	5	27
7	2	22
5	6	17
6	3	10
3	2	7
4	1	5
1	5	5
2	4	4

Figure 4.2 Line balancing by ranked positional weight method. (a) Ranked positional weight and its solution. (b) Material flow.

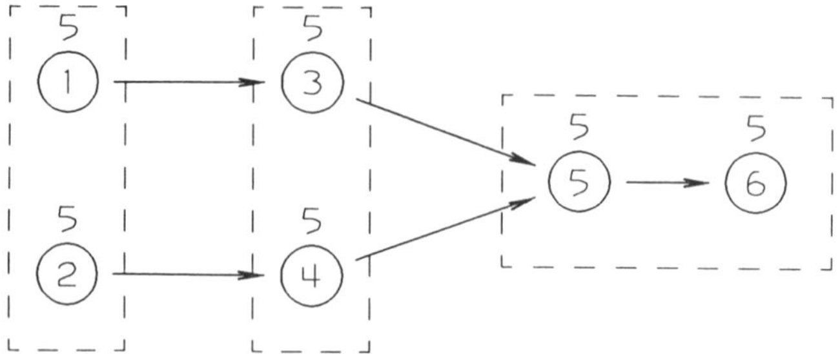

Figure 4.3 A balanced process with a poor material flow pattern.

different subassemblies. A better solution would be (1, 3), (2, 4), and (5, 6); that is, interrelated activities should be placed together. The next algorithm attempts to achieve this purpose.

Algorithm 4.2 (Largest Set Rule)

1. Select a cycle time, namely T. Let S be the set that consists of all elements and $k = 0$.
2. $k = k + 1$. (Consider a new operation.)
   Let $T_k = T$.
3. Compute the weight factor for each element in S. Find the element with the largest weight factor less than or equal to T_k, say element i. If no such element can be found, go to step 5. Otherwise, go to the next step.
4. Place element i and all its preceding elements in S into operation k. Delete all elements just placed from S. Reduce T_k by w_i, the weight factor of element i. If S is empty, a completed solution is found. Stop. Otherwise, go to step 3.
5. If $T_k = T$, then stop. The selected cycle time is too small and no feasible solution exists. If $T_k < T$, a new operation should be considered. Go to step 2.

Algorithm 4.2 also uses the weight factors, except that the weight factors are recomputed after each placement. Each placement involves a group of interrelated elements.

Example 4.2 Applying Algorithm 4.2 to the problem described in Figure 4.1, a different solution is obtained. Progressive solution steps are presented in Figure 4.4. The initial weight factors of nodes are computed and given by the numbers attached to individual nodes in

Figure 4.4(a). The largest factor that is less than or equal to the cycle time is $w_6 = 10$. Since the factor is the sum of t_1, t_3 and t_6, the first operation should include elements 1, 3 and 6. After having deleted these three nodes from the precedence diagram, Figure 4.4(b) is obtained. Then weight factors of unassigned elements are recomputed. Invoking the largest set rule again, elements 2 and 5 are assigned to the second operation. Finally the new value of w_9 becomes 10, and all remaining elements are included in the third operation. The completed solution by Algorithm 4.2 is (1, 3, 6), (2, 5) and (4, 7, 8, 9). In this case, we find a perfectly balanced solution and the balancing efficiency is 100%. □

Computer packages have been developed for line-balancing problems. These packages may enumerate and assess a large number of alternatives. A typical example is a solution package called COMSOAL, an acronym for Computer Method of Sequencing Operations for Assembly Lines. This method involves two steps: (i) to generate a feasible sequence of elements in accordance with their precedence relations, and (ii) to pack the elements into individual operations with a given cycle time in the order of the sequence. Once a packing sequence is defined, the packing itself becomes straightforward. It is the sequence generation method that needs to be developed. For a problem of n elements, there are $n!$ sequences. The number of feasible sequences is dependent on the structure of precedence relations. For example, the structure shown in Figure 4.5(a) has two feasible sequences: (1, 2, 3, 4) and (1, 3, 2, 4). However, the one in Figure 4.5(b) has six feasible sequences: (1,2,3,4), (1,3,2,4), (2,1,3,4), (2,3,1,4), (3,1,2,4), and (3,2,1,4).

Let p be the probability that a feasible sequence will lead to an optimal solution. The probability that an optimal solution can be covered by m feasible sequences is $q = 1-(1-p)^m$. Let $p = x/m$. Since $(1-x/m)^m \to e^{-x}$ as $m \to \infty$, $q \approx 1-e^{-x}$ for a large m. If $x = 5$, then $q \approx 0.9933$; that is, with a probability better than 0.99, at least one sequence is optimal. If $x = 10$, it is almost certain that an optimal solution will be found. In this case, $q = 0.9999546$. If p is known, the number of sequences that should be generated can be determined. For real-life problems, unfortunately, this information does not exist. Ad hoc rules have been suggested to increase the likelihood of reaching an optimal solution. Some of these rules may be illustrated by the following simple example.

Figure 4.6 is a precedence diagram with four elements. Three feasible sequences exist: (i) (1,2,3,4), (ii) (1,3,2,4), and (iii) (3,1,2,4). First, it can be seen that the first element in a feasible sequence must be either element 1 or element 3. For simplicity one may assume that both of them have

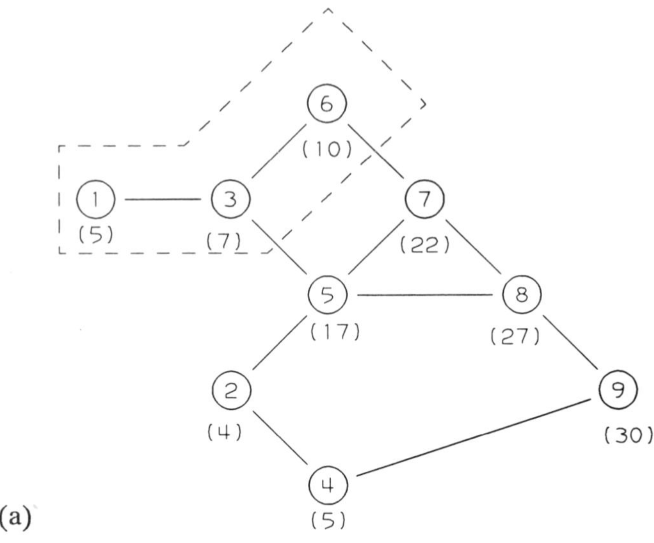

Figure 4.4 Line balancing under the largest set rule. (a) The first operation. (b) The second operation. (c) The third operation.

an equal chance to be selected. If element 3 is selected, the only possible sequence will be case (iii). Therefore the probability of generating case (iii) is 0.5. Suppose that element 1 is selected (with a probability 0.5), then either case (i) or (ii) will be obtained. The probability of either case will be 0.25. This means that some sequences have better chances to be generated as compared with others. To overcome this problem, a weight factor may be added to elements with a large number of followers. Next, it is known that one has a better chance to reduce the number of operations by packing the largest element first. A simple ad hoc rule is to add more weights to large elements. If the process times of elements 1 and 3 are 10 and 20 time units respectively, the probability of generating element 1 is $10/(10+20)$. A preference is given to element 3. Different rules may also be combined. For instance, the weight of element 1 is $10+15+5=30$ and of element 3 is 25. Then the probability of generating element 1 becomes 30/55.

Another enumeration scheme for line-balancing problems is the so-called *branch and bound* method. Instead of random generation, this method attempts to enumerate all feasible solutions. The enumeration process can be presented by a tree structure and starts at a root. First, all feasible combinations of elements for the first operation are enumerated. Each of these feasible combinations is represented by a branch emanating

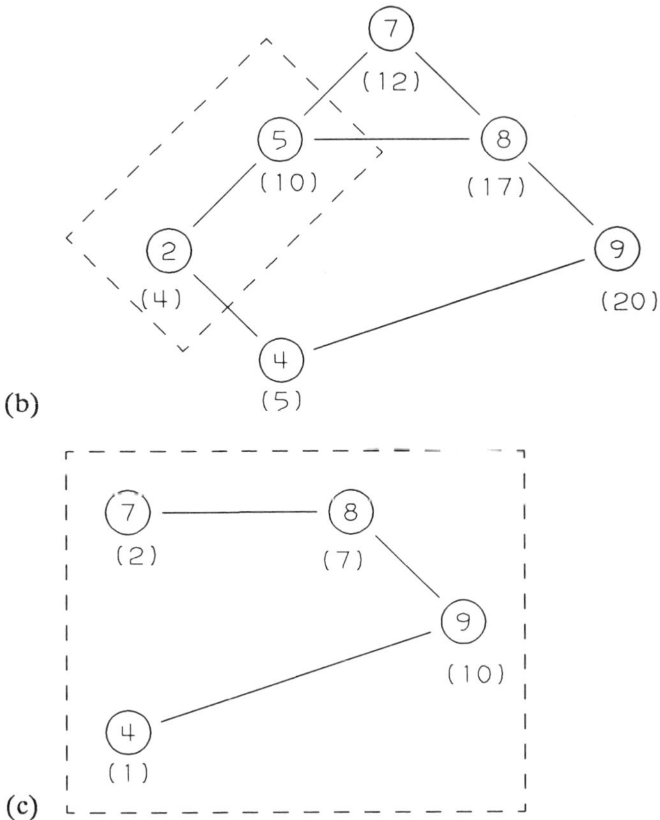

from the root. These branches may be called the first-level branches. Then for each first-level branch, a set of feasible combinations for the second operation is enumerated, using all unassigned elements. This process is carried out for as many levels as needed. Meanwhile, for each branch a lower bound of the total number of operations should be established. If the lower bound associated with a branch is no better than a solution already found, then the entire branch is discarded and no further enumeration from this branch is needed. If the known solution is close to optimal and lower bounds are tight, the branch and bound method can be very effective.

A simple lower bound is the least integer greater than or equal to the ratio of total process times of all unassigned elements to the selected cycle time. For instance, if $T = 10$ and $\{t_i\} = \{3,4,5,5,7,9\}$, then the minimal number of operations cannot be less than $(3+4+5+5+7+9)/10 = 3.3$. This implies that 4 is a lower bound for the minimal number of operations.

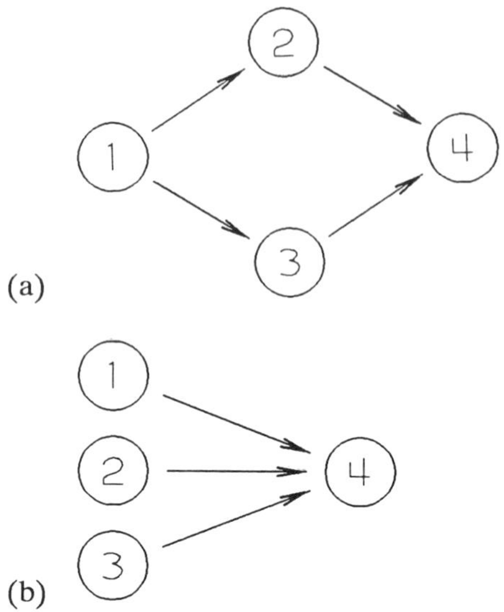

Figure 4.5 Precedence structure and the feasible sequence. (a) 4 nodes and 2 feasible sequences. (b) 4 nodes and 6 feasible sequences.

Suppose that a kth level branch is under consideration and S is the set containing all unassigned elements with respect to this branch. Then a lower bound associated this branch is given by $k + I[\sum_S t_i/T]$, where $I[x]$ is the least integer greater than or equal to x.

Intuitively, a branch with the smallest bound may have a better chance to cover an optimal solution. Therefore, a good solution may be obtained quickly by comparing the lower bounds of all branches. After having

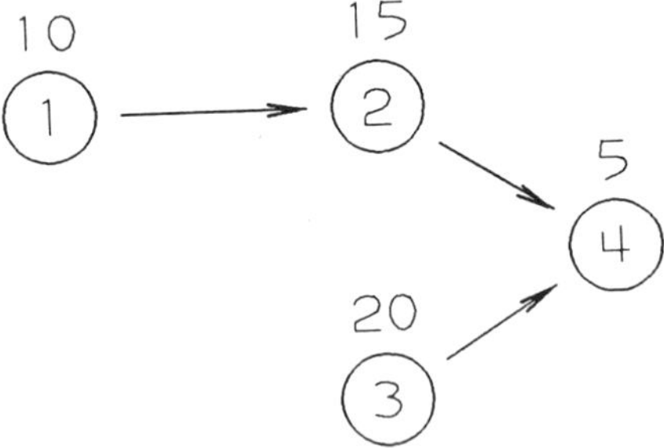

Figure 4.6 Generating feasible sequences with weight factors.

enumerated all first level branches, the one with the smallest bound is selected. From the selected branch, a set of second level branches and their lower bounds are created. Then the third level is under consideration. By repeating this procedure, an initial solution can be found. This solution may be replaced by a better solution during the branching process.

Unlike heuristic methods, the result from a branch and bound algorithm is the true optimal solution.

4.2 Other Considerations for Line Balancing

The problem of line balancing has variations which may add some complexity to the problem.

Multiple Products

First, a line-balancing problem becomes very complicated for assembly lines producing multiple products. Since assembly processes and process times may not be the same for different products, a single line cannot be balanced for all products. A popular approach is to balance a line by using average process times. In this case, however, the issue of scheduling different products must be considered. Small lot sizes may imply frequent tool changes and therefore high setup cost. On the other hand, since the line is not balanced for individual products, large lot sizes cause the line to be out of balance temporarily and create a high volume of work-in-process. To see this, consider two adjacent operations and two different products. The first operation has a process time of 3 minutes for product 1 and 7 minutes for product 2. At the second operation, process times for both products are 5 minutes. Assume that both products have the same demand. The line is balanced in an average sense. Suppose that the lot size is set to 1. The production cycles at operation 1 form an alternating process given by $(3, 7, 3, 7, \ldots)$, while the production cycles are identical at operation 2. For a buffer size of 1 between the operations, both operations can produce two pieces of product every 10 minutes. Hence, throughput is $2/10 = 0.2$. For a setup time of t minutes, the throughput is reduced to $2/(10 + 2t) = 0.2/(1 + 0.2t)$. This situation can be improved by increasing the lot size. Suppose that a lot size of 10 is chosen. The first operation will use $3 \times 10 = 30$ minutes to complete the first lot, then 70 minutes for the second. The throughput becomes $20/(100 + 2t) = 0.2/(1 + 0.02t)$. But when the first lot is sent to operation 2, the total work-in-process is increased to $10 + (10 - 30/5) = 14$.

To resolve the above dilemma, two approaches may be considered. One is called *group technology* and the other is *focused line*. The concept of group technology is to let the products of similar processes share the same line so that the line behaves as if there were a single product type. The focused line concept limits the line capacity so that each individual line deals with a small number of product types, e.g., one or two. Both approaches help to reduce setup times and work-in-process.

Variable Process Time

The second problem in line balancing is due to variable process times. Variability may take a number of different forms. The two most common ones result from human inconsistency found at manual operations and different reject conditions at test/inspection operations. It was mentioned in Chapter 2 that manual operation process time typically has a positively skewed distribution with a coefficient of variation between 0.1 and 0.65. (See Figure 2.9.) A test or an inspection operation may also cause variation problems, if a reject occurs before a complete test cycle. For instance, a complete test time is 10 minutes, but a bad product may be rejected in the first minute of the test time because a defect has been found. If the yield is high, say over 90%, then most test operations take a full cycle and a small percentage has considerably short test times. Consequently, the test operation time has a negative skewness index, i.e., a distribution with a left-hand tail.

Because of process time variation, a perfectly balanced line may not be fully utilized unless sufficient buffers are introduced between successive operations. The situation is similar to the case of multiple products described previously. Suppose that the average process time at an operation is 5 minutes, but the range of the process time is between 3 minutes and 13 minutes. A downstream operation has a constant process time of exactly 5 minutes. For a 3-minute cycle, the first operation works temporarily faster than does the second one. If no buffer is allowed between them, the first operation must wait 2 minutes before its next operation cycle begins. On the other hand, for a 10-minute cycle at the first operation, the second operation must wait 5 minutes before its next cycle. Consequently, for variable process times, the line-balancing problem becomes meaningless without considering the buffer-design problem.

Since a line is paced by its bottleneck, process time variation at the bottleneck area can be a severe problem. Figure 4.7 illustrates a simple example of four different operations: (i) short process time with low variability (ii) short process time with high variability, (iii) long process time with low variability and (iv) long process time with high variability. Suppose that the mean process times for the first two operations are S and for the last two are T. The process time distributions are shown by the four curves in Figure 4.7.

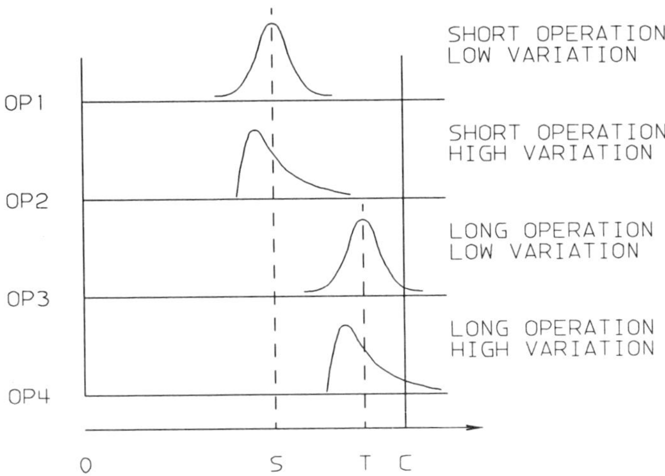

Figure 4.7 Process time and its variation.

A high variability implies that the distribution has more density (or more mass) at tail ends. One way of measuring line performance is to take the interval between two successive completion epochs as the line pace. This interval is called the *completion cycle.* If the mean process time is the only design parameter under consideration, one may conclude that the average completion cycle is close to T and line throughput will be $1/T$. Because of line detractors (e.g., material-handling delay, workstation availability, parts shortage, etc.), a small amount of contingency is normally planned. This makes a completion cycle, C, slightly larger than T. From a control point of view, a constant completion cycle is always desirable. If process times at an operation are consistently below C, then a constant completion cycle time is achievable. It can be seen from Figure 4.7 that none of the first three operations exceed the completion cycle time. A problem may arise at operation 4, where the probability that a process time exceeds C is significant. When an operation has a long process time and high variability, more work-in-process will be required to keep up the desired throughput; or else, a longer completion cycle and consequently a lower throughput can be expected. As a design guidance, a long process time should have low variability, and an irregular operation (i.e., high variability) should have a short process time.

Multiple Workstations

Another version of the line-balancing problem deals with multiple workstations. If the mean process time of an operation is larger than a planned completion cycle, multiple workstations are needed. In this case, the line is balanced at an operation level instead of at a workstation level. Since each operation may have multiple workstations, line sizing and balancing become interrelated problems. Compared to a small line, a large line tends to have a better balancing efficiency. On the other hand, a multiple-station line may complicate the material-handling system. Consider three adjacent operations: A, B and C. Operations A and C have only a single workstation, while operation B has two workstations, B1 and B2. The process flow is given by $A \rightarrow B \rightarrow C$. Since operation B has two workstations, the flow from A to B has two destinations while the flow from B to C has two origins. If B1 and B2 are placed in sequence, a flow from A to B2 and a flow from B1 to C would cross each other. Therefore, for each move, the material-handling system must (i) determine the origin and the destination, and (ii) prevent crossing flows from colliding.

Human Factors

Ergonomic concerns are also important in process design. Industrial experience has shown that repetition of the same motion pattern induces excessive muscle fatigue and may lead to body injury. This problem can be mitigated by mixing different motions or improving assembly tools and fixtures. These concerns sometimes can be more important than line balance.

As mentioned earlier, in Section 2.5, different operators may perform differently at the same operation. Line balancing becomes impossible if a poor job-assignment policy is implemented. On the other hand, assigning fast operators to bottleneck areas would help for balancing the line. This subject will be further discussed in Chapter 7.

Product Characteristics

The line designer does not always have the freedom to group elements any way he or she wants to achieve line balancing. In many cases an assembly is logically composed of a number of subassemblies. Tasks that belong to different subassemblies should not be assigned to the same operation. For instance, a product has a clean side and a dirty side. Assembly operations for the clean side must be kept in a clean room, while the dirty side can be exposed in ordinary factory air. In this case, separation becomes necessary. If the number of subassemblies is not many, the assembly process is often established by engineering judgment and economic reasons, such as the structure of assembly, tooling and fixture design, part usage and working space considerations.

Length of Cycle Time

Finally, it should be pointed out that the solution of a line-balancing problem is dependent on the selected cycle time, and that determination of cycle time is a complicated problem. As a mathematical problem, line balancing takes the selected cycle time as a constraint. A simple way to find the best cycle time is to try different cycles and compare their balancing efficiency. Normally, a large cycle time tends to permit a better solution. As a practical problem, however, many other factors must be taken into consideration. Further discussion of cycle time is given in the next section.

4.3 Assembly Cycle Time

The length of an assembly cycle time is a function of many parameters. Unfortunately, not all the parameters can be quantified, and therefore, no optimal solution is known. Decision is usually made by judgment. This section attempts to define the parameters that should be considered to reach a good decision. A list of such parameters and a summary of comparisons are given in Table 4.1.

A short (long) cycle often implies task simplicity (complexity). A short (long) cycle also means a relatively large (small) number of operations and a high (low) line throughput. For a given production demand, a short cycle usually results in fewer workstations per operation, as compared with a long cycle. For the same demand, a long-cycle line has more replicated tool sets and may need more work space. Furthermore, the complexity of a long task may make tool automation difficult. Thus, from a workstation-design viewpoint, a short cycle is more desirable.

Since a long cycle time will lead to fewer operations and more workstations per operation, the manufacturing line can be configured in either a vertical or horizontal fashion. A vertical integration policy tends to have a number of small (capacity) lines. Each line consists of a complete set of assembly operations. Horizontal integration, on the other hand, places all workstations of the same kind together, and the line is organized according to the function of each operation.

Uncertainty in product demand and major engineering changes often lead to line changes. For the same total capacity, a number of small lines has more flexibility than a single large line. If a production demand should be adjusted due to a demand change, one may add or delete (open or shut down) lines. For a small line size, this adjustment may follow the demand closely. For line modifications due to engineering changes, one can work

Table 4.1 Comparisons of short cycle and long cycle

Parameter	Short	Long	Comments
Workstation design			
Workstation cost	×		Fewer tools and spaces
Automation	×		Simplicity
Line flexibility			
Line organization		×	Vertical vs. horizontal
Capacity change		×	Demand uncertainty
Process change		×	Separate line
Line management			
Line scheduling		×	Focused line
Line inventory		×	Focused line
Dispatching		×	Fewer operators, fewer jobs
Routing complexity		×	Fewer workstations per operation
Quality			
Control		×	Problems identified
Handling damage		×	Fewer transports
Line efficiency			
Line balancing		×	Long cycle time
Material handling		×	Fewer operations
Line coordination		×	Short line, better throughput
Space utilization		×	Small line size
Human factors			
Job enrichment		×	Up to 10—12 minutes
Attainable pace			Short fatigue, long delay
Operator learning	×		Fast learning for short task
Group learning		×	Short line, better throughput

with one line at a time. Even in the case of line shut-down, it will not affect the entire production. Hence a long cycle time implies better line flexibility.

If a line is integrated horizontally (i.e., all workstations of the same kind placed together), capacity adjustment can be done by adding or deleting workstations. If a line is not well balanced and workstation cost is high, capacity adjustment under horizontal integration has a great cost benefit. Detailed discussion of line flexibility will be given in Chapter 8.

It has been mentioned that line balancing for multiple products is a difficult problem. This is also the case from a line-management point of view. Lot size and setup time are two major factors that affect line performance. The focused-line concept demands a number of small lines; each deals with one or two products. Hence both line-scheduling and line-inventory problems are simplified. Because there are fewer operators for fewer operations, job dispatching is relatively easy The job-routing problem becomes trivial if a material move between two successive operations has a single origin and a single destination.

A smaller number of operations means less material handling and, consequently, less handling damage. A long cycle time may further help to identify quality problems during an assembly process. Particularly, in a manual assembly environment, it is much easier to pinpoint the problem area if only a small number of operators participate in the assembly work.

Material-moving frequency is lower under a longer cycle assumption. Therefore, material-handling delay tends to decrease (see discussion in Chapter 5). The ratio of handling delay to cycle time is also smaller. A long cycle leads to a short line and facilitates line-coordination problems. (Communications between two people are far better than among twenty people.) Since the focused-line approach tends to have less line inventory and a small line size can fit demand better, a long cycle also means better utilization of floor space.

One of the major concerns in manual assembly is job enrichment. A long cycle usually implies a good variety and makes a job more interesting. From an ergonomic viewpoint, a long cycle design has many advantages. However, if an operation cycle is too long, the effective working speed may be reduced. An operator may not remember all assembly details and cause time waste on unnecessary work. Frequent tool changes make the work less productive. One empirical study in a telephone final assembly area (Buxey 1981) indicated that a change of cycle time from half a minute to 4.5 minutes greatly improved productivity. A feasible job duration can be as long as 12 minutes, according to a result from a domestic furnace assembly (Tuggle 1969).

Learning effect can be another major concern. The learning model, introduced in Chapter 2, suggests that the speed of an operator is a power function of the number of assembly cycles that have been completed. Therefore, a short cycle tends to have a high learning power. On the other hand, it is known that different operators have different learning speeds. Since the line throughput is dominated by the slowest operator, a large line tends to have a lower throughput. Thus the group learning speed is a decreasing function of the number of operators in the same line. From this aspect, a short cycle is not desirable.

Selection of cycle time is usually application-dependent and must be subject to the line-design objective. In the next section, a minimal cost method is discussed.

4.4 Design Optimization for Sequential Lines

In the previous sections, we have discussed the complexity of line-balancing problems and many concerns in the area of process design. If work elements in an assembly process can be arranged in a linear sequential order, it is possible to find a minimal cost solution by using a dynamic programming approach. (See Section 3.5.) From a practical viewpoint, a sequential line has several advantages. Since all material flows move in the same direction, line layout is straightforward. Material-handling and work-in-process management problems are simplified.

In the following paragraphs, an optimization procedure is given. This procedure considers explicitly three cost items: (i) tooling and fixture cost, (ii) space (installation) cost, and (iii) direct labor cost. For a given production demand the procedure generates a minimal cost solution, which includes cost estimation, operation definitions, the number of workstations per operation, workstation utilization, work-in-process level and labor requirement.

Consider an ordered list, $L = (l_1, l_2, \ldots, l_n)$, where l_i is the ith work element. The problem is to group these elements into a number of operations in accordance with their order in L. An operation that contains elements i, $i+1, \ldots, j$ is called i-j operation. The following notation is needed to facilitate our discussion:

D = production demand (number of assemblies per day)
TT = effective production hours per day
n_{ij} = number of workstations of i-j operation
t_i = mean process time of element i (hours)

Suppose that an i-j operation is established. The operation process time has a mean $t_{ij} = (t_i + t_{i+1} + \cdots + t_j)$. For a demand of D, the number of workstations, $n_{ij} = I[Dt_{ij}/TT]$. The incurred cost, $K(i,j)$, is the sum of tooling cost, space cost, and labor cost.

For a given sequential linear structure, the question is how to group the work elements into operations such that no two nonadjacent elements can be placed into the same operation. Figure 4.8 illustrates a set of 10 sequential tasks that have been grouped into six operations. The dynamic programming procedure starts with the first element and considers one

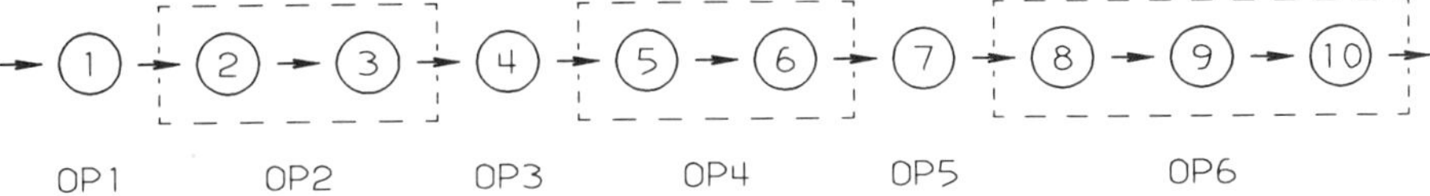

Figure 4.8 Group sequential tasks into operations.

element at a time. Let $S(j)$ be the minimal cost after having considered elements 1, 2, ..., j. For convenience, an initial condition for the dynamic programming procedure is defined by

$$S(0) = 0 \tag{4.1}$$

For a problem of n elements, the dynamic programming procedure has n stages. The minimal cost solution is defined by the following recurrent relation:

$$S(j) = \min_{1 \leq i \leq j} \{S(i-1) + K(i,j)\} \qquad \text{for } j = 1,2,\ldots,n \tag{4.2}$$

Equation (4.2) determines the minimal cost solution up to stage j by comparing j alternatives. The ith alternative suggests that elements i, $i + 1$, ..., j should be placed in the same operation with an incurred cost $K(i,j)$. The first $(i-1)$ elements have been considered before with the minimal cost solution $S(i-1)$.

When the nth stage is reached, the final answer is given by $S(n)$.

Example 4.3 Consider a four-element problem with constant process times. The daily production demand is 500 pieces. The element process times are 0.03, 0.04, 0.10, and 0.06 hours, respectively. Hence, the aggregated process times, $\{T_{ij}\}$, are given by the following triangular matrix:

$$T = \begin{bmatrix} 0.03 & 0.07 & 0.17 & 0.23 \\ & 0.04 & 0.14 & 0.20 \\ & & 0.10 & 0.16 \\ & & & 0.06 \end{bmatrix}$$

where the number in row i and column j is T_{ij} for $i < j$ and $T_{ii} = t_i$.

Assume that the effective production time is 21.5 hours per day. Then the number of workstations, n_{ij}, is given by $I[500 \times T_{ij}/21.5]$. The results are summarized by

$$N = \begin{bmatrix} 1 & 2 & 4 & 6 \\ & 1 & 4 & 5 \\ & & 3 & 4 \\ & & & 2 \end{bmatrix}$$

Again the number in row i and column j is n_{ij}.

To simplify the numerical evaluation process, only tooling replication cost is considered. For an i-j operation, the cost is given by the number located at the row i and column j of the following triangular matrix.

$$C = \begin{bmatrix} 10 & 15 & 30 & 60 \\ & 10 & 30 & 50 \\ & & 15 & 20 \\ & & & 20 \end{bmatrix}$$

Equation (4.2) is equivalent to

$$S(j) = \min_{1 \le i \le j} \left\{ S(i-1) + n_{ij} C_{ij} \right\}$$

For $i = 1$, $S(1) = S(0) + n_{11}C_{11} = 0 + 1 \times 10 = 10$.
For $i = 2$, there are two alternatives, i.e.,

$$S(2) = \min \begin{bmatrix} S(0) + n_{12}C_{12} = 0 + 2 \times 15 = 30 \\ S(1) + n_{22}C_{22} = 10 + 1 \times 10 = 20^* \end{bmatrix}$$

The first alternative combines elements 1 and 2 into one operation. The second alternative, however, demands both element 1 and element 2 to be standalone operations. Since the latter has a lower cost than the former, $S(2) = 20$. Continuing this process for $i = 3, 4$, we have

$$S(3) = \min \begin{bmatrix} S(0) + n_{13}C_{13} = 0 + 4 \times 30 = 120 \\ S(1) + n_{23}C_{23} = 10 + 4 \times 30 = 130 \\ S(2) + n_{33}C_{33} = 20 + 3 \times 15 = 65^* \end{bmatrix}$$

$$S(4) = \min \begin{bmatrix} S(0) + n_{14}C_{14} = 0 + 6 \times 60 = 360 \\ S(1) + n_{24}C_{24} = 10 + 5 \times 50 = 260 \\ S(2) + n_{34}C_{34} = 20 + 4 \times 20 = 100^* \\ S(3) + n_{44}C_{44} = 65 + 2 \times 20 = 105 \end{bmatrix}$$

The minimal cost is 100. By backtracking the computational process, it can be seen that the minimal cost solution requires 4 workstations for 3–4 operation. This is the second alternative at the last computational stage. After elements 3 and 4 have been assigned, the backtracking process goes to stage 2, where $S(2) = 20$. Element 2 should be a standalone operation and only one workstation is required. This leaves element 1 unassigned. Finally, the corresponding solution for $S(1)$ indicates that a single workstation for element 1 will sustain a demand of 500. The complete solution is given by $\{n_{11} = 1, n_{22} = 1, n_{34} = 4\}$. The total cost is $n_{11} \times c_{11} + n_{22} \times c_{22} + n_{34} \times c_{34} = 1 \times 10 + 1 \times 10 + 4 \times 20 = 100$, which is equal to $S(4)$. □

The computational procedure for real-life problems is more complex than illustrated by Example 4.3, because of cost structure and possible constraints. The tooling cost usually consists of two parts. The design and development cost for a new tool is charged only for the first tool set. This cost includes engineering cost, laboratory expenditures and software development expenses. The replication cost is mainly hardware and debugging cost. Space cost consists of installation cost and overhead. The installation cost is a one-time charge, and the overhead includes utility and maintenance. If a production facility will be used for five years, the total overhead should cover five years' expenses. Similarly, labor cost is computed by multiplying the number of person-years required and the average annual cost per person.

Line constraints may be considered during the computational process. Examples observed from existing lines are discussed as follows.

1. Workstation size The optimization procedure, defined by (4.2), attempts to combine work elements into an operation. If the combined operation requires a very large workstation, the solution may become infeasible. For example, in a manual workstation, an operator will have an ergonomic problem if the reaching distance is too far.

2. Nonintegrable operations Sometimes two adjacent work elements cannot be combined because of different working environmental requirements. In the electronics industry, for example, a clean-room operation must not be mixed with nonclean-room operations.

3. Utilization and effective hours Although productivity is directly related to workstation utilization, it is impractical to plan for 100% utilization. A manual operation should allow time for lunch and breaks, personal fatigue and delay, training, education and meetings. Tooling and fixtures in a workstation have both scheduled and unscheduled downtime. The unscheduled downtime is basically due to failures. Calibration, engineering changes and preventive maintenance are examples of scheduled downtime. Because of workstation downtime and operator allowance, the effective working time is less than 8 hours per shift. This fact can be reflected in the procedure by specifying the effective hours per shift (or per day) and placing an upper bound on workstation utilization.

4. Start factor When an operation has a yield problem, the number of assemblies started at the beginning of the line is usually higher than the number of finished ones. For an isolated operation, if the yield is 0.9, then on the average 1.111 products must be assembled in order to produce a good one. The average number of starts is called the *start factor*. If a line has multiple operations, the start factor of each operation is not necessarily identical to the reciprocal of the yield. In most cases, start factors can be obtained by solving simultaneous linear equations. The formulation of linear equations is identical to that defined by Equation (3.26).

Example 4.4 Computation of start factors may be illustrated by a numerical example. Figure 4.9 shows a simplified assembly process from a real assembly line, which consists of four assembly operations (P1, P2, P3 and P4) and two rework operations (P5 and P6). The yields of the assembly operations are 1.0, 0.97, 0.99 and 1.00, respectively. Let

v_i = start factor of operation i
q_i = number of assemblies scheduled at operation i
r_{ij} = probability that a product leaves operation i for j

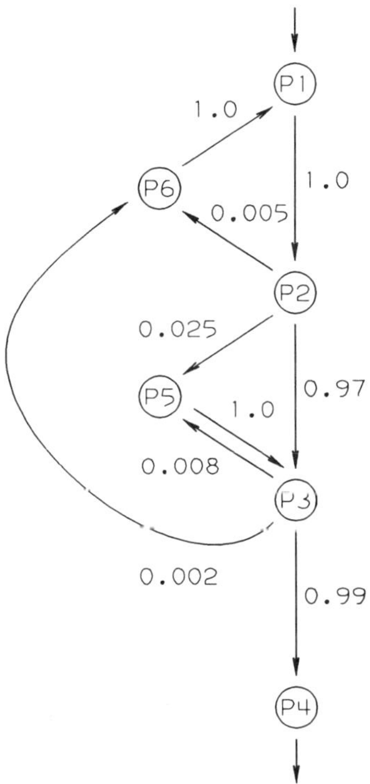

Figure 4.9 A simplified assembly process.

Since the number of starts is equal to the total number of arrivals at an operation, it follows that

$$v_j = (\sum r_{ij} v_i) + q_j \qquad \text{for all } j \tag{4.3}$$

The linear system becomes

$$v_1 = (v_6) + 1$$
$$v_2 = (v_1) + 0$$
$$v_3 = (0.97 v_2 + v_5) + 0$$
$$v_4 = (0.99 v_3) + 0$$

$$v_5 = (0.025v_2 + 0.008v_3) + 0$$

$$v_6 = (0.005v_2 + 0.002v_3) + 0$$

The solution is given by $V = (v_1, \ldots, v_6) = (1.0434, 1.0071, 1.0071, 1.1010, 1.0000, 0.0333, 0.0071)$.

All operations are manual and the line is planned for three shifts a day. The time detractors during a shift are:

Tool calibration	0.20 (hours)
Setup	0.20
Two 15-minute breaks	0.50
Workstation clean-up	0.15
Department meetings	0.15
Total	1.20 (hours)

Consequently, the total effective hours per day is $TT = 3 \times (8 - 1.2) = 20.4$. The *yield capacity* of a workstation is given by

$$g_i = TT/(v_i \times T_i)$$

where T_i is the average process time of operation i.

Suppose that the process times (in minutes) $T = (T_1, T_2, \ldots, T_6) = (4.0, 4.5, 3.8, 4.2, 4.3, 9.0, 8.5)$. The yield capacities of workstations are given by $G = (296, 273, 323, 268, 287, 4124, 20481)$. The estimated line capacity is given by the lowest value in G, that is, $g_4 = 268$ per day.

For a daily demand $D = 1150$, $n_i = I[1150/g_i]$, and the expected workstation utilization is given by $u_i = 1150/(g_i n_i)$. Hence, $N = (4, 5, 4, 5, 5, 1, 1)$ and $U = (0.971, 0.843, 0.890, 0.860, 0.800, 0.279, 0.056)$.

Because time detractors, yields and operation process times are variable, the estimated line capacity should be regarded as an average value. Actual daily production quantity may be higher in one day and lower in another. For this reason, a line designer may add a 5% contingency to stabilized production volume. This implies that utilization should be consistently below 95%. Consequently, the new value of n_i is $I[D/(1-e)g_i]$, where e is the contingency factor added to operations.

Since u_1 is the only utilization factor exceeding 0.95, the new values of n_1 and u_1 are 5 and 0.777, respectively. □

Because of rework operations, the assembly process illustrated in Figure 4.9 is not sequential. Therefore the recurrent relation (4.2) cannot be used directly. There are two ways to deal with rework, contingent on the rework-detection mechanism. If a bad assembly is found at the same operation that causes defects, the extra rework time may be considered as a part of operation process time. For example, if a new build time and a rework process time are 5 and 9 minutes, respectively, then for a 0.95 yield the expected operation process time is $5 + 0.05 \times 7/0.95 = 5.37$. If a bad assembly is found downstream (typically by an inspection or test operation), a separate rework operation is usually required. In this case, the separate rework operation is not allowed to be combined with new build operations, and can be treated as a nonintegrable operation.

The previous discussion is summarized in Algorithm 4.3.

Algorithm 4.3

1. For a given set of work elements, compute their start factors, $\{v_i\}$, by (4.3).
2. Determine the effective production hours per day, TT, and the expected product life-time in years, Y.
3. Let $S(0) = 0$ and $j = 0$.
4. Let $j = j + 1$. (Consider the next element.)
5. Compute $K(i,j)$ for $i = 1, 2, \ldots, j$.

   a. Yield process time

$$\bar{t}_{ij} = \sum_{k=i}^{j} (t_k v_k)$$

   b. Number of workstations

$$n_{ij} = I[D\bar{t}_{ij}/TT(1-e)]$$

   c. Tooling cost

$$C_{ij}^{t} = d_{ij} + n_{ij} c_{ij}$$

where d_{ij} is the design and development cost and c_{ij} is the replication cost for operation i-j

d. Space cost

$$C_{ij}^{s} = n_{ij} s_{ij} (a + b)$$

where a is the unit installation cost, b the total overhead charge per unit area over Y years, and s_{ij} the space for an i-j workstation.

e. Labor cost

$$C_{ij}^{l} = n_{ij} l_{ij}$$

where l_{ij} is the total labor cost for an i-j workstation over Y years.

Then $K(i,j) = C_{ij}^{t} + C_{ij}^{s} + C_{ij}^{l}$. If operations i, $i + 1$, ..., j are not integrable due to ergonomic or environmental constraints, then let $K(i,j) = \infty$.

6. Invoke relation (4.2) to obtain $S(j)$.
7. If $j = n$, the optimal solution is given by $S(n)$. The algorithm stops. Otherwise, go to step 4.

This algorithm can also be used to improve an existing line. A real-life case study is presented in the next section.

4.5 A Case Study

Consider a subassembly area of an existing line. This area consists of 10 operations in a clean-room environment. The space cost is $500 per square foot. Since all necessary tools have been installed, no design and development cost is included. To illustrate the results with different objectives, the following four cases will be compared:

1. Business as usual (BAU)
2. Minimal tooling cost only (MTO)
3. Minimal space cost only (MSO)
4. Minimal tooling and space costs (MTS)

Table 4.2 Input Data to Sequential Process Design

Operation	Cost (k$)	Start Factor	Space (sq. ft.)	Time (hrs.)
OP-1	53.2	1.10	120	0.0078
OP-2	3.5	1.10	100	0.0561
OP-3	1.5	1.10	100	0.0934
OP-4	16.2	1.10	120	0.0110
OP-5	1.8	1.10	100	0.0110
OP-6	8.2	1.06	120	0.0833
OP-7	25.2	1.08	60	0.0900
OP-8	13.0	1.06	100	0.0171
OP-9	1.2	1.03	120	0.0094
OP-10	1.2	1.03	120	0.0084

The replication costs, start factors, space requirements, and average process times are given in Table 4.2.

The assembly line is running two shifts a day with an effective time of 15.2 hours. The production demand is 236 pieces per day. A team of three engineers (a cost engineer and two manufacturing engineers) has worked out the cost and space requirement under different process design alternatives. The values of $\{c_{ij}\}$ and $\{s_{ij}\}$ are given respectively in Tables 4.3 and 4.4, where i is an index for row number and j for column number. In Table 4.3, for example, the entry at row 1 and column 3 means that operations 1, 2 and 3 may be combined into a single operation at a cost of $56,200. However, operation 4 is for oven cure, and must be a standalone operation. All entries in column 4, except c_{44}, correspond to infeasible cases and are indicated by minus signs.

The estimated cost for BAU is shown in Table 4.5, where operation number, the number of workstations per operation (NWS), tooling cost per workstation type (TOOL/$k), and space requirement (SPACE) in square feet are given. Results of the other three cases (i.e., MTO, MSO and MTS) are obtained by invoking Algorithm 4.3, and are shown in Tables 4.6, 4.7, and 4.8, respectively. The total number of workstations, the total tooling cost and total space requirements are given at the bottom of each table. The first column of each table indicates the result of executing Algorithm 4.3. In the MTO case, for example, operations 5 and 6 are combined into one operation, and operation 9 and 10 are merged with operation 8. Consequently, only 7 operations are shown in Table 4.6.

The results from Tables 4.5 through 4.8 are summarized in Table 4.9. It can be seen that MTS has the lowest cost. The saving mainly comes from

Table 4.3 Replication Costs for Operation Merge (Thousands of Dollars)

53.2	55.7	56.2	—	—	—	—	—	—	—
	3.5	4.0	—	—	—	—	—	—	—
		1.5	—	—	—	—	—	—	—
			16.2	—	—	—	—	—	—
				1.8	9.0	—	—	—	—
					8.2	—	—	—	—
						25.2	—	—	—
							13.0	13.2	13.2
								1.2	13.2
									1.2

Table 4.4 Space Requirements for Operation Merge (Square Feet)

120	120	120	—	—	—	—	—	—	—
	100	120	—	—	—	—	—	—	—
		100	—	—	—	—	—	—	—
			120	—	—	—	—	—	—
				100	120	—	—	—	—
					120	—	—	—	—
						60	—	—	—
							100	120	120
								120	120
									120

Table 4.5 Line Cost for BAU Case

OP	NWS	TOOL/$k	SPACE
1	1	53.2	120
2	2	7.0	200
3	2	3.0	200
4	1	16.2	120
5	1	1.8	100
6	2	16.4	240
7	2	50.4	120
8	1	13.0	100
9	1	1.2	120
10	1	1.2	120
Total	14	163.4	1440

Table 4.6 Line Cost for MTO Case

OP	NWS	TOOL/$k	SPACE
1	1	53.2	120
2	2	7.0	200
3	2	3.0	200
4	1	16.2	120
5/6	2	18.0	240
7	2	50.4	120
8/9/10	1	13.2	120
Total	11	161.0	1120

Table 4.7 Line Cost for MSO Case

OP	NWS	TOOL/$k	SPACE
1/2/3	3	168.6	360
4	1	16.2	120
5/6	2	18.0	240
7	2	50.4	120
8/9/10	1	13.0	120
Total	9	266.2	960

Table 4.8 Line Cost for MTS Case

OP	NWS	TOOL/$k	SPACE
1	1	53.2	120
2/3	3	12.0	360
4	1	16.2	120
5/6	2	18.0	240
7	2	50.4	120
8/9/10	1	13.0	120
Total	10	162.7	1080

Table 4.9 Comparison of Different Objective Values

Case	No. of Sta's	Tool Cost($k)	Space (sq. ft.)	Space Cost($k)	Total Cost($k)	Saving (%)
BAU	14	163.4	1440	720	883.4	—
MTO	11	161.0	1120	560	721.0	18
MSO	9	266.2	960	480	746.2	16
MTS	10	162.7	1080	540	702.7	20

better space utilization. The maximal saving in tooling cost is only $2,400 in the MTO case. On the other hand, the maximal space cost saving is found to be $240,000 (= 720k–480k) in the MSO case.

4.6 Test/Inspection Strategy

To assure product quality, testing and inspection may be necessary. In some cases, the cost of test equipment can be over one million dollars and a test operation may last a day or more. The complexity of a test function is dependent on product technology, complexity, and expected performance. A line designer must decide what to test, how to test, and when to test. A good methodology should help to reduce manufacturing cost by carefully planning test strategy. In principle, a test operation should be performed as early as possible in an assembly line so that defects can be quickly found. On the other hand, since an assembly line is progressive, a test function should be installed after all interrelated components have been placed together. Early detection of defects often implies frequent test installations and therefore high cost.

Optimization of test/inspection strategy can be a very complex problem. This section considers a relatively simple case, and presents a computational method and the structure of a cost model. Figure 4.10 shows a flow diagram, where circles represent assembly operations, the triangle stands for a test/inspection operation, and squares for reworks. It is assumed that:

1. a test/inspection operation, H, occurs after the k-th assembly operation, A_k
2. all defects will be found at the test/inspection station
3. defects are reworkable and will be sent to rework stations, $\{R_i\}$

At the test/inspection operation, i.e., operation H, causes of defects are determined. Bad assemblies are sent for rework. A defect caused by A_i is called a type-i defect. A type-i defect will be handled by rework station R_i. The function of a rework operation is to disassemble the product. If a defect caused by A_i is found after A_k for some $k > i$, then the product must go through $R_k, R_{k-1}, \ldots, R_i$ for rework. After rework, the product is reintroduced into the line at A_i. If product defects are caused by two or more assembly operations, the product must be sent back to each of the corresponding rework operations and is reintroduced to the line. Since it is assumed that defects are reworkable, each assembly scheduled at the beginning of the line will eventually get through the final stage as long as the overall yield is nonzero. Consequently, if an input flow of value one is applied to the network shown in Figure 4.10, the output flow should have a value of one, too. To formulate the problem, let

y_i = yield factor at A_i

$d_i = 1 - y_i$ = defect rate at A_i

v_i = amount of flow leaving A_i

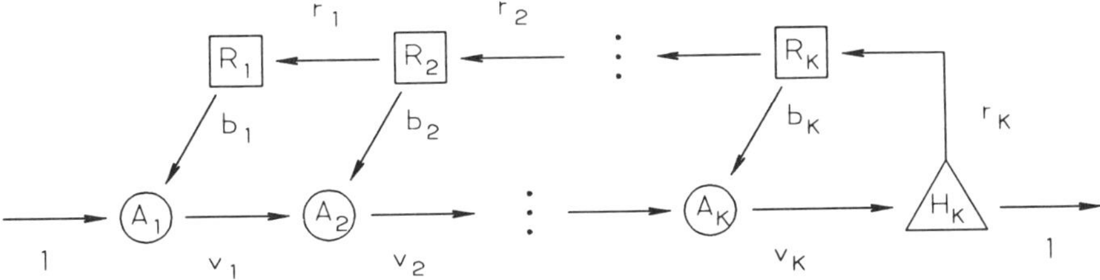

Figure 4.10 An assembly process with test and rework operations.

b_i = amount of flow from R_i to A_i

r_i = amount of flow entering R_i

To analyze process flow, it suffices to consider an input flow of a unit value. For instance, if the production demand is 150 pieces a day, the amount of flow leaving A_i will be $150v_i$. Under a unit demand, the total amount of flow entering (or leaving) an operation is equivalent to the start factor of the operation.

For a given set of yield factors, $\{y_i\}$, the flow values can be obtained by induction. At A_1, the law of flow conservation leads to

$$b_1 + 1 = v_1 \tag{4.4}$$

Since the defect rate is d_1, it follows that

$$v_1 d_1 = b_1 \tag{4.5}$$

Solving (4.4) and (4.5), we have

$$v_1 = 1/y_1 \qquad \text{and} \qquad b_1 = d_1/y_1$$

Considering the flows at R_1, r_1 is given by

$$r_1 = d_1/y_1$$

v_1 consists of two parts: good assemblies and bad assemblies. Since assemblies with defects will eventually be sent to R_1 and reintroduced to A_1, the amount of flow carrying defective products is equal to b_1. The amount of flow for the good assemblies is then equal to $v_1 - b_1 = 1$. The throughput flow at A_2 is composed of four different types: (i) good assemblies, (ii) assemblies with defects caused by A_1 only, (iii) assemblies with defects caused by A_2 only, and (iv) assemblies with defects caused by both A_1 and A_2. The flow values of these four types are, respectively:

1. $(1 + b_2)y_2$
2. $b_1 y_2$
3. $(1 + b_2)d_2$
4. $b_1 d_2$

Types 2 and 4 are always returned to R_1, regardless of defect types. Type 3 will be handled by R_2 only. Thus, $b_2 = (1 + b_2)d_2$. This implies that $b_2 = d_2/y_2$, and the flow value of type 1 becomes $(1 + d_2/y_2)y_2 = 1$. Thus $r_2 = r_1 + b_2 = b_1 + b_2 = d_1/y_1 + d_2/y_2$, and $v_2 = v_1 + b_2 = 1 + d_1/y_1 + d_2/y_2$.

By induction, it is straightforward to show that

$$b_j = d_j/y_j \tag{4.6}$$

$$r_j = \sum_{i=1}^{j}(d_i/y_i) \tag{4.7}$$

$$v_j = 1 + \sum_{i-1}^{j}(d_i/y_i) \tag{4.8}$$

The value of v_j consists of a good product and b_i bad products that should returned to stage i, for $i = 1, 2, \ldots, j$. After a test operation at H, all defective products are sent back for rework and the flow value becomes one again. It is assumed that a final test operation is installed so that the amount of flow leaving the line is always identical to the amount of flow scheduled at the beginning of the line.

Two types of costs should be considered. By setting up many test operations along a line, it will indeed reduce the total amount of rework—but will increase the testing cost. Therefore, a trade-off between testing cost and rework cost exists. Intuitively, a test operation can be discarded if the yield factor is high. For a more accurate answer, a quantitative method should be used.

Different objectives may be considered in problem formulation. Examples are (i) minimal total process time, (ii) maximal throughput, and (iii) minimal cost. However, their computational principles are similar. The system shown in Figure 4.10 can be regarded as the basic working structure in the development of test strategy. Suppose that two test stations are installed after A_i and A_j with no other test operations between them. Then the amount of flow entering operation $i + 1$ and the amount of flow leaving operation j are both equal to one. The flow structure between A_{i+1} and A_j is similar to the basic structure shown in Figure 4.10. For a line with k test operations, there are k basic structures. Each contains a number of assembly operations, a number of rework operations, and a test operation. The assembly and the rework operations are in one-to-one correspondence, as illustrated in Figure 4.10. Each basic structure can be evaluated independently by the same method. This property simplifies the searching procedure for the optimal strategy. For convenience, a basic structure between test operations just after A_{i+1} and A_j is denoted by $B(i,j)$. Figure 4.11 illustrates an assembly process with three such structures. The problem is to partition an assembly process into a number of basic structures to achieve a predefined objective. For mathematical analysis, the following notation is adopted:

T_i = mean process time at A_i
S_i = mean process time at R_i
U_i = mean test time if the test is performed just after A_i
g_i = (workstation) cost of A_i
w_i = (rework station) cost of R_i
h_i = cost of the test station just after A_i
z = labor cost per time unit
TA_i = total available time of A_i each day
TR_i = total available time of R_i each day
TH_i = total available time of H_i (the test just after A_i) each day
Q = daily production demand

If a line is operating three shifts a day, the value of TT is usually less than 24 hours. Time detractors may include workstation maintenance, machine

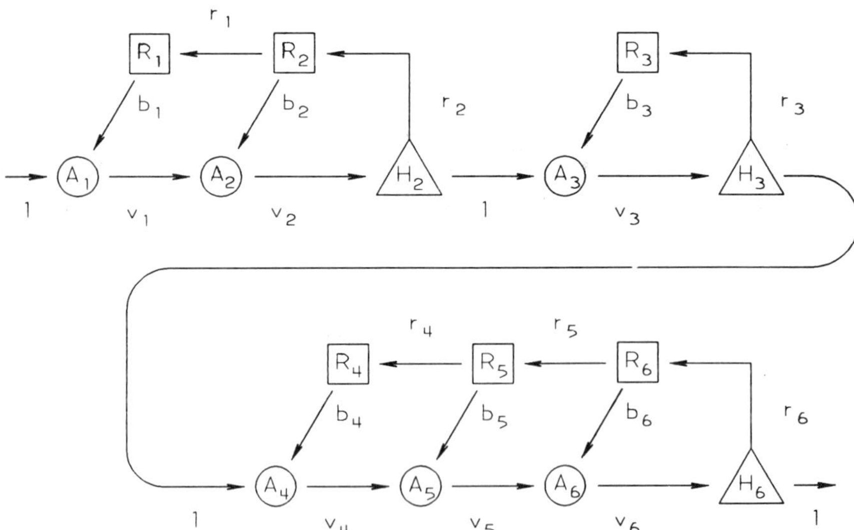

Figure 4.11 Basic structures in an assembly process.

calibration, setup, operator training, lunch time, break, meetings, and so on. (See Example 4.4, for instance.)

Minimal Total Process Time

Let $w(i,j)$ be the total process time in $B(i,j)$. Since the throughput at A_i is v_i, the assembly time for v_i products at A_i is $v_i T_i$. Likewise, the rework time at R_i is $r_i S_i$. Hence,

$$w(i,j) = \sum_{k=i}^{j} v_k T_k + \sum_{k=i}^{j} r_k S_k + v_j U_j \tag{4.9}$$

where r_k and v_k are given by (4.7) and (4.8), respectively.

Let $F(i)$ be the minimal total process time for the first i operations (including the first i rework stations), given that a test operation exists after operation i. A dynamic programming procedure, illustrated in Algorithm 4.4, will find the optimal test strategy.

Algorithm 4.4

1. Let $F(0) = 0$ and $j = 0$.
2. $j = j + 1$. (Proceed to the next stage.)
3. Compute

$$F(j) = \min_{1 \leq i \leq j} \{F(i) + w(i + 1, j)\}$$

   Record the index i^o such that $F(j) = F(i^o) + w(i^o + 1, j)$.
4. If $j < n$ (the last operation number), go to step 2. Otherwise, stop.

The index set, recorded in step 3, reports the locations of test operations under the optimal strategy. The minimal total process time is given by $F(n)$.

Maximal Throughput

Let $w(i,j)$ be the minimal yield capacity of all operations contained in $B(i,j)$. The yield capacity is equal to the total available time per day divided by the product of the start factor and the mean process time. $w(i,j)$ is the minimum of the following:

1. $TA_k/(v_k T_k)$ for all k in $B(i,j)$

2. $TR_k/(r_kS_k)$ for all k in $B(i,j)$
3. $TH_j/(v_jU_j)$

Let $F(i)$ be the maximal throughput for a line consisting of the first i operations, including the corresponding rework stations and the last test operation.

Algorithm 4.5

1. Let $F(0) = 0$ and $j = 0$.
2. $j = j + 1$. (Proceed to the next stage.)
3. Compute

$$F(j) = \max_{1 \leq i \leq j} \{\min[F(i), w(i+1,j)]\}$$

Record the index i^o such that $F(j) = \min\{F(i^o), w(i^o + 1,j)\}$.
4. If $j < n$ (the last operation number), go to step 2. Otherwise, stop.

In step 3 the smaller value of $F(i)$ and $w(i+1,j)$ is the throughput of the bottleneck before operation j if the last two test operations are placed after i and j, respectively. The index set, recorded in step 3, reports the locations of test operations under the optimal strategy. The maximal throughput is given by $F(n)$.

Minimal Cost

For a trade-off analysis, two major cost items should be considered. One is the workstation cost and the other is labor. If the daily demand is Q, the numbers of workstations for assembly, rework, and test operations are respectively given by

1. $NA_k = I[Qv_kT_k/TA_k]$
2. $NR_k = I[Qr_kS_k/TR_k]$
3. $NH_k = I[Qv_kU_k/TH_k]$

where $I[x]$ is the least integer greater than or equal to x.

Workstation cost may include tooling cost and space cost. Once the number of workstations has been determined, the workstation cost for each operation can be computed in the same way as described in steps 5c and 5d of Algorithm 4.3.

The labor cost is dependent on the total workload encountered at each workstation that requires operators, and is usually proportional to the product of the start factor and the mean process time. For example, if operation i is a manual operation, the average daily workload is $v_i T_i Q$. Dividing the workload by the available time per day, the expected number of operators, m_i, is obtained. If each operator costs x dollars during the product life, the labor cost associated with the operation is xm_i.

Let $w(i,j)$ be the incurred cost for all operations contained in $B(i,j)$, including both workstation cost and labor cost. Let $F(i)$ be the minimal cost for a line consisting of the first i operations, including the corresponding rework stations and the last test operation.

Algorithm 4.6

1. Let $F(0) = 0$ and $j = 0$.
2. $j = j + 1$. (Proceed to the next stage.)
3. Compute

$$F(j) = \min_{1 \le i \le j} \{F(i) + w(i+1,j)\}$$

   Record the index i^o such that $F(j) = F(i^o) + w(i^o + 1,j)$.
4. If $j < n$ (the last operation number), go to step 2. Otherwise, stop.

The index set, recorded in step 3, reports the locations of test operations under the optimal strategy. The minimal cost is given by $F(n)$.

It can be seen that Algorithms 4.4, 4.5, and 4.6 are all based on the same concept, except that their objectives are different. To illustrate these algorithms, let us consider a simple example.

Example 4.5 Consider a line with four assembly operations, as shown in Figure 4.12. The average assembly process times are 4.3, 5, 3.8, and 4.7 minutes, respectively. A defective product is sent to rework where an assembly is disassembled and reintroduced to the line. The disassembly time is the same as the corresponding assembly time, i.e., $S_i = T_i$ for all i. A test operation may be placed after each assembly operation. The average test times after operations 1, 2, 3 and 4, are 10, 12, 15, and 20 minutes, respectively. The yield factors at these four operations are 0.995, 0.9, 0.95 and 0.99, respectively. Hence, $B = (b_1, b_2, b_3, b_4) = (0.0050, 0.1111, 0.0526, 0.0101)$.

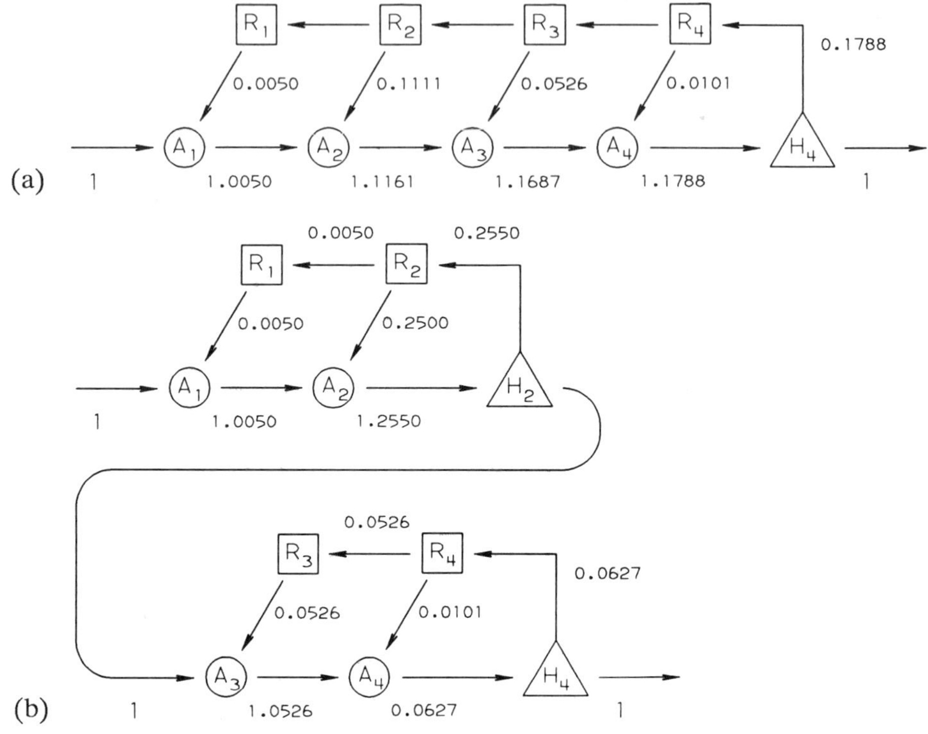

Figure 4.12 Optimal test strategies. (a) A long test time case. (b) A short test time case.

Our problem is to find the best test strategy such that the total process time is minimal. Let $F(i)$ be the total process time after having considered the first i operations. Algorithm 4.4 gives the following results:

$$F(1) = (1 + b_1)T_1 + b_1 S_1 + (1 + b_1)U_1$$

$$= 1.005 \times 4.3 + 0.005 \times 4.3 + 1.005 \times 10 = 14.39$$

$$F(2) = \min \begin{bmatrix} (1 + b_1)T_1 + (1 + b_1 + b_2)T_2 + b_1 S_1 + (b_1 + b_2)S_2 \\ \quad + (1 + b_1 + b_2)U_2 = 23.90^* \\ F(1) + (1 + b_2)T_2 + b_2 + (1 + b_2)U_2 = 33.83 \end{bmatrix}$$

$$F(3) = \min \begin{bmatrix} (1+b_1)T_1 + (1+b_1+b_2)T_2 + (1+b_1+b_2+b_3)T_3 \\ \quad + \cdots + (1+b_1+b_2+b_3)U_3 = 33.12^* \\ F(1) + (1+b_2)T_2 + (1+b_2+b_3)T_3 + \cdots \\ \quad + (1+b_2+b_3)U_3 = 43.00 \\ F(2) + (1+b_3)T_3 + b_3S_3 + (1+b_3)U_3 = 43.88 \end{bmatrix}$$

$$F(4) = \min \begin{bmatrix} (1+b_1)T_1 + (1+b_1+b_2)T_2 + \cdots \\ \quad + (1+b_1+b_2+b_3+b_4)U_4 = 45.55^* \\ F(1) + (1+b_2)T_2 + (1+b_2+b_3)T_3 + \cdots \\ \quad + (1+b_2+b_3+b_4)U_4 = 55.36 \\ F(2) + (1+b_3)T_3 + (1+b_3+b_4)T_4 + \cdots \\ \quad + (1+b_3+b_4)U_4 = 54.64 \\ F(3) + (1+b_4)T_4 + b_4S_4 + (1+b_4)U_4 = 58.12 \end{bmatrix}$$

By inspecting the result for $F(4)$, it is found that the optimal test strategy is to place a test operation at the final stage only. Since the yield at each operation is high and test times are much longer than assembly and rework times, it will be more appropriate to have a fewer number of test operations as possible. The flow process under the optimal strategy is given in Figure 4.12(a).

Suppose that $y_2 = 0.8$ and test times become 2, 2.4, 3, and 4 minutes, respectively. The other parameters remain unchanged. Hence, $B = (0.0050, 0.2500, 0.0526, 0.0101)$. The computational results are

$$F(1) = (1+b_1)T_1 + b_1S_i + (1+b_1)U_1$$

$$= 1.005 \times 4.3 + 0.005 \times 4.3 + 1.005 \times 2 = 6.35$$

$$F(2) = \min \begin{bmatrix} (1+b_1)T_1 + (1+b_1+b_2)T_2 + b_1S_1 + (b_1+b_2)S_2 \\ \quad + (1+b_1+b_2)U_2 = 14.9^* \\ F(1) + (1+b_2)T_2 + b_2 + (1+b_2)U_2 = 16.85 \end{bmatrix}$$

$$F(3) = \min \begin{bmatrix} (1+b_1)T_1 + (1+b_1+b_2)T_2 + (1+b_1+b_2+b_3)T_3 \\ \quad + \cdots + (1+b_1+b_2+b_3)U_3 = 21.95^* \\ F(1) + (1+b_2)T_2 + (1+b_2+b_3)T_3 + \cdots \\ \quad + (1+b_2+b_3)U_3 = 23.86 \\ F(2) + (1+b_3)T_3 + b_3S_3 + (1+b_3)U_3 = 22.26 \end{bmatrix}$$

$$F(4) = \min \begin{bmatrix} (1+b_1)T_1 + (1+b_1+b_2)T_2 + \cdots \\ \quad + (1+b_1+b_2+b_3+b_4)U_4 = 30.99 \\ F(1) + (1+b_2)T_2 + (1+b_2+b_3)T_3 + \cdots \\ \quad + (1+b_2+b_3+b_4)U_4 = 32.84 \\ F(2) + (1+b_3)T_3 + (1+b_3+b_4)T_4 + \cdots \\ \quad + (1+b_3+b_4)U_4 = 28.64^* \\ F(3) + (1+b_4)T_4 + b_4S_4 + (1+b_4)U_4 = 30.79 \end{bmatrix}$$

The optimal strategy now is to place test operations after operations 2 and 4. This is because operation 2 has a relatively poor yield and test times are smaller than assembly times. The flow process is given by Figure 4.12(b). □

So far rework operations have been considered as a part of an assembly process. How to perform rework operations can be a major concern in line design. Different strategies are discussed in the following section.

4.7 Rework Strategies

If defects are found, an assembly may be sent for rework so that assembly work and material can be partially saved. However, rework may also present problems. Rework flow in a line usually deviates from the normal assembly process flow. Line operation therefore may be disrupted. From a logistics point of view, rework creates additional types of work-in-process and therefore complicates flow control. Rework may lead to a line discipline problem, too. Often working on a bad product is much more difficult than assembling a fresh product. For this reason, if one's merit is measured simply by output, most people would prefer new build to rework. As a consequence, a lot of bad products may accumulate in the line, waiting for rework.

If product cost is low and yield is high, it may be less expensive to scrap a bad product than to rework it. A trade-off analysis can be done by comparing rework cost with scrap cost. In Figure 4.12(a), for instance, since A_1 has a very high yield, the rework operation R_1 may be eliminated from the process so that 0.5% ($= b_1$) of the total number of products become scrap. Saving from this elimination includes the workstation cost of R_1 and its associated labor cost. If this saving exceeds 0.5% of a product cost, then R_1 may be eliminated.

To reduce product cost and understand a yield problem, rework may become essential. Rework strategy therefore is an important subject in process design. At least three different rework strategies, illustrated in Figure 4.13, are worth further discussion.

1. *Dedicated rework* All rework is sent to a separate, dedicated production line that does only rework.

2. *Integrated rework* Rework is removed from the line, disassembled at some rework station, and reintroduced to the line. The rework stations may be located at an area separated from the main assembly line.

3. *In-line rework* Once a bad assembly is found, it is sent back to the previous operations in the line, i.e., the flow is reversed. The assembly is torn down and reassembled.

• DEFECT OCCURS

x DEFECT FOUND

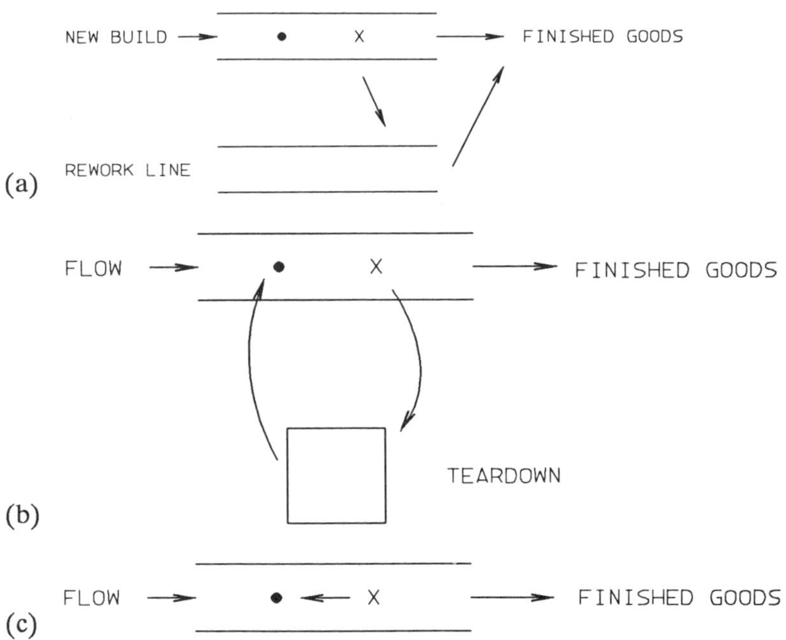

Figure 4.13 Three rework strategies. (a) Dedicated rework. (b) Integrated rework. (c) In line rework.

These three strategies may be compared from different aspects. A summary comparison is given in Table 4.11, and detailed discussion follow.

A dedicated rework strategy is feasible only for a high volume and high yield condition. Since a dedicated rework line must have a complete tool set, poor utilization is expected if the production demand is low. On the other hand, low yield simply implies high cost. For a 50% yield, each product has to be processed twice on the average. Consequently, the product cost is approximately doubled, as compared to a 100% yield. A good production strategy should produce a small quantity while yield is low. Bad products should be sent back to the same line and reworked by the same operators so that the problems can be corrected and avoided in the future. Quick problem feedback is a key to reduce rework effort in many cases. Therefore, this strategy may not be suitable for a line at its early production stage.

Since all reworks are removed to a separated rework area, line disruption on the new-build line is very slight. However, this does not mean that everything moves smoothly. If daily yield is variable, workload balance on the rework line can be a problem. The data reported in Section 2.6 show that daily yield can deviate from its mean by nearly 10%. If the average yield is 0.9, then the rework-line capacity should be approximately 10% of the new-build capacity. On the other hand, even under a normal condition, yield may be 10% lower than the average on given a day. This 10% change on the new-build line generates nearly 100% additional workload on the rework line and causes extra WIP. If the yield on another day is 10% higher than the average, then the rework line will not have enough work to do.

Table 4.11 Comparison of Different Rework Strategies

Factors	Dedicated	Integrated	In-line
Feasibility	High yield only	Feasible	Tool-dependent
Defect feedback	Slow	Automatic	Automatic, fast
Yield effect	High variation	Self-adjusting	Self-adjusting
Degree of line disruption	Little effect on new build	Remove defect insert rework	Reverse flow, reduce pace
WIP level	High	Medium	Low
Tool cost	High	Medium	Low
Space cost	High	Medium	Low
Labor cost	High	Medium	Low

The second strategy attempts to integrate rework operations with new build. Normally separate workstations are needed for disassembling bad products. The disassembly tools may or may not be identical to the assembly tools. After having been disassembled, the product is sent to the operation that caused the defects. In this case, the operation will handle both new build and rework. A high priority should be given to the rework so that necessary corrections can be made and future yield improved. Since the rework products are always fed back to the same line, a low yield simply slows down the throughput. No extra WIP will be created. The key problem here is line discipline. The bad products should always receive immediate attention. Otherwise, both WIP and yield problems will occur.

The in-line strategy has advantages such as defect information feedback, low WIP, and self-adjustment to yield fluctuation. Since no separate workstations are used, both space and labor costs are relatively low. Two problems should be considered. First, the feasibility of this strategy is tool-dependent. If an assembly tool and a disassembly tool are different, it may not be possible to install them in the same workstation due to ergonomic reasons. For example, it is unsafe to have a manual disassembly operation too close to a robotic assembly station. The second problem is that the rework flow is opposite to the new-build flow. Therefore, the line disruption may cause throughput reduction. This problem is usually related to the material-handling system. A one-way conveyor obviously cannot support the in-line strategy.

Finally, the reader should be reminded that there is no reason why a mixed strategy cannot be adopted. Rework can be done in a separated area in one stage of the assembly line while in another stage it is integrated into the new-build operations. For example, an electronic manufacturing line may elect a mixed strategy. In the early production phase, the major cause of rework is workmanship. An integrated rework strategy allows quick feedback. Therefore, operators can correct their own mistakes. When this phase is over, the dominate factor of product yield become very complex such as component quality and product design problems. Usually, specialists may be needed to perform failure analysis to determine the causes of product failure. In this case, a dedicated rework strategy may be appropriate.

4.8 Remarks

Many articles have discussed the problem of line balancing. The two heuristic methods introduced in this chapter can be found in Helgeson (1961) and Agrawal (1985), respectively. Discussion on COMSOAL is given by Arcus (1966). A branch-and-bound method is presented in Johnson

(1983). Other works in assembly line balancing can be found in Ignall (1965), Mastor (1970), Dar-El (1975), Sharp (1977), Pinto (1983), and Baybars (1986).

Process-design optimization is a very complex problem. The work described in Section 4.3 considers only sequential lines and is by no means complete. In many cases, an assembly process may have a tree structure such that each branch of the tree corresponds to a process for a subassembly. Then the method for sequential lines can be applied to each branch process. Sometimes, in an assembly process a feedback loop may exist for rework. In this case, the method can be used by first ignoring rework. After having defined the optimal solution, one may put all rework operations back. This approach normally is acceptable if yield is high enough that rework operations have relatively small start factors.

Test strategy is one of the key functions for many electronics manufacturing lines. This area is closely related to yield management policy as illustrated by Example 4.5. Real-life problems can be more complex than the one considered in Section 4.5. A simplified assumption that has been used is that a test operation is able to detect all defects. This may not be the case in reality. If the probability of finding a defect is known, a different dynamic programming procedure may still be established for the optimal solution. It appears that the optimization procedure is application-dependent. Development of a general algorithm is not a simple task.

Two important subjects in the area of process design have not been discussed in this chapter—automatic assembly and ergonomics. From a technical point of view, automatic assembly is not impossible. But design for automation still has many problems. One common mistake is to install automation without a long-term strategy. First, to facilitate automatic assembly, the product must be designed with that in mind. Second, the assembly process must be also designed for automation. If a product life is short and frequent engineering changes are anticipated, automation may not be economical. In some cases, automation may not be feasible if the assembly task is too complex. For a new product, one may first install a manual line with short operation process times. (But each operator may handle more than one operation for the sake of job enrichment.) After the process has become stable, automatic workstations can be gradually introduced. Short process times will force task simplicity, and therefore increase feasibility for automation and workstation reliability. Boothroyd (1982) has given a systematic treatment on this subject.

For a manual operation, on the other hand, job content is considered as a key parameter in process design. A short job tends to have a monotonous and restricted motion pattern. Repeating such a pattern continuously would lead to muscle fatigue and consequently slows down the working pace. If job duration is too long, slowing of pace may also occur due to loss of motor skill, frequent tool changing, forgetting and fumbling. This phenomenon is called *short-task fatigue* and *long-task delay*. A case study of a TV chassis assembly by Kilbridge (1962) shows that the maximal working speed was attained when the job duration is ranged from 0.5 to 1 minutes. From a line performance point of view, however, this may not be optimal. If the total assembly work lasts 60 minutes, a one-minute operation time means 60 operations. This will make line coordination very difficult. A typical department has a size of about 20 operators. If a line has 60 operators, it ought to be partitioned into three departments such that time for communication, negotiation and problem tracking can be reduced. Although no general agreement has been reached, many authors believe job enrichment is important for productivity, and it is achievable by increasing job content per operation or allowing job rotation.

For manual assembly, the reader may find useful information regarding job design and related subjects in Kilbridge (1961a), Kilbridge (1961b), Prenting (1964), van Beek (1964), Smith (1968), Tuggle (1969), and Buxey (1981). A review of reported experiments in job design is given in Birchall (1973). Extensive studies in the ergonomics area have been reported in two volumes published by the Eastman Kodak company (1983, 1986). The first volume discusses workstation design and the second job design. Both are excellent reference books.

REFERENCES

Agrawal, P. K. (1985). The Related Activity Concept in Assembly Line Balancing, *International Journal of Production Research*, v. 23, pp. 403–421.

Arcus, A. L. (1966). COMSOAL—A Computer Method of Sequencing Operations for Assembly Lines, *International Journal of Production Research*, v. 4, pp. 259–277.

Baybars, I (1986). An Efficient Heuristics Method for the Simple Assembly Line Balancing Problem, *International Journal of Production Research*, v. 24, pp. 149–166.

Birchall, D., and R. Wild (1973). Job Restructuring amongst Blue-Collar Workers, *Personnel Review*, v. 2, pp. 40–56.

Boothroyd, G., C. Poli, and L. E. Murch (1982). *Automatic Assembly*, Marcel Dekker, New York.

Buxey, G. M., and J. J. Owens (1981). The Operation of a Conveyor System Supplying Unit Build Assemblies, *International Journal of Production Research*, v. 19, pp. 123–137.

Dar-El, E. M. (1975). Solving Large Single-Model Assembly Line Balancing Problems—A Comparative Study, *AIIE Transactions*, v. 7, pp. 302–310.

Eastman Kodak Company (1983). *Ergonomic Design for People at Work, Volume 1*, Van Nostrand Reinhold, New York.

Eastman Kodak Company (1986). *Ergonomic Design for People at Work, Volume 2*, Van Nostrand Reinhold, New York.

Helgeson, W. P., and D. P. Birnie (1961). Assembly Line Balancing Using Ranked Positional Weight Techniques, *Journal of Industrial Engineering*, v. 12, pp. 394–198.

Ignall, E. J. (1965). A Review of Assembly Line Balancing, *Journal of Industrial Engineering*, v. 16, pp. 244–254.

Johnson, R. V. (1983). A Branch and Bound Algorithm for Assembly Line Balancing Problems with Formulation Irregularity, *Management Science*, v. 29, pp. 1309–1324.

Kilbridge, M. (1961a). Nonproductive Work as a Factor in the Economic Division of Labor, *Journal of Industrial Engineering*, v. 12, pp. 155–159.

Kilbridge, M. (1961b). Turnover, Absence, and Transfer Rates as Indicators of Employee Dissatisfaction with Repetitive Work, *Industry and Labor Relations Review*, v. 14, pp.21–32.

Kilbridge, M. (1962). A Model for Industrial Learning Costs, *Management Science*, v. 8, pp. 516–527.

Mastor, A. A. (1970). An Experimental Investigation and Comparative Evaluation of Production Line Balancing Techniques, *Management Science*, v. 16, pp. 728–746.

Pinto, P. A., D. G. Dannenbring, and B. M. Khumawala (1983). Assembly Line Balancing with Processing Alternatives: An Application, *Management Science*, v. 29, pp. 817–830.

Prenting, T. O. (1964). Better Selection for Repetitive Work, *Personnel*, v. 41, pp.26–31.

Sharp, W. I. (1977). Assembly Line Balancing Techniques, *Society of Manufacturing Engineers*, Paper MS77-313, Dearborn, Michigan.

Smith, D. M. (1968). Job Design: From Research to Application, *Journal of Industrial Engineering*, v. 19, pp. 177–482.

Tuggle, G. (1969). Job Enlargement: An Assault on Assembly Line Inefficiencies, *Industrial Engineering*, v.1, pp. 26–31.

van Beek, H. G. (1964). The Influence of Assembly Line Organization on Output, Quality and Morale, *Occupational Psychology*, v. 38, pp. 161–172.

5

Material Handling Systems

In an assembly line, a material handling system (MHS) usually serves two functions: storage and transport. Although these two functions do not directly add any value to the product, the design problems of MHS cannot be overlooked. A good system should be able to deliver the *right* amount of the *right* material to the *right* place at the *right* time under the *right* conditions. This chapter discusses the role and the basic types of material handling systems, design considerations and performance characteristics. Two case studies of real-life problems are also presented.

5.1 The Role of Material Handling Systems

The MHS can be regarded as a service system that provides certain functions to facilitate assembly jobs. Therefore the MHS should be integrated into line design. Compared to other manufacturing operations such as assembly, inspection and testing, material handling is relatively simple and mechanical. Therefore automated material handling systems are very popular in manufacturing lines. Past experience proves that an automated system can handle material more safely, more economically and more accurately. The design complexity of a material handling system comes from its interaction with other manufacturing components.

First of all, the MHS must be compatible with product characteristics, such as size, weight, fragility and other factors. For instance, a manual delivery and handling method will not be appropriate for a heavy or bulky product. An automated system would help reduce product handling damage. In some semiconductor factories, product quality is a function of cleanliness. Consequently, assembly operations must be performed in a clean room environment, where both the number and the size of particles in the air must be controlled. In this case, the MHS must be clean-room-compatible.

Next, the interface between the MHS and workstations can be important. In some cases the material handling device may directly interact with a workstation; in other cases an additional local device is needed. A parts-feeding device at an automated workstation is a typical example of a local device. They may also be used for local storage, transport, or machine loading and unloading.

Third, material handling complexity is a function of material handling frequency and the duration of handling time. Therefore the manufacturing process and line layout must be taken into consideration for MHS design. For a given product, a line with longer (shorter) operation process times means a smaller (larger) number of operations and less (more) material handling effort. For a given manufacturing process, line layout is very dependent on the choice of MHS. If the selected MHS has direct access capability (e.g., an automatic guided vehicle or simply a cart pushed by an operator), then line layout need not necessarily match with manufacturing process flow. The line designer has flexibility to try different line layouts. For a sequential MHS (such as a conveyor), on the other hand, line layout should be coordinated with the manufacturing process for efficient line performance. If the line designer anticipates process changes, layout flexibility could be a more important factor than material handling performance. If the process is stabilized, flexibility may not be of any concern.

Finally, material handling delay and storage capacity should be kept at proper levels so that the line can be operated smoothly. Material handling delay is determined by the handling speed, physical flow pattern and production volume. Usually the handling delay can be analyzed by a queuing model. Line internal storage is justified because of time variation. An assembly line consists of a sequence of operations. At each operation, two successive completion time epochs constitute an operation cycle. A cycle time is equal to an operation process time plus any delay due to machine failure, parts shortage and operator delay and fatigue, etc. Synchronization of operation cycles is difficult. A buffer may be installed between operations to absorb cycle variation and keep the line running at a desirable throughput level. Consequently, the buffer storage capability should be considered as a design parameter for the MHS.

In the subsequent sections, we shall discuss different types of MHS with an emphasis on design integration. Since the number of MHS types is enormous, it is impossible to deal with every single one of them in this book. For convenience, they may be classified into two basic types: *discrete* and *continuous*. An automated storage and retrieval system (AS/RS) is a discrete MHS; it handles one object at a time. A conveyor system, on the other hand, sequentially moves many items at the same time, and therefore is regarded as a continuous type. In this chapter only these two types of systems are considered. Other systems may be treated in a similar fashion or considered as cases between these two basic types. For instance, an automatic guided vehicle system (AGVS) may be analyzed by using a model similar to that proposed for AS/RS in the following section. If an AGVS has multiple vehicles that follow the same move pattern, such a system can be viewed as an intermediate case between an AS/RS and a conveyor.

5.2 Automated Storage and Retrieval Systems

An automated storage and retrieval system is a robot-like MHS composed of (i) a direct access handler (DAH), (ii) a structure for material storage, and (iii) a control system. The DAH performs physical material handling such as pick, place, load, unload, reach and move. Usually a rack is installed for material storage. The DAH and the rack together serve as random access storage. The access pattern and the speed of the DAH are guided by the system control unit. When an AS/RS is employed for assembly lines, the DAH may interact with assembly workstations. The storage rack can be used for internal storage.

The control system may have two separate components, a local controller and an on-board controller. Usually the former is handled by a personal computer placed in a control room, and the latter by a micro-computer residing in the DAH. The on-board controller accepts commands from the local controller and conducts physical material handling motions. These commands are derived from line operation requirements.

The DAH consists of (i) an x-carriage that moves along a linear track (x-axis), (ii) a y-carriage that moves up and down along a mast placed on the x-carriage (y-axis), (iii) a z-carriage, attached to the y-carriage, serves as a pick device and can move back and forth (z-axis), and (iv) a θ-carriage that can perform a circular motion (θ-axis). All directional moves, except z-axis, can be performed simultaneously.

A DAH and a possible line layout with an AS/RS are illustrated in Figures 5.1 and 5.2, respectively. In Figure 5.2, workstations are placed on both sides of the linear track. The interface between the DAH and a workstation is through a pick-and-delivery (P/D) port. A vertical rack is used for work-in-process storage.

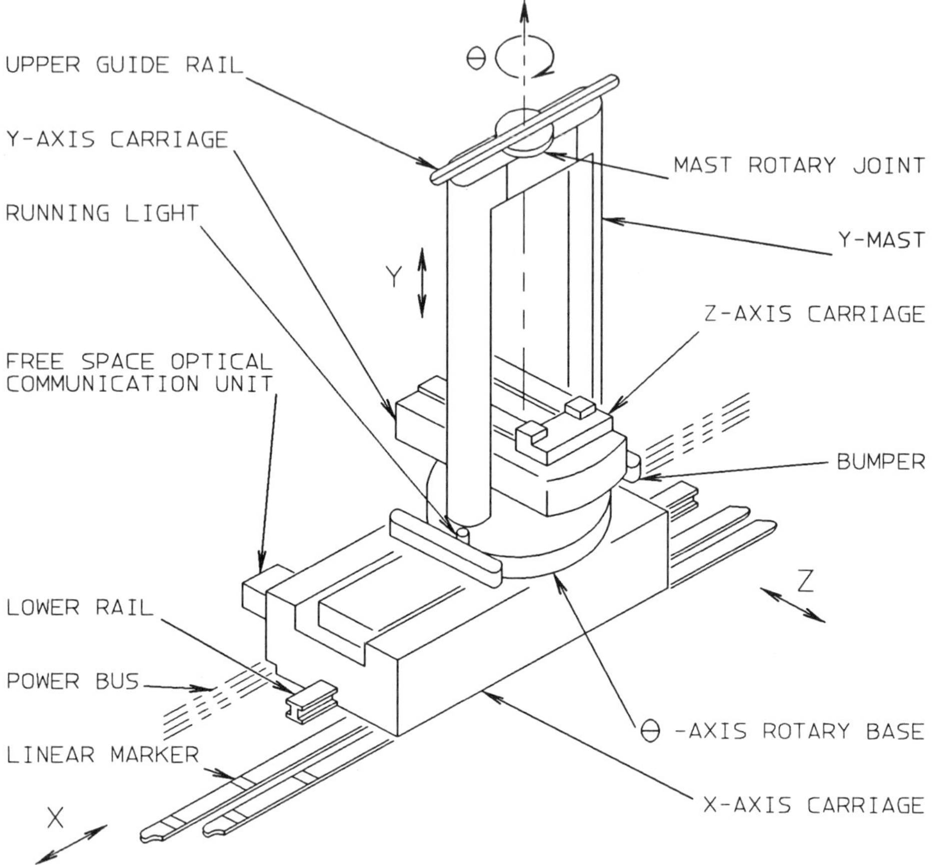

Figure 5.1 A DAH configuration.

5.2.1 An AS/RS Performance Model

A service request involves a pair of pick and place. If a request is to move objects from location i to location j, it is then called an (i,j) request. There are three types of requests: (i) workstation to workstation, (ii) workstation to storage, and (iii) storage to workstation. Depending on the origin and the destination, a motion sequence can be established. The duration of time to complete the motion sequence is the service time. For a given service request pattern and a request rate, MHS performance can be analyzed by an M/G/1 queuing model. (See Section 3.3.) The input data to this model include arrival rate and service time distribution.

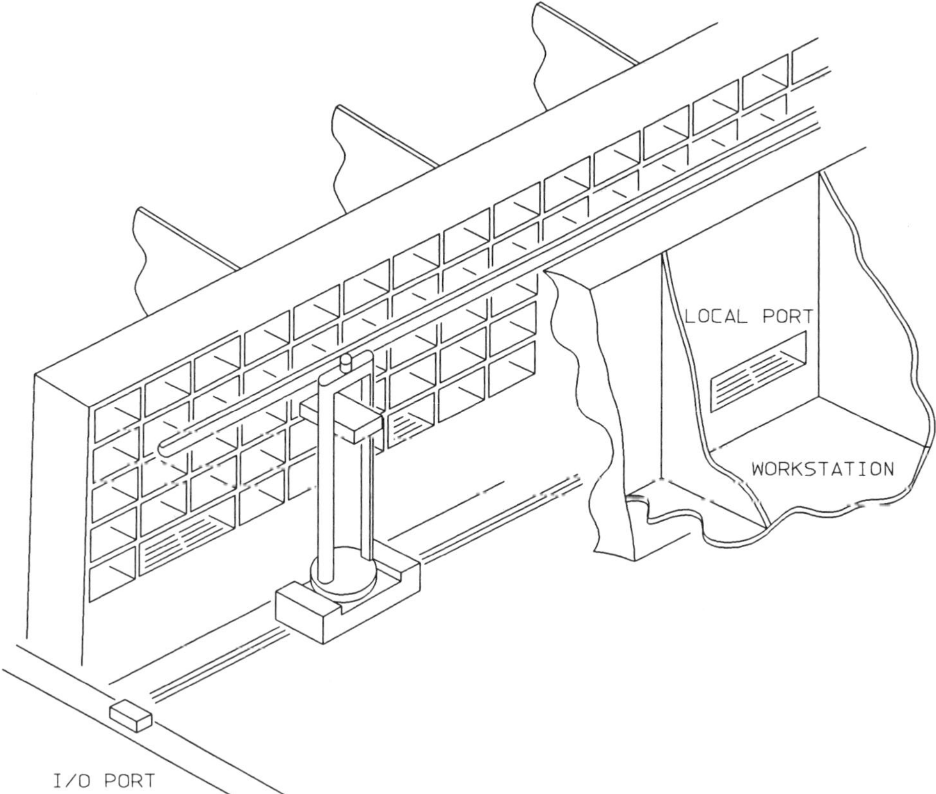

Figure 5.2 A DAH layout.

Service Request Rate

The arrival rate is a function of production volume, the number of operations, and the probability that a work unit must be placed in the storage rack. If an AS/RS serves a line of n operations which produces v assemblies per day, then each assembly will be handled nv times. If the receiving workstation is temporarily out of space for an incoming work unit, the AS/RS may have to place the unit into the storage rack. Usually the rack has multiple levels and can hold a large amount of storage. The probability of an additional storage operation, g, can be estimated by a queuing model. Suppose that the mean process time at a workstation is s hours and the line is operated in TT hours a day. Then the probability that the workstation is busy is expressed by $u = sv/TT$. If the working bench does not have any space for work-in-process, then $g = u$. If the working bench can hold up to m units of work-in-process, $g = P[\text{queue length} \geq m]$. Using an M/M/1 queuing model, it follows that $g = u^m$. If an operation has k identical workstations, then an M/M/k model or an M/G/k model may be used to estimate g. For operation i, denote this probability by g_i. Since v assemblies will be delivered to operation i, the expected number of additional storage operations is vg_i. Consequently, the arrival rate seen by the AS/RS is

$$\lambda = (n + g_1 + \cdots + g_n)v/TT \qquad \text{(per hour)} \tag{5.1}$$

Service Time

A service usually involves three locations: the current location of the DAH, the origin (or the sending location), and the destination (or the receiving location). Denote these three locations by i, j and k, respectively, and the corresponding service time by S_{ijk}. A service is typically composed of the following motions:

1. Move the DAH from location i (the current location) to location j (the origin). This move may involve x, y and θ motions. A y-motion is normally triggered by a storage access. For instance, a service request may remove a work unit from the rack to a workstation. Since the size of an assembly is usually considerably less than the workstation size, the travel time along the x-axis is much longer than that along the y-axis.
2. Pick the work unit from location j. This consists of a sequence of motions in z-axis: (i) forward z-carriage, (ii) lift the object and (iii) retract the z-carriage to its home position.
3. Move the DAH to location k (the destination).
4. Place the work unit at location k. This consists of a sequence of motions in z-axis: (i) forward z-carriage, (ii) release the object and (iii) retract the z-carriage to its home position.

Letting h be the travel distance, a the acceleration and b the maximal speed, the travel time is given by

$$t = \begin{cases} h/b + b/a & \text{if } h \geq b^2/a \\ 2\sqrt{h/a} & \text{if } h < b^2/a \end{cases} \tag{5.2}$$

Denote tx, ty, $t\theta$, u and v as travel times in x, y, θ axes, pick time and place time, respectively. Then the service time is given by

$$S_{ijk} = \max(tx_{ij}, ty_{ij}, t\theta_{ij}) + u + \max(tx_{jk}, ty_{jk}, t\theta_{jk}) + v \tag{5.3}$$

where the subscripts indicate the locations.

Usually tx is larger than ty. Therefore, the exact location in the y-axis need not be considered. On the other hand, the travel time in the θ axis, $t\theta$ is either 0 (if the starting and the ending locations are on the same side of the linear track) or equal to a 180-degree rotation time (if the two locations are on opposite sides). Sometimes a service request is to retrieve a work unit from the vertical rack to the receiving workstation. In this case, tx_{jk} and $t\theta_{jk}$ are both zero, but ty is positive.

Given that there exists a service request, the conditional probability that it is an (i,j) request is proportional to the expected amount of flow from location i to j, f_{ij} and is defined by

$$r_{ij} = f_{ij}/F \qquad i,j = 1,2,\ldots,n \tag{5.4}$$

where F is the expected total amount of flows, i.e., $F = \sum_{ij} f_{ij}$.

If a first-come, first-served dispatching rule is implemented, the request to be served is independent of the DAH's current position. Therefore the probability that the DAH is at location i just before a service begins is

$$q_i = \sum_{j=1}^{n} f_{ji}/F \qquad i = 1,2,\ldots,n \tag{5.5}$$

Consequently, the probability that a service time is S_{ijk} becomes

$$p_{ijk} = r_{jk} q_i \qquad i,j,k = 1,2,\ldots,n \tag{5.6}$$

Thus the service time, S, has a distribution defined by $\{S_{ijk}\}$ and $\{p_{ijk}\}$. The rth moment of the service time is given by

$$E[S^r] = \sum_{ijk} (S_{ijk})^r p_{ijk} \tag{5.7}$$

Using the results from M/G/1 queue in Section 3.3, we have the following DAH utilization:

$$\rho = \lambda E[S] \tag{5.8}$$

Average material handling delay

The average material handling delay is the expected time interval between a service request time epoch and its completion time epoch.

$$E[W] = E[S] + \frac{E[S]\rho}{2(1-\rho)}(1+C^2) \tag{5.9}$$

where C is the coefficient of variation of S.

Average number of incomplete requests

$$E[N] = \rho + \frac{\rho^2}{2(1-\rho)}(1+C^2) \tag{5.10}$$

It should be pointed out that the M/G/1 queuing model assumes that all service times are independent. In a real problem, however, this is not the case. Consider successive service requests generated from workstations deployed along the linear track. Since the service time, S, is dependent on the travel distance in the x-axis, a request involving workstations at either end of the track tends to have a relatively long service time. In other words, a long service time implies a high probability that the receiving station is near one end of the track. Since a previous receiving location becomes the starting point of the next DAH service, a long service time also means a relatively high probability that the next service time is also long. Hence the successive service times are not strictly independent. The proposed M/G/1 queuing model only gives an approximation. On the other hand, a service consists of two x-motions and a pair of pick/place operations. This dependency between successive service times is usually rather weak. The proposed model should be a good approximation.

Higher moments of material handling delay can also be obtained by the model. The following results for an M/G/1 queue are given without proof:

$$Var[W] = Var[S] + \frac{\rho}{1-\rho}\frac{E[S^3]}{3E[S]} + (\frac{\rho}{1-\rho}\frac{E[S^2]}{2E[S]})^2 \tag{5.11}$$

$$Var[N] = \lambda^2 Var[W] + E[N] \tag{5.12}$$

In the last two equations, the moments of S can be evaluated by (5.3)–(5.7) and λ is given by (5.1). AS/RS performance is a function of (i) DAH hardware characteristics, (ii) assembly process, (iii) production volume, and (iv) workstation layout. Different performance measures have different emphases. From a capital investment viewpoint, it is customary to measure system utilization. From a workstation viewpoint, material handling delay is more important. If storage space is of concern, the measure for service queue becomes relevant.

Previous discussions can be summarized into a computational algorithm for AS/RS performance evaluation. In the following, it is assumed that an assembly process, a layout, a production volume during a period of TT, and a DAH hardware are given.

Algorithm 5.1

1. Determine the flow volume between locations i and j for every pair of (i,j). Compute the total number of service requests during TT. This is given by

$$F = \sum_i \sum_j f_{ij}$$

2. Compute the service request rate, $\lambda = F/TT$.
3. Compute $\{r_{ij}\}$ by (5.4), $\{q_i\}$ by (5.5), and $\{p_{ijk}\}$ by (5.6). Note that p_{ijk} is the probability that the DAH is at location i but is going to pick an object at j and to deliver it to k.
4. For each service defined by (i,j,k), compute its service time S_{ijk} by (5.2) and (5.3). The service time distribution of DAH is then defined by $\{S_{ijk}\}$ and $\{p_{ijk}\}$.
5. Using (5.7) and the results obtained from steps 3 and 4, calculate the first three moments of DAH service time.
6. Compute ρ, $E[W]$, $E[N]$, $Var[W]$ and $Var[N]$ by (5.8), (5.9) and (5.10), respectively.

This algorithm is useful for design study of an AS/RS, hardware trade-off analysis and line configuration with an AS/RS. Because of its simplicity, the line designer may conduct a large number of model experiments in a very short time. Experimental results and detailed performance characteristics will be discussed in the next section.

5.2.2 Model Experiments and Operating Characteristics

Assume that an assembly line has 10 workstations, each 12 feet wide. One end of the linear track is designated as an input/output (I/O) port. All parts come to this port, are picked by the DAH and distributed to workstations. Completed assemblies are delivered to the port and taken to a shipping area. Figure 5.3 shows a top view of two possible layouts. The first layout places all stations on one side of a linear track; the second one distributes them onto both sides.

The line is well balanced so that all stations are almost always busy. A delivery is sent to a buffer and then removed to the working bench. For each station, the buffer is about three feet away and is large enough to prevent buffer overflow.

Parts are placed in trays for delivery. The part density is expressed in terms of the number of assemblies being made by the parts from the same tray. The workload is characterized by (i) a total of 500 assemblies to be made, (ii) part density varying from 2 to 8 assemblies per tray, and (iii) up to four types of workstations. For two assemblies per tray and two types of stations, as an example, we have 5 type-1 and 5 type-2 stations. The number

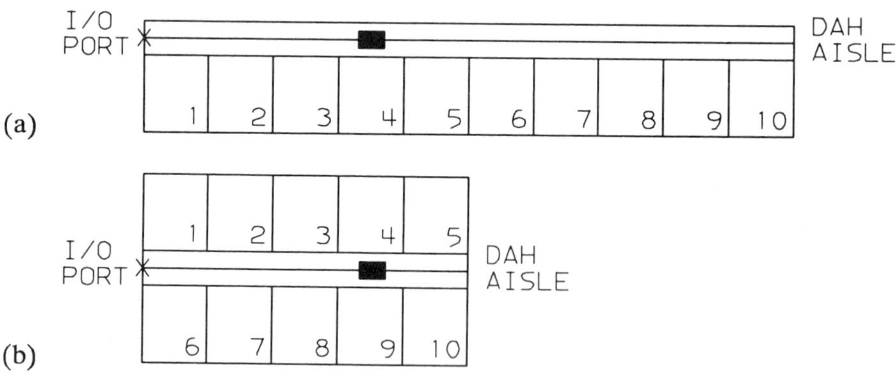

Figure 5.3 10-station layout with AS/RS. (a) Single-side. (b) Double-side.

of full trays (with parts) is 500/2 = 250. Flows between buffers, workstations and the I/O port are as follows:

From	To	Flow	Content
I/O port	Each buffer	25	Parts
Each buffer	Each station	25	Parts
Type-1 station	Type-2 buffer	100	Subassemblies
Type-2 buffer	Type-2 station	100	Subassemblies
Type-2 station	I/O port	100	Completed assemblies
Each station	I/O port	25	Empty trays

The total flow volume is 2250. If the AS/RS works 20 hours a day, the service request rate becomes $\lambda = 2250/(20 \times 3600) = 0.03125$ per second.

Typical DAH hardware characteristics found in the existing technology are assumed:

Acceleration in x-axis	2 ft/sec/sec
Top speed in x-axis	6 ft/sec
Angular speed	60 degree/sec
Pick time	6 sec
Place time	6 sec

For example, since the distance from the I/O port to the buffer of the second stations in Figure 5.3(b) is 21 feet, the travel time in the x-axis is $21/6+6/2 = 6.5$ seconds. In the meantime, the circular motion for 90 degrees takes $90/60 = 1.5$ seconds. Therefore the travel time from the I/O port to the buffer is 6.5 seconds. Let the locations i, j and k represent station 8, the I/O port and the buffer of station 2, respectively. For an (i,j,k) request,

$$S_{ijk} = (30/6 + 6/2) + 6 + (21/6 + 6/2) + 6 = 25.5 \text{ sec}$$

Since station 8 assembles 100 units a day, it also receives 100 assemblies a day, $q_i = 100/2250$. Station 2 should receive 25 part trays a day, therefore $r_{jk} = 25/2250$, and $p_{ijk} = 2500/(2250 \times 2250) = 0.0004938$. For each (i,j,k)

combination, the service duration and its probability can be obtained in the same manner. By (5.8)–(5.12), performance measures follow.

Performance Behavior

By varying the parts packing density and the number of workstation types, a number of cases can be analyzed. Figure 5.4 shows the relations between ρ, $E[W]$ and $E[N]$ for both of the layouts illustrated in Figure 5.3. The dots represent the cases of serving one side (Figure 5.3(a)) while the squares are for the cases of serving two sides. The relations between DAH utilization and the average material handling delay are fitted by two dashed lines, one for each layout. As one would expect, the average delay is an increasing function of ρ. A DAH with rotation capability will reduce the average delay by at least 20 seconds for $\rho > 0.6$. Both curves increase very rapidly once ρ exceeds 0.75. On the other hand, one can hardly discern whether the DAH is serving one side or two merely by inspecting the average incomplete number of requests. This situation is reflected by the single solid line.

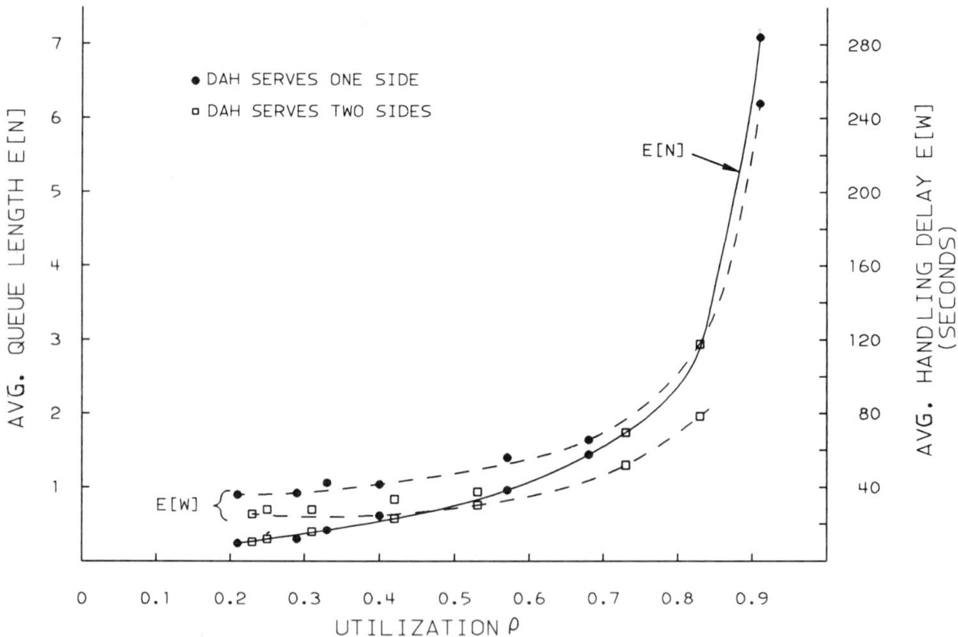

Figure 5.4 DAH utilization, queue length and handling delay.

Note that $E[N]$ also bends upward quickly when $\rho > 0.75$. If the DAH is utilized over 75%, any small additional workload will degrade performance drastically. This means that each service call may have a long waiting time, that a workstation has a long operation cycle, and that the line throughput may decrease. Consequently, a desirable operating range may be confined to $\{\rho \,|\, 0.6 \leq \rho \leq 0.7\}$.

The difference between the two layouts is that the track length of the single-side case is twice as long as that of the double-side. Since the total number of requests per day remains the same, the difference in DAH performance behavior is due to the service time. The results given in Figure 5.4 indicate that, compared to the average number of incomplete requests, the average material handling delay is more sensitive to the service time. A service time is composed of a deterministic part and a stochastic part. The former is equal to a pick time plus a place time. The latter is the sum of two travel times, from locations i to j and from j to k. Since a significant portion is due to the deterministic part, the service variability is small. In each numerical case presented in Figure 5.4, the coefficient of variation of DAH service time, C, is between 0.2 and 0.55. The last term on the right-hand side of Equation (5.10) has a factor of $(1 + C^2)$. If $C < 0.55$, this factor can be altered no more than 30%. Furthermore, if the utilization factor is kept between 0.6 and 0.7, it can be shown by using (5.10) that the value of $E[N]$ may vary at most 15% due to different C. This implies that the average number of incomplete requests is dominated by the utilization factor in the recommended operating range.

If the variation effect due to C is negligible, it can be seen from (5.10) that $E[N]$ is uniquely determined by ρ. On the other hand, Equation (5.9) shows that the average material handling delay is a function of ρ and $E[S]$. Since $\rho = \lambda E[S]$, one could simultaneously change the average service time and service request rate so that ρ remains the same, but $E[S]$ becomes less. As a result, the average material handling delay can be reduced even if the expected number of incomplete requests does not change much. The average service time may be reduced by upgrading the DAH hardware or improving the DAH dispatching rule (other than first-come, first-served). The latter approach is discussed in the next section. Hardware issues are considered first.

Hardware Consideration

The speed of a DAH can be improved by:

1. Decreasing pick/place times

2. Increasing acceleration/deceleration
3. Increasing the top speed

The distance and the time required to reach the top speed are, respectively, $d = b^2/2a$ and $s = b/a$, where b is the top speed and a the acceleration/deceleration. Suppose that the travel distance $h > 2d$. The travel time then becomes $t = (h - 2d)/b + 2s = h/b + b/a$. Increase a and b by the same percentage. The travel times are respectively

$$t_a = \frac{h}{b} + \frac{b}{a(1+\delta)}$$

$$t_b = \frac{h}{b(1+\delta)} + \frac{b(1+\delta)}{a}$$

It is easy to show that for all $\delta \geq 0$

$$t_a \leq t_b \qquad \text{if and only if } h \leq (2+\delta)b^2/a$$

If $a = 2$ and $b = 6$, the above condition becomes $h \leq 36$ (at $\delta = 0$). If the travel distance is less than 36 feet, the marginal gain from increasing the acceleration/deceleration is larger than that from improving the top speed.

For each delivery, the DAH involves two stops and a pair of pick and place motions. The service time is approximately (rotation is ignored)

$$S = \left(\frac{h_1}{b} + \frac{b}{a}\right) + u + \left(\frac{h_2}{b} + \frac{b}{a}\right) + v$$

Denote the service times by S_1, S_2 and S_3, respectively, for the cases of decreased pick/place times, increased acceleration/deceleration and increased top speed. Let $h = (h_1 + h_2)/2$ and $w = (u + v)/2$. Then

$$S_1 \leq S_2 \qquad \text{if and only if } (b/a(1+\delta)) \leq w$$

$$S_1 \leq S_3 \qquad \text{if and only if } (h/b(1+\delta)) - b/a \leq w$$

For $a = 2$, $b = 6$ and $h = 36$, these two conditions (at $\delta = 0$) are reduced to $w \geq 3$. This means that the pick and place times are dominant factors if $h \leq 36$ and $w \geq 3$.

From our discussion it is clear that selection of an AS/RS should be based on individual applications. Design optimization of such a system is achievable only after a careful analysis of its cost and performance. For an existing AS/RS, however, performance may be improved by invoking a different dispatching rule, which is our next subject.

5.2.3 Dispatching Rules and Delay Reduction

Whenever the DAH becomes available for multiple service requests, the sequence of service must be determined. The previous discussions assume a first-come, first-served (FCFS) rule, i.e., the order of service is identical to the order of arrival. This rule does not necessarily lead to the minimal material handling delay. Determination of the optimal DAH dispatching rule is equivalent to finding the optimal queuing discipline when the AS/RS is viewed as a queuing system. As discussed in Section 3.3, if there are multiple jobs in the queue, a simple rule to reduce the total waiting time is to serve the shortest job first. Unfortunately, this rule cannot be implemented directly for the DAH. Difficulties may come from two sources. First, successive service times are not independent. The exact service duration for each request cannot be determined unless the initial position of the DAH is known. For a given set of requests, the problem is to find the order of service such that the total travel time for all the service is minimal. (This is recognized as the *traveling salesman problem.*) An effective solution technique is not yet available. Second, the problem is compounded by its stochastic nature, as the requests are generated by workstations in a random fashion. For these reasons, only heuristic rules are considered.

In addition to FCFS, two other rules are worth consideration. One is called *nearest-first* (NRFS) and the other *shortest-first* (STFS). Under NRFS, the DAH always serves the nearest station where a request has been generated. After each service completion, the dispatcher simply compares the distances between the current DAH position and the locations of all sending stations. The nearest station will be served next. If STFS is invoked, both the sending and the receiving stations must be considered. The request that has the least total travel time is selected. The total travel time is the sum of travel time from the current position to the sending station and the time from the sending station to the receiving station. Again the service request is picked after each service completion. Both of NRFS and STFS are one-stage-look-ahead policies. Figure 5.5 shows three travel patterns under FCFS, NRFS and STFS, respectively.

Between the completion of the previous service and the beginning of the next service, a small delay may exist due to the software control process

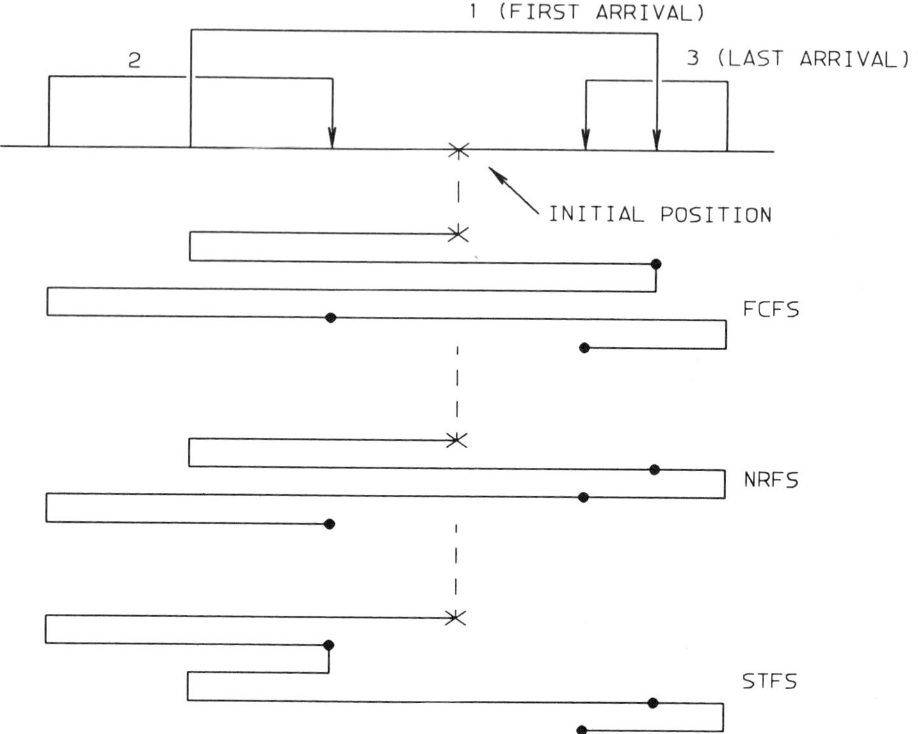

Figure 5.5 Traveling path under different dispatching rules.

time. Under FCFS, incoming requests may be stacked in a *first-in, first-out* fashion. Selection of the next service is simply to read out from this stack; therefore the control process delay is negligible. For NRFS and STFS, determination of the next service requires arithmetic operations and comparisons. Contingent on DAH utilization, the control process delay may or may not be significant. One way to resolve the problem is to start the control process early enough so that the next service can be chosen before the DAH becomes idle. But once the process is initiated, no incoming requests are allowed to compete for the DAH.

Since the determination of service times are state-dependent, a queuing analysis becomes very inefficient even for a small number of stations. For NRFS and STFS, DAH performance is analyzed by simulation. Two cases are considered here.

Case	1	2
Number of workstations	10	10
Station width (ft)	8	12
Top speed (ft/sec)	6	6
Acceleration (ft/sec/sec)	2	2
Angular speed (degree/sec)	60	60
Pick/Place time (sec)	6	4
Hourly request rate	90	90

The difference between these two cases is that the pick/place time in case 1 play a relatively important role. Model solutions of these two cases are summarized in Tables 5.1 and 5.2. In addition to simulation results, the analytic queuing solution under FCFS is also shown. Performance measures include DAH utilization, material handling delay and queue length. For each measure, both mean and standard deviation are estimated. These two numbers are separated by a slash in the tables.

First note that the analytic results under FCFS are very close to those from simulation. The former are approximated by using independent service times, while the latter are obtained by correctly simulating the

Table 5.1 Performance of DAH: Case 1

Rule	Utilization	Handling Delay	Queue Length
FCFS/A	0.65	51.68/34.11	1.29/1.43
FCFS/S	0.66	54.37/40.23	1.37/1.59
NRFS	0.65	47.42/34.14	1.21/1.29
STFS	0.63	45.95/38.14	1.16/0.82

Table 5.2 Performance of DAH: Case 2

Rule	Utilization	Handling Delay	Queue Length
FCFS/A	0.67	56.77/39.74	1.42/1.55
FCFS/S	0.66	56.36/41.14	1.40/1.58
NRFS	0.65	47.85/34.45	1.23/1.32
STFS	0.62	44.58/37.00	1.10/0.75

sequence of services and their respective travel times. This validates the M/G/1 model introduced in the previous section.

Table 5.1 does not show significant differences in DAH performance under the three dispatching rules. In terms of average material handling delay, STFS is about 10% better than FCFS. NRFS provides an intermediate case. The mean service time under FCFS is about 26 seconds of which 12 seconds are pick and place times. The average travel time is 14 seconds. The largest average queue length among all three dispatching rules is 1.37, which is found under FCFS. Note that the mean plus three standard deviations is $1.37 + 3 \times 1.59 = 6.14$. One could expect that the number of incomplete service requests very rarely exceeds 7.

The modeling results in case 2 are not much different from those in case 1, except that the difference between FCFS and STFS becomes larger. The pick and the place times are reduced, and enlarged workstations lead to relatively long travel time. Consequently, DAH performance becomes sensitive to dispatching rules. The handling delay difference between FCFS and STFS is 20 to 25%. However, differences in utilization are consistently small. Therefore average material handling delay is a better measure for the comparison of dispatching rules.

Sometimes a small handling delay may seriously degrade throughput of an assembly line. For instance, consider a case in which the bottleneck operation has a cycle time of 5 minutes, and the theoretical throughput is 12 assemblies per hour. If the material handling delay is 60 seconds, then the operation cycle becomes 6 minutes and the line throughput is reduced by about 17%. In this case, a priority dispatching rule may be considered. A high priority is given to the requests related to the bottleneck operations. This will reduce handling delay at the critical operations but prolong the delay at the nonbottleneck area. One may define as many priority classes as required. For each class, the request rate can be estimated by applying (5.1) to those operations that belong to the same priority class. Similarly, S_{ijk} and p_{ijk} can be obtained by the procedure in Section 5.2.1. Once the arrival rate and the first two moments of service time are known, the results from M/G/1 nonpreemptive priority queues can be applied directly. (See Section 3.3.)

5.2.4 Implementation

The previous two sections have discussed how to improve AS/RS performance by altering hardware configuration and dispatching rules. A desirable operating range is defined based on the relation between the average material handling delay and DAH utilization. However, this range may not be the same for different hardware. Table 5.3 lists three different

Table 5.3 Different DAH Hardware Configurations

Case	1	2	3
Top Speed (ft/sec)	3.3	6.0	6.6
Acceleration (ft/sec/sec)	1.6	2.0	3.3
Rotation Speed (degree/sec)	0.0	60.0	60.0
Pick time (sec)	7.5	6.0	3.8
Place time (sec)	9.0	6.0	3.8

DAH hardware configurations under the existing technology. Case 1 does not have rotation capability; all its workstations must be placed on the same side of the linear track.

A number of experiments have been conducted by varying the request rate, flow pattern and workstation layout. A model based on an M/G/1 queue under FCFS is employed. Figure 5.6 shows three sets of numerical results indicated by discrete points; each corresponds to an individual case given in Table 5.3. Three smooth curves, derived from the best fit, are used for comparison. These curves can be constructed by least-square models (discussed in Chapter 2) or simply hand-sketched.

The vertical and horizontal axes in Figure 5.6 are respectively the average material handling delay and DAH utilization. Clearly, for a high performance DAH, the curve should appear in the lower right corner. For the same average delay, a high performance DAH results in higher utilization. It is seen that all three curves have very steep slopes if the average material handling delay is greater than 60 seconds. Thus, for a range of DAH hardware under the existing technology, a stable performance region is defined by looking at the average material handling delay. When the delay is kept at 60 seconds, DAH utilizations for cases 1, 2 and 3 are about 0.65, 0.74 and 0.82 respectively. In other words, at the same service level, case 3 has about 26% more useful capacity than case 1 does.

In addition to material delivery, if a DAH also performs load and unload functions at automated workstations, then the pick/place time may be considerably longer, e.g., 30 seconds to 2 minutes. In this case, the 60-second bound is no longer valid for the stable region. The line designer should reconstruct the performance curve and determine a new bound. The same M/G/1 model, but with a different mean service time, can be used. Empirical results show that the 60-second bound may be considered valid as long as the sum of the pick and the place times is below 18 seconds.

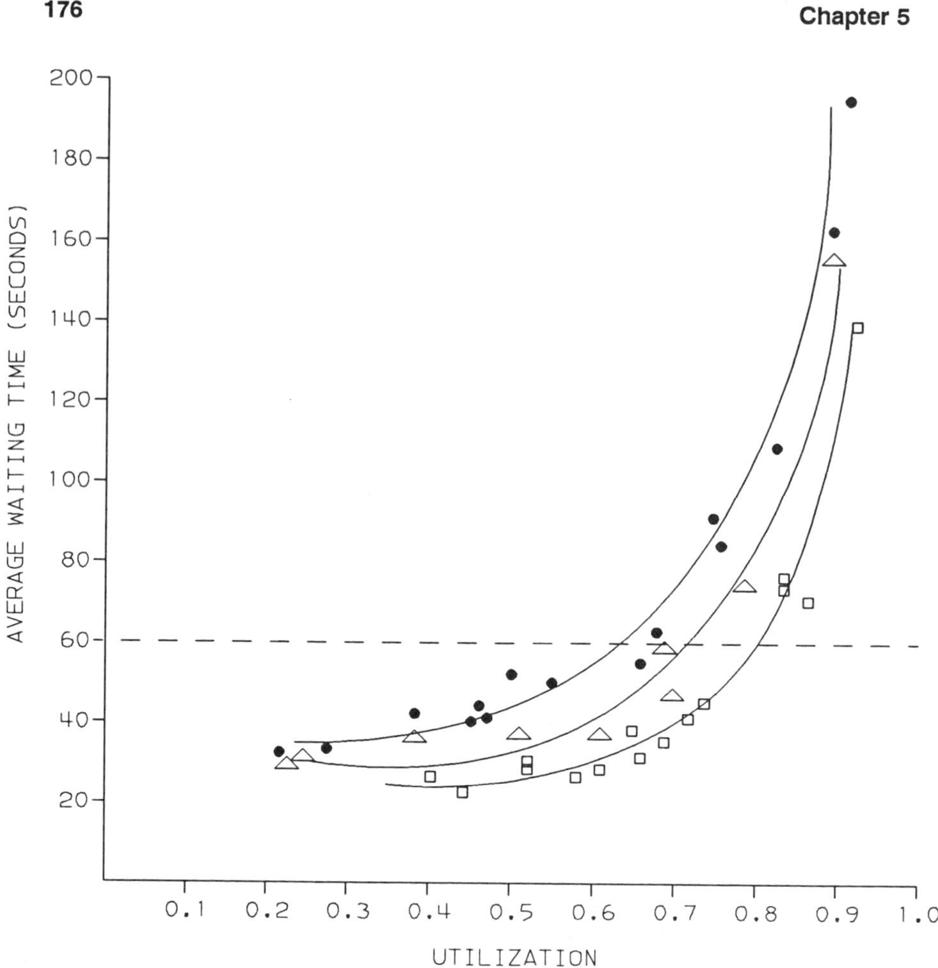

Figure 5.6 Operating characteristics of AS/RS.

If a workstation has to wait for material handling service after a job completion, then the waiting time becomes a part of operation cycle (a cycle is the interval between two successive completions). From a line performance viewpoint, a long cycle means low throughput. In this case, a trade-off evaluation is necessary. On one hand, it is desirable to keep the DAH busy. On the other hand, material handling delay should be kept at a minimum level. One should assess both cost and line performance. Unfortunately this cannot be done without considering the overall line structure. An effective local solution is introduced in the following.

Between each workstation and the DAH, an interface may be installed to hold a work unit. Removal of the work unit to the assembly work bench triggers a request for a new unit. If the material handling delay for the new request does not exceed the assembly cycle time, then the delay becomes transparent to the workstation. The assembly cycle is equal to an operation process time plus a removal time. The latter is dependent on the weight, size and fragility of the object and may range from 5 to 20 seconds. If the average DAH delay is 60 seconds, the saving is 40 to 55 seconds. This is achievable at the cost of an extra work unit and an interface device.

By changing the service request rate and the service pattern, different distributions of material handling delay are obtained by simulation, as shown in Figure 5.7. All these distributions are positively skewed and the 90-percentile is approximately equal to the mean plus 1.5 standard deviation. In real-life applications, this number may be regarded as the

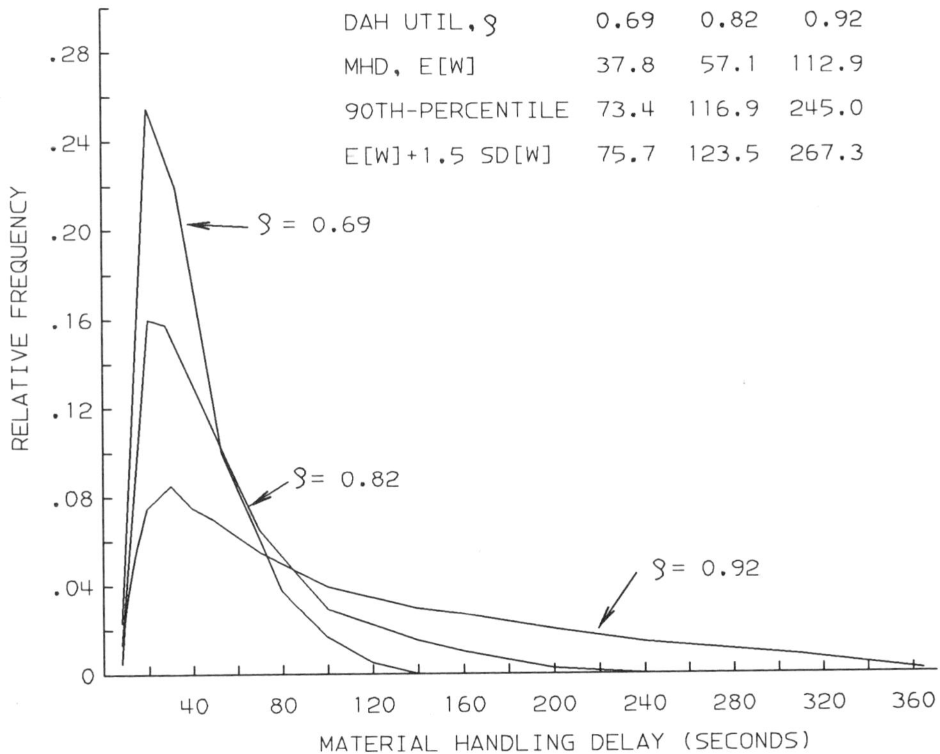

Figure 5.7 Material handling delay distribution of DAH.

maximal material handling delay. Under FCFS, the standard deviation can be estimated by Equation (5.11).

Based on the above discussions, the service limitation of an AS/RS is given by two principles:

1. The average material handling delay should be kept below 60 seconds. This will guarantee performance stability.
2. The mean plus 1.5 standard deviation of handling delay should never exceed the workstation process time. Therefore the AS/RS can be free from being the line bottleneck.

A line configuration based on AS/RS will be introduced in Chapter 8.

5.3 Conveyor System

There is a vast body of literature dealing with different kinds of conveyors. Some have accumulation capability; some do not. Some use discrete buckets to hold workpieces; some are equipped with belts. Some have powered rollers; some are not powered.

An assembly line normally consists of a number of operations arranged sequentially. After completion at an operation, a workpiece continues to its next operation at a subsequent workstation. If stations of the next operation are busy, the workpiece usually has to wait in a queue until a workstation becomes free. If a conveyor does not have an accumulation capability, the workpiece must bypass the next operation. The incomplete workpiece may re-enter the missed operation through a return path, but this complicates flow control and line discipline problems.

From a workstation point of view, the time interval between successive arrivals should be kept below the operation process time. For an automated workstation, a major concern is workstation failure. This problem may be resolved by stopping the conveyor. For a manual operation, pacing may or may not be a good approach. If the assembly operation is very simple, line pacing may work well. But a simple repetitive task may not be attractive to many operators. For a complex operation, pacing may cause problems in both quality and productivity. For a long feeding cycle, some workstation idle time is inevitable. On the other hand, if the cycle is short, the operator must respond to incoming works quickly, and therefore, the quality of work may deteriorate. Since manual assembly time is variable, workpiece misses may occur. A miss means extra waiting time for the next arrival and extra handling effort for the missing piece.

If the missing workpieces are re-introduced into the conveyor line, either the conveyor or workstations must have an intelligence to identify the right workpiece for the right operation. For an automated line, vision devices such as bar code scanners may be installed. Flow control is handled by complex software systems. It is not uncommon that several man-years are required for system development. For a manual assembly operation, identification may be done by an operator. Therefore, discipline is important. If a line is poorly managed, missing workpieces may stay on the conveyor for a long time and the amount of work-in-process can be very high.

For these reasons, a conveyor without accumulation capability is more suitable for workstations that perform identical operations. A typical example is that each workstation does the complete assembly work. In this case, the duration of an operation cycle should be carefully studied. If the cycle is longer than 20 or 30 minutes, both learning and productivity may present problems.

For an assembly line with multiple operations, local storage is often required for smooth production flow. The manufacturing concept derived from Toyota's Kanban system (discussed in Chapter 6) uses local storage to control work-in-process, to regulate production schedules and to detect line problems. This concept can be directly implemented for a conveyor line with accumulation capability. The relation between a conveyor and workstations is illustrated in Figure 5.8. The conveyor space between two adjacent workstations may be used for storage. The length of a conveyor should be determined by the storage space requirement. Insufficient space usually leads to significant reduction in line throughput. Determination of the minimal storage space for a given throughput is an important design problem, and will be discussed in Chapter 6. Analysis of a given conveyor configuration and a flow requirement is presented in the following.

Three different approaches may be used for analysis. The first one views workpieces on a conveyor as vehicles on a highway. Traffic engineering principles can be applied directly. The second approach treats material flow on a conveyor system as flow in a network. By solving a maximal flow problem, the capacity of the conveyor can be analyzed. This approach leads to a well-defined problem that is solvable by effective algorithms, but the stochastic nature of material flow is not considered. The last approach is

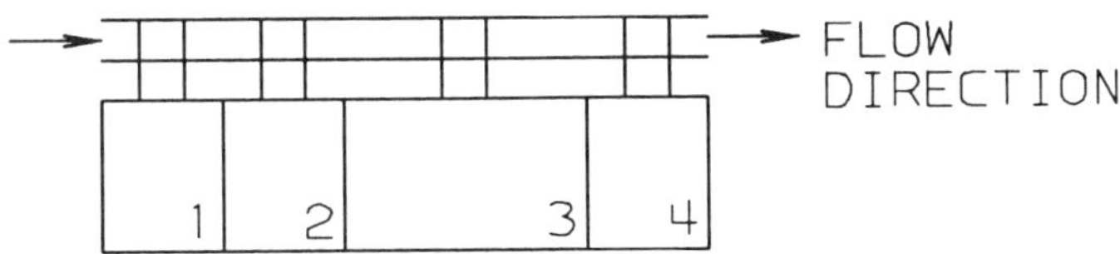

Figure 5.8 A conveyor line.

to assess a conveyor system by a stochastic model, e.g., a queuing model may be employed. However, simplified assumptions must be used in many cases because of mathematical intractability. All three approaches will be discussed under the assumption of accumulation.

5.3.1 Traffic Model

Consider that two workstations are connected by a conveyor. Let s be the average size of workpieces, h be the minimal distance headway between workpieces on the conveyor, and v be the conveyor speed. (See Figure 5.9) The conveyor capacity (i.e., the maximal number of workpieces that may pass through in a unit of time) is $c = v/h$. If the traffic flow requirement exceeds the capacity, the conveyor becomes a bottleneck of the production system. If the flow volume is fixed and below the conveyor capacity, then the moving speed of a workpiece is v. The material handling delay is simply equal to the travel distance divided by v. The case of interest is that the flow requirement may be changed dynamically. If the requirement temporarily exceeds the capacity, excessive workpieces may be either blocked from entering the conveyor (if accumulation is not allowed) or accumulated on the conveyor. The former case can be handled in a straightforward manner. It is the latter case that needs our attention.

Suppose that a conveyor system has accumulation capability. If a section of the conveyor temporarily becomes a bottleneck, workpieces may have to be accumulated in its upstream section. This situation is illustrated in Figure 5.10(a). If the upstream section has a length of l, the maximal number of workpieces that can be accumulated is $r = l/s$. The speed in the congested area may be dropped to zero while the speed in the free flow area is still v. The speed of accumulation is dependent of the flow rate in the free flow area. For simplicity, let us treat conveyor flow as continuous fluid. When workpieces are accumulated, the boundary between the congested area and the free flow area moves in an opposite direction against the conveyor flow. In Figure 5.10(b), this is the frontage of the shaded area. Let x be the speed of the moving frontage. The free flow speed is simply the conveyor speed. Since the flow direction and the frontage direction are

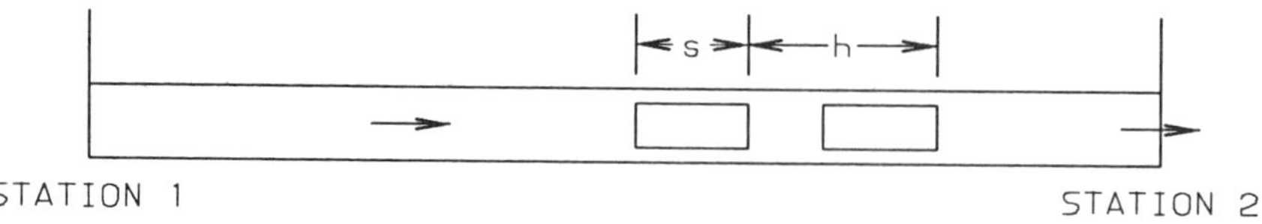

Figure 5.9 Work units on a conveyor.

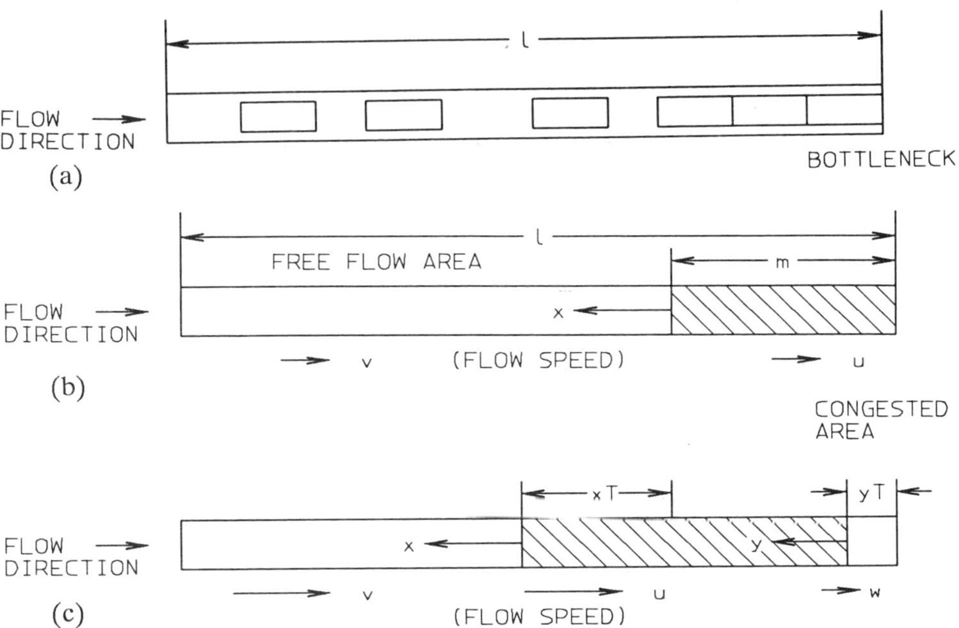

Figure 5.10 Conveyor flow under a congestion condition. (a) Accumulation. (b) Shock wave. (c) Recovery wave.

opposite, the flow moving into this frontage has a relative speed $(v + x)$. Similarly, the flow leaving the frontage is $(u + x)$, where u is the average flow speed in the congested area. Normally $u = 0$. Assume that the average headway in the free flow area is g. The densities in both free flow and the congested areas are $1/g$ and $1/s$, respectively. Because the flow volume is equal to the product of the density and the flow speed, and because the flow volume entering the moving frontage must be identical to the volume leaving the frontage, it follows that

$$(v + x)/g = (u + x)/s$$

Solving for x, we have

$$x = \frac{v/g - u/s}{1/s - 1/g} \tag{5.13}$$

The above equation defines the speed of accumulation on a conveyor section. In fluid dynamics, this is referred to as the *shock wave* speed. In t time units, the accumulated workpieces will cover an area of size xt with xt/s pieces. If the workpieces are completely stopped, i.e., $u = 0$, the number of accumulated pieces becomes

$$q = vt/(g - s) \tag{5.14}$$

If later the bottleneck condition is removed, then a downstream free flow area is formed. In Figure 5.10(c), a recovery region is shown by the unshadowed area downstream of the congested area. Again the boundary of the recovery area is moving against the traffic flow. Let w be the average flow speed in the recovery region and k be the average headway. Then the moving speed of the recovery wave is given by

$$y = \frac{w/k - u/s}{1/s - 1/k} \tag{5.15}$$

If $y > x$, then eventually all accumulated workpieces will be cleared out. Since a conveyor usually has a constant speed, $w = v$. Comparing (5.13) with (5.15), the condition that $y > x$ is equivalent to that $k < g$. This simply says that the discharge rate from a congested area must be greater than the arrival rate to the congested area. Once the bottleneck condition is removed, the discharge rate is usually equal to the maximal flow rate, at which the headway attains its minimum, i.e., $k = h$.

If a bottleneck condition has been imposed on the conveyor for t time units, the recovery time becomes $t' = xt/(y - x)$. If $u = 0$, then

$$t' = \frac{(k - s)t}{g - k} \tag{5.16}$$

During the recovery period, the congested area is diminishing and moving upstream. The vanishing point is located at $x(t + t')$ upstream of the bottleneck.

The traffic quality can be measured by looking at the average speed, density, flow rate, travel time and occupancy. Assume that at time 0 the length of the congested area is m as indicated in Figure 5.10(b). The free flow area is $l - m$. Since the speed of the congestion wave is x, after T time units the wave frontage arrives at $m + xT$. Meanwhile a recovery wave is formed and moves at a speed of y. The entire conveyor section is partitioned into three regions: (i) a free flow area of length $l-(m+xT)$, (ii) a congested area of $m-(y-x)T$, and (iii) a recovery area of yT. The average headways, speeds and occupancies in these three regions are summarized in Table 5.4.

The total number of workpieces in a region is equal to the region length divided by the headway. Let n_1, n_2 and n_3 be the numbers in the regions of free flow, congestion and recovery, respectively. Let l_1, l_2 and l_3 be the average lengths of these three regions during $(0,T)$. Then

$$\begin{aligned}
l_1 &= [(l-m)+l-(m+xT)]/2 = (l-m)-xT/2 \\
l_2 &= [m+m-(y-x)T]/2 = m-(y-x)T/2 \\
l_3 &= yT/2 \\
n_1 &= l_1/g \\
n_2 &= l_2/s \\
n_3 &= l_3/k
\end{aligned} \tag{5.17}$$

The average speed over the entire section of the conveyor is given by

$$\bar{v} = (n_1 v + n_2 u + n_3 w)/n \tag{5.18}$$

where $n = n_1 + n_2 + n_3$

The average density (i.e., the number of workpieces per unit length) is

$$\bar{d} = n/l \tag{5.19}$$

Table 5.4 Traffic Quality on Conveyor

Region	Size at 0	Size at T	Mean Headway	Average Speed	Occupancy
Free Flow	$l-m$	$l-(m+xT)$	g	v	s/g
Congestion	m	$m-(y-x)T$	s	u	1
Recovery	0	yT	k	w	s/k

The average traffic flow rate (i.e., the expected throughput) is

$$\bar{q} = \bar{v}\bar{d} \\ = (n_1 v + n_2 u + n_3 w)/l \tag{5.20}$$

The average occupancy is

$$\overline{occ} = sn/l \tag{5.21}$$

Finally, the average travel time during time $(0, T)$ is given by

$$\bar{t} = l/\bar{v} \tag{5.22}$$

The above discussion provides a basis for flow analysis, if the flow pattern and traffic demand are known. The study period can be divided into a number of slices such that the flow pattern and the demand in each slice is fixed. The traffic condition at the end of a time slice is used as the initial condition for the next slice. Algorithm 5.2 gives a stepwise computational procedure. This algorithm considers a single conveyor section. At the beginning of each time slice, it is assumed that accumulated workpieces are all gathered at the end of the section. This assumption is not essential for traffic evaluation, but simplifies the data processing effort and reduces computational time.

Algorithm 5.2

1. Let T be the duration of a time slice and $i = 0$.
2. $i = i + 1$ (A new time slice is under consideration).
3. Determine the downstream traffic condition, that is, the departure rate from the conveyor. (Usually this rate is the downstream process rate.) Denote this rate by b.
4. Based on the results from the previous time slice, determine the size of the congested area. If there is no accumulated workpiece left over from the previous time slice, the size is zero.
5. Determine the traffic volume, a. If the demand is greater than the capacity, i.e., $d > c$, then $(d - c)T$ workpieces will be delayed to the next time slice. (In this case, the average headway is minimal, i.e., h.) The flow volume is $a = \min(d, c)$.

The next step can be one of three mutually exclusive cases described in steps 6, 7 and 8, respectively.

6. If no congested area exists and $a \leq b$, the conveyor has enough capacity for the passing flow. The material handling delay (i.e., the travel time) is l/v, where l is the length of the conveyor. The average headway is given by $g = v/a$. The average number of workpieces on the conveyor is l/g and the average occupancy s/g. Go to step 2.
7. If no congested area exists but $a > b$, compute x by (5.13) with $g = v/a$ and $u = bs$. This is illustrated by Figure 5.10(b). Compute the average travel time, the average flow speed, occupancy, the average headway and flow rate by (5.17)–(5.22) with $m = 0$ and $y = 0$, i.e.no recovery wave exists. Go to step 2.
8. If a congested area exists in accordance with the result from step 4, then compute (i) the speed of the recovery wave, y, by (5.15) with $w = v$ and $k = v/b$, and (ii) the propagation speed of accumulation, x, by (5.13) with $g = v/a$ and $u = 0$. This case is shown in Figure 5.10(c). Compute the average travel time, the average flow speed, occupancy, the average headway and flow rate by (5.17)–(5.22). Go to step 2.

Let t_1 be the time when the congestion wave reaches the upstream end of the conveyor and t_2 be the time when the recovery wave catches up with the congestion wave, i.e., the congestion area vanishes. To execute the last step of Algorithm 5.2, five different cases should be considered, contingent on the magnitudes of T, t_1 and t_2.

8a. $T \leq t_1$ and $T \leq t_2$

In this case, m, x and y are all positive. Relations (5.17)–(5.22) can be applied directly.

8b. $T > t_1$ but $T \leq t_2$

The time period, $(0,T)$, can be partitioned into two smaller intervals: (i) During $(0,t_1)$ use (5.17)–(5.22) with T replaced by t_1. (ii) During (t_1,T) the upstream portion of the conveyor has become congested and no additional incoming traffic is allowed. Consequently, $x = 0$.

8c. $T \leq t_1$ but $T > t_2$

This is the case when the recovery region takes over the congested area. Again during $(0,t_2)$ (5.17)–(5.22) can be employed by replacing T by t_2. Since the conveyor usually has a constant speed, $w = v$. The free flow area and the recovery area, however, may have different mean headways. The boundary between these two areas moves in the same direction as the traffic flow at a speed of v. Thus, after yt_2/v time units, this boundary will reach the downstream end of the conveyor.

8d. $T > t_1 > t_2$
Since t_2 comes first, this case is identical to case 8c.

8e. $T > t_2 > t_1$
Traffic quality may be evaluated by looking at $(0,t_1)$, (t_1,t_2) and (t_2,T) separately. Traffic flow during the first interval can be evaluated by (5.17)–(5.22) with T replaced by t_1. The same approach can be applied to the second interval, except that there is no free flow region and the congested area stops growing. During the last interval, the recovery area is diminishing and is taken over by free flow traffic.

The traffic flow model closely describes the physical flow behavior. If the number of conveyor sections is large and the flow pattern changes very frequently, computational complexity may present a problem. Next we consider a network flow model for conveyor analysis.

5.3.2 Network Flow Model

The configuration of a conveyor often lends itself to a network flow analysis. A network can be constructed in a two-dimensional space. Figure 5.11 shows a network of 20 nodes with four workstations (the horizontal axis) and five time increments (the vertical axis). A node is denoted by (i,j) for workstation j and increment i, $i = 1, \ldots, 5$ and $j = 1, \ldots, 4$. The link between nodes reflects the travel time between the two nodes. For example, since the travel time between workstations 1 and 2 is one time unit, a link is shown from $(i,1)$ to $(i+1,2)$. Similarly, a link from $(i,2)$ to $(i+2,3)$ implies that the travel time from station 2 to station 3 is two time units. If the study period consists of five time units, a wraparound structure may be used; that is, the sixth and the first time units are equivalent. This situation is illustrated in the figure. The link capacity is given by $c = v/h$, where v is the conveyor speed and h is the minimal headway. To complete the network, a pair of source and sink may be added and denoted by a and b, respectively. For each node associated with workstation 1, there is a direct link from the source to that node. For each node with workstation 4, there is a direct link from that node to the sink. All links connecting the source or the sink have infinite capacities.

If a conveyor has accumulation capability, then workpieces may be kept on a conveyor section. This can be modeled by introducing additional links. For each workstation in Figure 5.12, additional links are placed between (i,j) to $(i+1,j)$. These links form a cycle with respect to j, i.e., following the

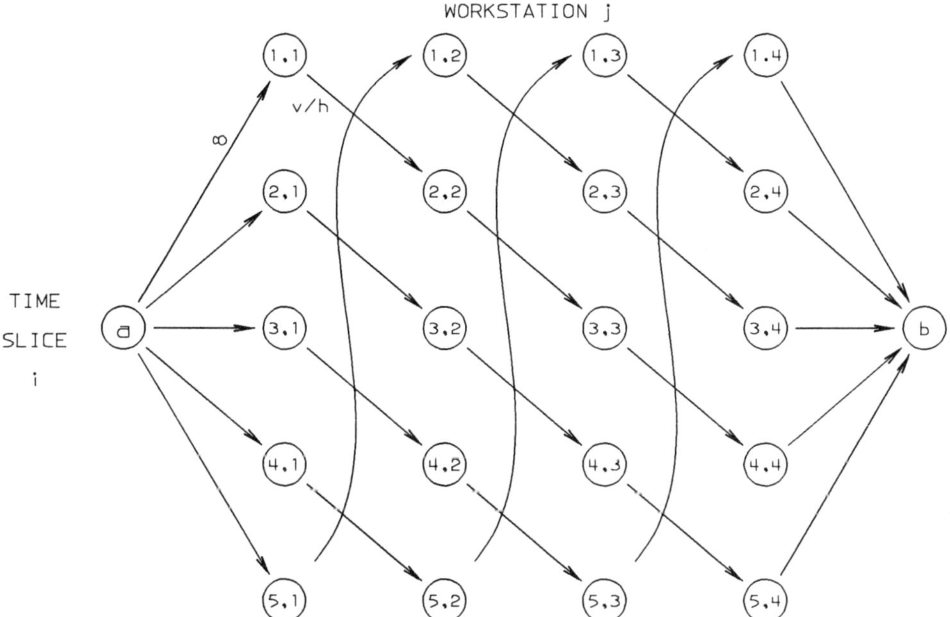

Figure 5.11 A network flow model for a conveyor.

last time unit is again the first unit. Since flows on a cycle never actually leave a workstation, this is equivalent to the accumulation effect. The capacity of an accumulation link is equal to its accumulation capacity, i.e., the maximal number of workpieces that can be stored on a conveyor section. If l is the length of the section and s the average size of workpieces, the link capacity for accumulation is given by $c = l/s$.

The network model may also include workstation capacities. This is done by replacing a workstation node by a pair of nodes with a direct link between. As illustrated in Figure 5.13, both workstations 2 and 3 are represented by node pairs. The capacity of a link between a node pair is the workstation capacity during a time unit. To reduce the size of a network, nodes representing the first and the last workstations need not be changed into node pairs. The workstation capacity, however, may be handled by limiting the capacity of each link connecting the source or the sink. For instance, the link capacity between the source and node (2,1) is the maximal number of assemblies that can be completed by workstation 1 during the second time unit. Similarly, the capacity of the link from node (3,4) to the sink is the capacity of workstation 4 during the third time unit. For a given

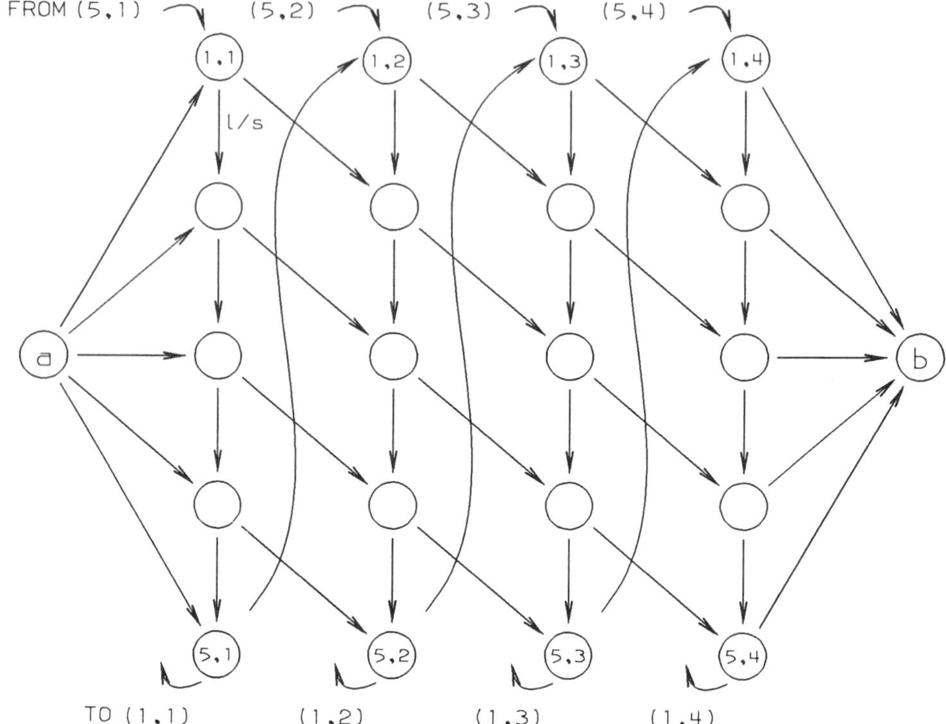

Figure 5.12 A network flow model with accumulation links.

conveyor configuration and its speed, and given workstation capacities as functions of time, a maximal flow model can be used to estimate the line capacity. A simple illustration given in the following example.

Example 5.1 Consider a conveyor serving three workstations, as shown in Figure 5.14(a). Workpieces of a uniform size are sent over the conveyor from station 1 to station 2 and then to station 3. The conveyor is equipped with transfer ports, one for each workstation. The function of a transfer port is to control material flows into and out of its associated workstation. The size of a transfer port is identical to that of a workpiece. Both are 2 feet. The distance between port 1 and port 2 is 8 feet, and the distance between ports 2 and 3 is 16 feet. Therefore, the accumulation capacity between ports 1 and 2 is $8/2 = 4$ and between ports 2 and 3 is $16/2 = 8$.

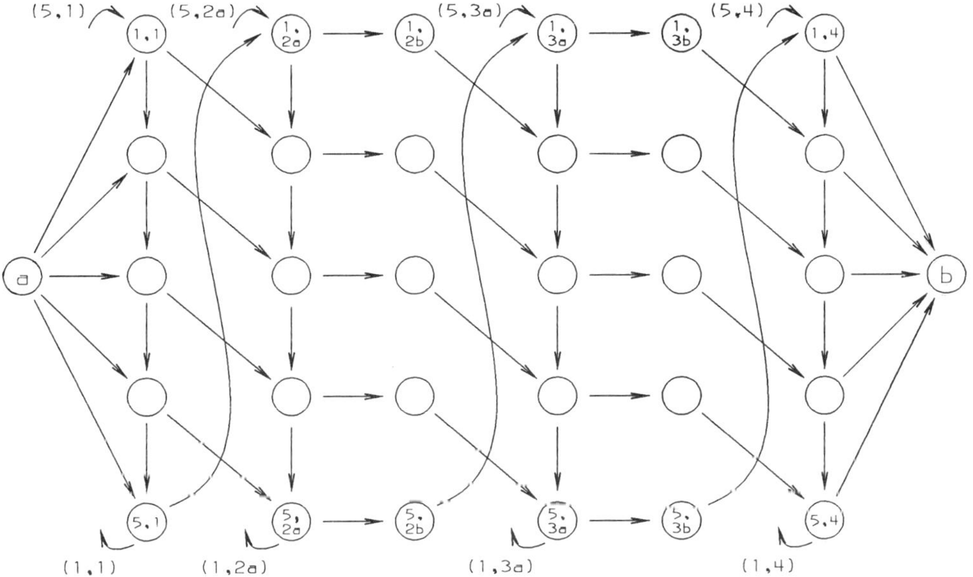

Figure 5.13 A network flow model with workstation links.

The time unit may be selected by taking the greatest common divisor of travels times between adjacent ports. The assembly is made of fragile material. Therefore, the conveyor speed is kept at 0.1 foot per second and the minimal distance headway is 2.5 feet. The travel times between adjacent ports are 80 seconds and 160 seconds, respectively. Therefore, a time unit is chosen to be 80 seconds. Consequently, the maximal flow volume on the conveyor is $80 \times (0.1/2.5) = 3.2$ per time unit, and an eight-hour shift consists of 360 time units.

A network with 360 time units and four columns (two columns for station 2 and one column for stations 1 and 3, respectively) is given in Figure 5.14(b), where $u(i,j)$ is the capacity of workstation j during the ith time unit. This capacity may be changed from time to time because of break, workstation failure, varying work speed of an operator, different operators on the station, or simply a scheduled downtime such as preventive maintenance.

Once a network has been constructed, the solution can be obtained by invoking Algorithm 3.2. □

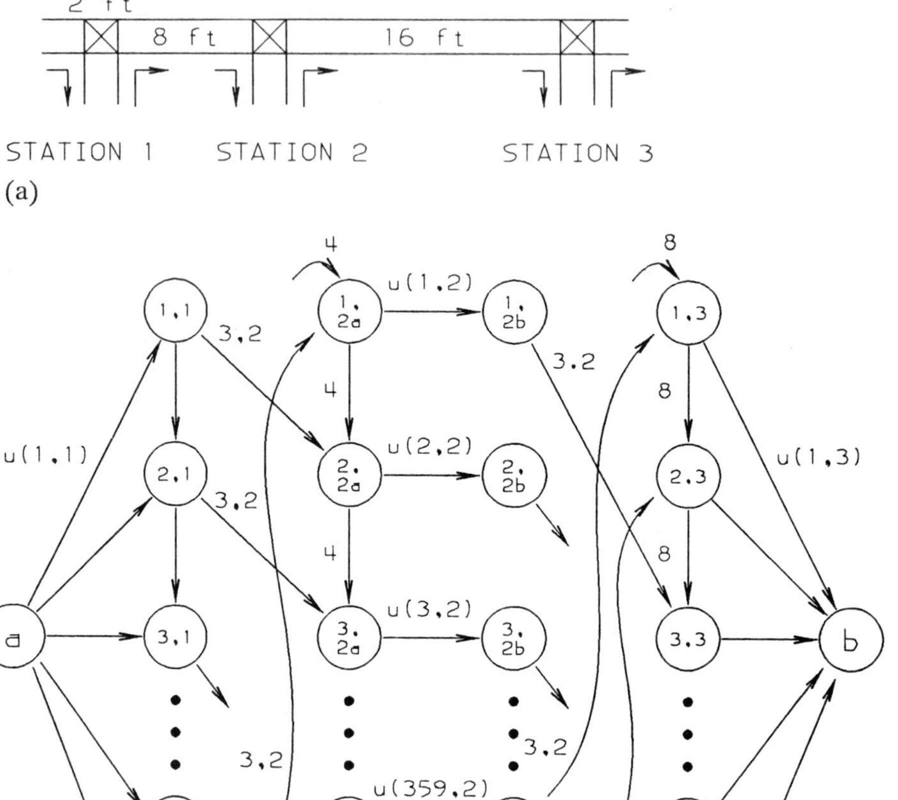

Figure 5.14 A network flow model for a conveyor with 3 workstations. (a) A conveyor with 3 workstations. (b) A network flow model.

The computational method for a network flow model can be summarized as follows:

Algorithm 5.3

1. Determine the number of conveyor sections. (Usually a section is defined as a piece of conveyor that links two workstations.)
2. Compute travel times of all conveyor sections. Let l_k be the length of section k and v be the conveyor speed. Then the travel time over section k is $t_k = l_k/v$.
3. Let d be the time unit equal to the greatest common divisor of $\{t_k\}$.
4. Determine the length of the study period, T. For convenience, T is so chosen that $m = T/d$ is an integer. Thus, m is the total number of time units considered in the model.
5. If workstation capacities should be included in the model, then for each station introduce a pair of nodes, and place a direct link between the pair. Otherwise, each workstation is represented by a single node.
6. Expand the network size by duplicating each node or node pair m times. Number all nodes or node pairs in accordance with their physical orders and time orders. A node is denoted by (i,j), if it corresponds to the ith time unit and the jth workstation. A node pair has two nodes denoted by (i,ja) and (i,jb), respectively, in the same fashion.
7. Define workstation links. The capacity of the link between (i,ja) and (i,jb) is equal to $d \times \mu(i,j)$, where $\mu(i,j)$ is the process rate of station j during the ith time unit. Repeat this step for all workstations, except the first and the last.
8. Add a source node and a sink node. Place links from the source to $(i,1)$ or $(i,1a)$ for $i = 1, 2, \ldots, m$. Put direct links from (i,n) or (i,nb) to the sink. Capacities of these links are identical to their correspondent workstation capacities during the ith time unit.
9. Convert travel time $\{t_k\}$ into time units. Since d is a common divisor of all travel times, $x_k = t_k/d$ must be an integer. Repeat this for all conveyor sections.
10. Define transport links. If section k connects stations j and $j+1$, place a direct link from node (i,j) to node $(y,j+1)$

where $y = i + x_k$ if $i + x_k \leq m$ and $y = i + x_k - m$ otherwise. If workstation capacity should be considered, then place the direct link from (i,jb) to $(y,(j+1)a)$. Repeat this for all sections.

11. Define accumulation links. If output from workstation i is transported over section k, place a link from node (i,j) to node $(i+1,j)$ with a capacity of l_k/s, where s is the size of workpiece. If workstation capacity is under consideration, a link is placed between node (i,ja) and node $(i+1,ja)$. Repeat this for all workstations.
12. Define wraparound links. If, after m time units, the process flow starts over again, then place a direct link from node (m,j) to node $(1,j)$ or from (m,ja) to $(1,ja)$. Repeat this for all sections.
13. Solve the network problem by Algorithm 3.2. The flows carried by transport links are flow volumes on conveyor sections. Flow values on accumulation links or wraparound links are the amount of accumulated workpieces. Values on the workstation links are the workstation throughputs.

The network flow approach has some practical problems. It can be seen from the above example that there are $4 \times 360 = 1440$ nodes. It is not uncommon that an assembly job takes one hour to finish but the assembly process is so designed that the mean operation process time is only 5 to 6 minutes. This implies a line of 10 to 12 workstations. For a 24-hour analysis, there will be $3 \times 360 \times 10 \times 2 = 21{,}600$ nodes. For an automated station, it is desirable to keep its operation process time below one minute for the sake of simplicity and reliability. Then the total number of nodes becomes five times as many, that is, 108,000! This certainly causes a computational problem.

Another problem of using the network flow model involves flow branching. Suppose that an assembly operation has a yield problem, with a defect rate of 0.05. All defective pieces are sent to a rework area and reintroduced into the assembly line later. Good assemblies are sent to downstream operations for completion. The flow stream from the operation is split into two substreams. This case may be represented by using two links from the operation; one downstream and the other to the rework area. But the model fails to take branching probabilities into consideration.

If a conveyor has sufficient space for accumulation, then an assembly line can be analyzed by using a queuing network model. Neither the problem size nor the branching probability may cause difficulty. Such a model is discussed in the next section.

5.3.3 Queuing Model

If workpieces in an assembly line are regarded as jobs in a queuing system, both workstations and material handling devices can be regarded as servers. They have no basic difference. Typically a workstation is equivalent to a single server. The service time is identical to its operation process time. An operation of multiple workstations can be modeled by a service center of multiple servers.

A conveyor section can also be considered as a service center. Since a conveyor may serve a number of jobs simultaneously, it is convenient to represent a conveyor section by infinitely many servers. The conveyor with three workstations discussed in Example 5.1 may be modeled by a series of queues, as illustrated in Figure 5.15. The three single servers are workstations and the two service centers with infinitely many servers are the two conveyor sections. A service center, corresponding to a conveyor section, has a constant service time which is equal to the travel time through the section.

Suppose that workstation 3 performs a test operation. Assemblies with defects will be sent back to workstation 1 for rework. Assume that the defect rate is 0.05. The line configuration and its queuing model are given in Figure 5.16. The numbers, associated with the two paths from workstation 3, are the branching probabilities derived from the defect rate. If the arrival process is Poisson and the service time at each single server (i.e., the process time of a workstation) is exponential, then each queue has an $M \Rightarrow M$ property. The entire queuing network has a product form solution, as discussed in Section 3.4.

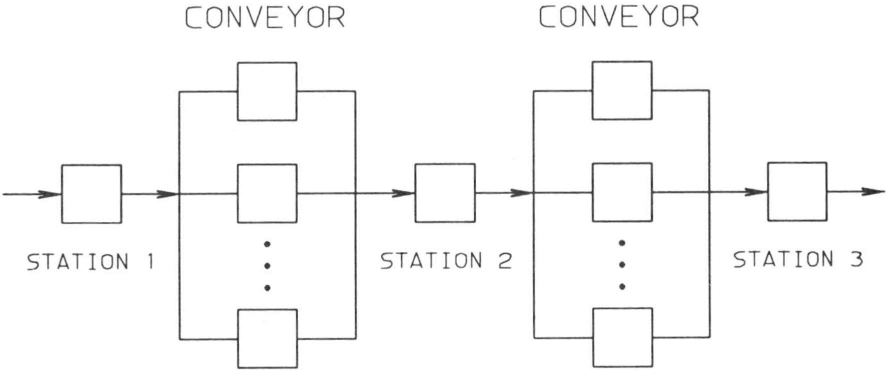

Figure 5.15 A queuing model for a conveyor.

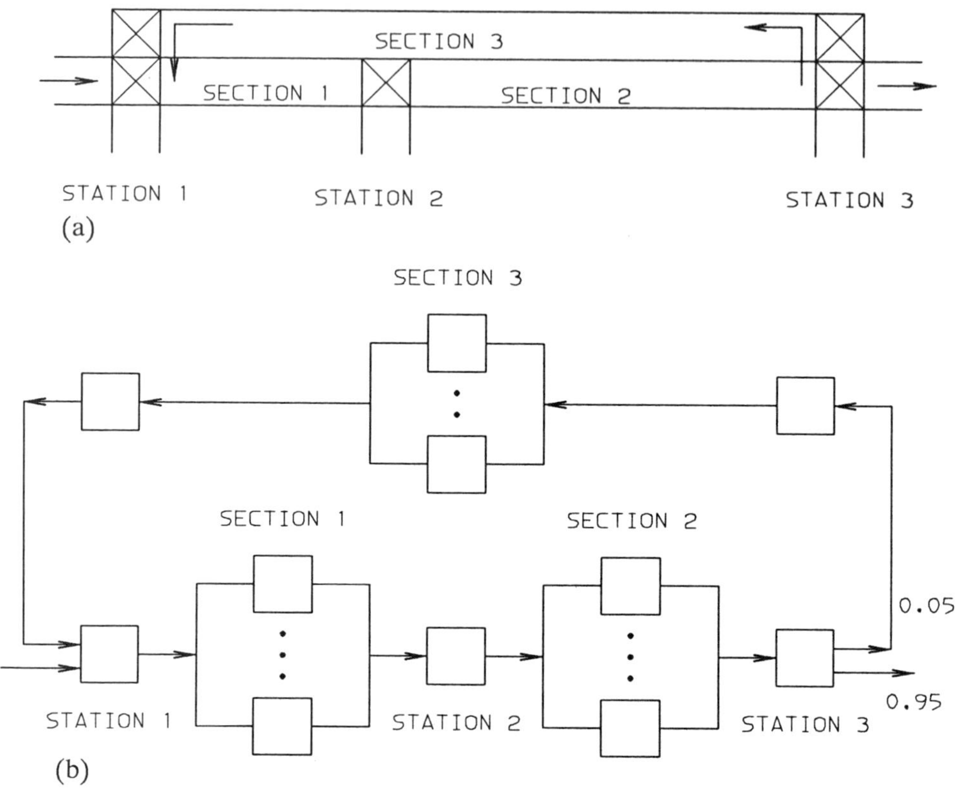

Figure 5.16 (a,b) A conveyor loop and its queuing model.

However, two problems may arise: First, the service time of a workstation may not be exponential, and second, insufficient storage space may block upstream operations. An exponential random variable has a coefficient of variation equal to one. This value is greater than what one would expect from reality. From our previous discussion, in Section 2.5, it is known that the coefficient of variation of a manual process time is typically between 0.1 and 0.65. For an automated operation, the process time is close to a constant, i.e., the coefficient is near zero. To resolve this problem, one may treat workstations as independent M/G/1 queues. Let ρ be workstation utilization and C be the coefficient of variation of a process time. The former can be obtained by taking the product of throughput and the average process time (see Example 3.3). Throughputs can be estimated by solving the simultaneous linear equations defined in (3.26). Let v_i be the expected number of times that a job visits queue i before leaving the queuing network and r_{ij} be the routing probability from queue i to queue j. The exogenous arrival rate to queue i is denoted by λ_i. For given $\{\lambda_i\}$ and $\{r_{ij}\}$, $\{v_i\}$ can be obtained by solving

$$v_j = \sum_i p_{ij} v_i + \lambda_j \qquad \text{for all } j \tag{5.23}$$

Let S_i be the process time at queue i. By using the results from Example 3.3, it follows that

$$\rho_i = v_i E[S_i]$$

The mean queue length is given by Equation (3.11), i.e.,

$$L_i = \rho_i + \frac{\rho_i^2}{2(1-\rho_i)}(1 + C_i^2) \tag{5.24}$$

Since an M/G/1 queue does not have an $M \Rightarrow M$ property, the arrival process, seen by each queue, is no longer Poisson. Insertion of M/G/1 queues can only lead to an approximate solution.

The second problem is blocking. Consider a conveyor section that connects two workstations. The output from the upstream station is sent over the conveyor to the downstream station. If the downstream station temporarily becomes a bottleneck, workpieces may be accumulated on the conveyor. If this happens at a time the upstream station has just completed an operation cycle and finds no available storage space on the conveyor, the station will stop its operation. This situation is called *blocking*. Unfortunately, a queuing problem with blocking is very difficult; no simple solution exists. If the probability of blocking is small (say, no greater than 0.05), an approximate solution may be obtained by "discounting" the service rate of the blocked station. Let b be the blocking probability and μ the service rate. The discounted rate is $(1-b)\mu$.

The blocking probability may be estimated by first solving the queuing network with infinite accumulation capacity, i.e., a no-blocking case. Suppose that workstations k and $k+1$ are connected by a conveyor section of length l and that the size of a workpiece is s. Let p_i be the probability of

i workpieces in the conveyor queue and q_j be the probability that j workpieces are found in the queue of station $k+1$. Then the blocking condition corresponds to $\{i+j \geq x\}$, where x is the least integer greater than l/s. The probability that station k is blocked is approximated by

$$b_k = \sum_{i=x}^{\infty} \sum_{j=0}^{\infty} p_i q_j + \sum_{i=0}^{x} \sum_{j=x-i}^{\infty} p_i q_j \tag{5.25}$$

If a workstation is modeled by an M/G/1 queue, its distribution of queue length, $\{q_j\}$, can be approximated by relation (3.12). For a small b_k, (5.25) usually gives a good approximation. Algorithm 5.4 defines a computational method for a conveyor serving a sequence of workstations.

Algorithm 5.4

1. Construct a network of queues. For each conveyor section (i.e., a piece of conveyor between two adjacent workstations), a queue of infinitely many servers is defined such that service time is identical to the travel time over the section. For each workstation, an M/G/1 queue is defined such that service time is equal to operation process time.
2. The job routing pattern in the network of queues is the same as the assembly process flow pattern. Number workstations by 1, 2, ..., n such that station 1 performs the first operation and station n does the last. Compute $\{v_i\}$ by (5.23). Let a conveyor section between stations i and $i+1$ be the section i.
3. Let $k = n$ (start from the last workstation).
4. Compute the queue length distribution at workstation k by relation (3.12). This distribution is a function of throughput, v_k, and the first two moments of the process time, S_k.
5. Compute the queue length distribution for conveyor section $k-1$ by (3.14). This distribution is a function of flow rate on the conveyor and travel time.
6. Use Equation (5.25) to estimate b_{k-1}, the blocking probability of workstation $k-1$. Discount the service rate of station $k-1$; that is, replace $E[S_{k-1}]$ by $E[S_{k-1}]/(1-b_{k-1})$.
7. If $k = 1$, all stations have been analyzed. Stop. Otherwise, let $k = k-1$ and go to step 4.

If a conveyor connects work centers of multiple workstations, then step 4 in the above algorithm should be modified. A work center with c workstations can be represented by an M/G/c model, an approximated queue length distribution is given by (3.15)–(3.17).

Numerical examples of conveyor queuing models are presented in the next section.

5.4 Comparison of AS/RS and Conveyor

Discussed in the previous sections are two basic types of material handling systems: AS/RS as a discrete system and a conveyor as a continuous one. This section attempts to compare these two systems from several different aspects. Our comparison is limited to the material handling systems only. No workstation performance is under consideration.

From a service capacity point of view, the DAH of an AS/RS can only serve one request at a time, while a conveyor can move multiple items simultaneously. On the other hand, with proper design a DAH may carry a workpiece at a much higher speed than does a conveyor. To make a meaningful comparison, a common manufacturing environment must be assumed. Workstations are either connected by a conveyor loop or placed on both sides of a DAH aisle, as illustrated in Figure 5.17. The conveyor system consists of transfer ports where objects can be loaded or unloaded. This arrangement is comparable to the case where a DAH is equipped with a rotation arm and serves stations on both sides of the DAH track. The interface between the DAH and a workstation is a pick-and-delivery (P/D) port, which corresponds to a transfer port on the conveyor. No matter which system is chosen, an I/O port of infinite storage size is always assumed.

The AS/RS is treated as an M/G/1 queuing system as described in Section 5.2, while the conveyor is modeled by a queuing network, discussed in Section 5.3.3, except that each workstation is replaced by a transfer port. For each transfer port, there are two types of jobs: (i) a job to forward an object through the port, and (ii) a job to transfer an object so that the moving direction of the object is changed. The latter is to load workpieces from workstations to the conveyor, to unload from the conveyor to workstations, or to transfer workpieces from one conveyor section to another. A queue with infinitely many servers is used to represent a conveyor section, while each transfer port is modeled by a single-server queue. A piece of conveyor and its queuing model are illustrated in Figure 5.18, where each single server has a constant service time, identical to travel time over a transfer port. The flow requirement at each section and at each port is given by $\{\alpha_i\}$ and $\{\beta_i\}$. The following data are assumed for numerical comparison:

AS/RS Data Set		Conveyor Data Set	
Top speed	6 ft/sec	Port Size	2 ft
Acceleration	2 ft/sec/sec	Speed	0.5 ft/sec
Pick/place time	6 sec	Transfer time	5 sec
Angular speed	60 degree/sec	Object size	2×2 ft^2

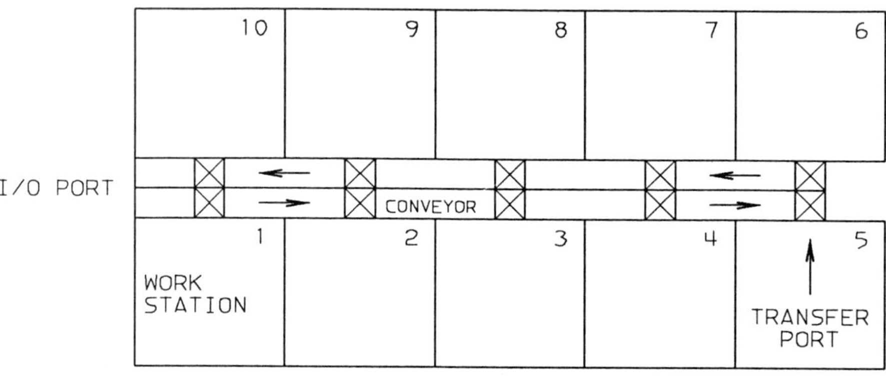

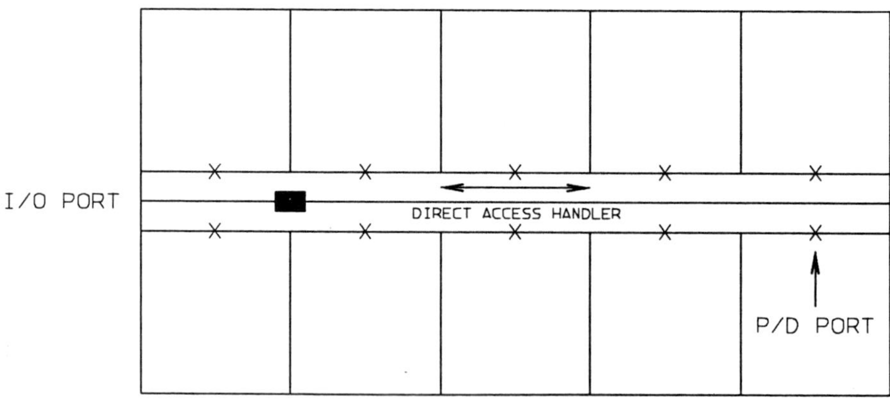

Figure 5.17 A conveyor layout and a DAH layout.

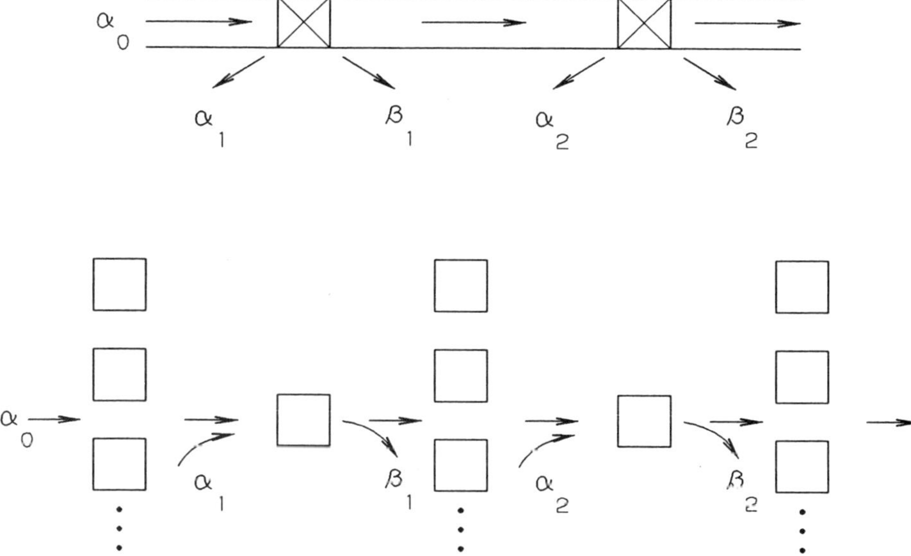

Figure 5.18 A queuing model for a conveyor.

Suppose that m assemblies are produced by two different types of workstations in T time units. Each type has five identical stations. For simplicity it is assumed that arrival materials can be accepted by stations without using intermediate buffer storage. The width of each station is 12 feet with a P/D port or a transfer port at the center. The parts packing density is such that each full tray provides two sets of assemblies. Material flow is evenly distributed among the workstations. The flow pattern is summarized as follows:

Station	Arrival Flows	Departure Flows
Type 1	Initial m subassemblies $m/2$ full trays of parts	m completed ones to type 2 $m/2$ empty trays
Type 2	m subassemblies from type 1 $m/2$ full trays of parts	m completed assemblies $m/2$ empty trays

The numerical results are presented in Table 5.5 for different traffic conditions during 20 hours. Average queue length, average material handling delay (in seconds), utilization, the average and the maximal blocking probabilities among all ports are reported. The queue length on the conveyor represents the total number of items that are moving on the conveyor or waiting for an available transfer port.

It is clearly shown that if the total flow volume is low, the AS/RS is superior to the conveyor due to fast speed. As the flow volume becomes higher and higher, eventually the conveyor catches up with the AS/RS. Before m reaches 1000, the AS/RS has already run out of capacity, but the conveyor can still provide a reasonable service quality with a steady material handling delay. The low blocking probability, b, and small queue length indicate that the transfer ports are under light traffic conditions. In fact, occupancies of transfer ports are consistently less than 0.09 when $m = 650$, and never exceed 0.14 when $m = 1000$. The approximations by the discount factor should not give us any serious problems.

Table 5.5 also reconfirms that for an AS/RS the average material handling delay should be kept below 60 seconds. This can be seen by plotting the average delay against utilization. From a practical point of view, if operation cycle times are less than 2 minutes, AS/RS may not be a proper choice for material handling. The tabulated results also indicate that the maximal service request rate that can be handled by a single DAH is about 120 per hour. If the rate exceeds this limit, either additional DAHs are needed or a conveyor should be considered.

Although a conveyor has a much larger service capacity because of its continuous flow, a sequential access method must be invoked. This means a relatively complex flow control mechanism. In order to select the right workpiece from a conveyor, each workstation must identify every passing

Table 5.5 Performance Comparison of AS/RS and Conveyor

		AS/RS			Conveyor		
ASM QTY	*MTRL* FLOW	QL	MHD	Util.	QL	MHD	b(AVE/MAX)
150	750	0.27	25.4	0.23	0.62	59.7	0.001/0.007
300	1500	0.66	31.8	0.46	1.25	60.0	0.001/0.013
450	2250	1.48	47.4	0.69	1.88	60.3	0.002/0.020
500	2500	2.06	59.4	0.76	1.94	60.5	0.002/0.022
650	3250	71.60	1585.6	0.99	2.70	60.8	0.003/0.029
1000	5000		Overload		4.29	61.7	0.005/0.044

item. Therefore, for each transfer port at least one detector (e.g., bar code scanner) is required. If an AS/RS is installed, it is easy to establish a point-to-point control method. A single detector, installed on the DAH, will perform the same function. Moreover, due to its direct access capability, an AS/RS may invoke a priority rule.

Since an AS/RS may offer a very large vertical storage space, the work-in-process control logic is nearly independent of physical storage. This can be valuable if the control logic has to be modified during the life of an assembly line. Typically a modification follows changes in assembly process, product yield, production demand or operation management strategy. On the other hand, since a conveyor has limited storage space, it is relatively difficult to accommodate these changes.

Installation of a storage buffer between operations is justified by process time variation. In general, the higher the variation, the more buffer space will be required. Therefore, an AS/RS is more suitable if operation process times have high coefficients of variation.

In many cases, line flexibility is desirable. It may be economical to adjust line capacity if production demand is changed. Because of the direct access method, a layout of an DAH line does not have to be consistent with the process flow. Therefore, a line expansion can be accomplished by simply adding workstations at the end of the DAH aisle. Similarly, an assembly line can be reconfigured without major layout change. This means a minimum

Table 5.6 An Overall Comparison of AS/RS and Conveyor

Factors	AS/RS	Conveyor
Duration of process time	long (> 2 min)	short (< 1 min)
Variation of process time	high	low
Moving speed	high	low
Capacity (service volume)	low (120/hour)	high
Flow control method	end-to-end	sequential
Flow direction	bi-directional	uni-directional
Storage structure	vertical	horizontal
Storage size	large	limited, fixed
Storage access method	direct	sequential
Work-in-process management	flexible	constrained
Line configuration	flexible	constrained
Process/engineering change	less impact	more impact

line disruption. For a new product, it may take a number of engineering or process changes to develop a good manufacturing process.

Finally, from a maintenance point of view, job complexity is a function of technology, design and quality. A conveyor system tends to have more parts than does an AS/RS. For instance, a power roller conveyor may have over one thousand belts, one for each roller. Belt replacement therefore can be very time-consuming. For this reason, an AS/RS may have an advantage.

The above discussion is summarized in Table 5.6.

Selection of a material handling system is one of the major problems in line design. In addition to minimal cost, there are other objectives that can be derived from a given manufacturing environment. The next two sections present two real-life case studies.

5.5 Case Study I

A manufacturing line produces high-technology products. Because of competition, a product lifetime of no more than five years is expected. Accurate demand forecast is difficult. Engineering changes may be frequent during the first two years. This means that both manufacturing process and production demand are subject to change. Since different processes mean different material flow patterns, the line may have to be reconfigured for each manufacturing process change. This will disrupt the line operation and sometimes cause a line shutdown. Furthermore, demand change may lead to line capacity adjustment and alter the total number of workstations. Both insertion or deletion of workstations will again disturb the material flow and the line operation. The above problems may be alleviated if the material handling system has multi-dimensional movement capability and can support a flexible line.

The manufacturing line consists of 10 operations. Some are automated operations and their workstations are subject to failures. Workstation downtime ranges from 5 minutes to 8 hours, and the average downtime is at least one order of magnitude longer than the average operation process time. The process times of manual assembly operations lie between 5 to 20 minutes and their coefficients of variation are within (0.2, 0.6). Daily yield can be regarded as a random variable ranging from 0.75 to 0.95. Most bad products are reworkable. Rework is scheduled in a batch mode, that is, no rework is done until there are enough bad products in a batch, and batch size is determined by the line manager.

Workstation reliability, variable process time and product yield are three major performance detractors. Work-in-process is therefore introduced to smooth production flow and to increase line throughput.

The assembly line is operated in a clean room, where high air quality must be maintained. Both installation and operation cost of the clean room are very expensive and contribute a significant part of total manufacturing cost. Therefore, space utilization becomes a major concern. Material handling aisle, storage space and workstation size should be kept at a minimal level. Furthermore, any selected material handling system should be clean-room compatible.

Workstation cost varies from an order of \$100,000 to \$1,000,000. The total workstation cost counts as the major portion of the line installation cost. To fully utilize workstations, material handling delay should be short.

The product's size is about $(2 \times 2 \times 2)$ cubic feet and it weighs slightly less than 150 pounds. The product is fragile. Careless handling may cause product damage. The actual demand is unknown. But the line designer believes that a demand of 220 pieces a day is a good estimate. For three shifts a day, the total productive time is targeted for 19 hours. Since there are 10 operations, the material handling request rate is $220 \times 10/(19 \times 60) \approx 2$ per minute.

Based on the above information, the selection criteria can be summarized as follows:

cost
flexibility
multi-dimensional movement
space requirement
material handling delay
product compatibility
environment compatibility

Considered are three alternatives: (i) an AS/RS, (ii) a power-roller conveyor, and (iii) manual delivery. To reduce travel distance, the line is configured in a U-shape. If an AS/RS is chosen, the DAH aisle is placed between two rows of workstations. Thus, rotation capability is assumed. For the second case, a conveyor loop should be installed. These two cases have layouts similar to those illustrated in Figure 5.16. A typical workstation size is 12×10 square feet. Therefore, the DAH aisle is $5 \times 12 = 60$ feet long, while the conveyor length is 120 linear feet.

Algorithm 5.1 is invoked to analyze the AS/RS performance. The computed mean service time of the DAH is about 20 seconds. Therefore, utilization is $\lambda E[S] = 2 \times 20/60 = 0.67$. The average handling delay is estimated to be 40 seconds by an M/G/1 queuing model. The multi-dimensional movement capability of the AS/RS offers a high degree of flexibility in line capacity adjustment. Additional workstations can always be placed at the end of the DAH aisle. A process flow change does not affect line operation very much. Vertical storage has good floor space utilization and reduces clean room cost. Handling damage almost never becomes a problem. The cost range of an AS/RS is between $80,00 and $150,000.

A 120-foot conveyor covers 10 workstations. When a delicate product travels over rollers, the speed cannot be too fast. Normal speed is set at 0.5 feet per second. The queuing model described in Section 5.3.3 is used to characterize conveyor performance. The average handling delay is about 60 seconds. The major problem of using a conveyor is its flexibility. Once a line is laid out, addition or insertion of workstations is difficult without reconfiguration. One may have to shut down the line for a new layout. Space is not economically utilized because of horizontal storage. Due to contamination problems, stainless steel rollers are used. The unit cost of the conveyor ranges from $600 to $1,000 per linear foot. Hence the total cost is between $72,000 and $120,000.

A manual delivery system can support any line layout and has the maximal flexibility. However, there are three difficulties: product weight, handling speed and space utilization. From an ergonomic point of view, anything over 50 pounds requires at least two persons to perform load/unload operations. When pushing a cart with a heavy load, an operator's travel speed is expected to be between 2 and 3 feet per second. If k operators are available, their performance may be analyzed by an M/G/k queuing model (see Section 3.3). For the same traffic condition and $k = 2$, the average handling delay is nearly 120 seconds and the utilization factor is 0.8. If the line is scheduled for three shifts a day, then six operators are needed. For a five-year product life, the total cost is $\$30{,}000 \times 6 \times 5 = \$900{,}000$.

Comparison of these three systems is summarized in Table 5.7. Since the AS/RS is less expensive than the manual system and more flexible than the conveyor, in this situation the AS/RS is clearly a better choice.

Table 5.7 Comparison of Three Alternatives

	AS/RS	Conveyor	Manual
Cost (thousands of dollars)	80–150	70–120	900
Flexibility	good	fair	excellent
Multi-dimensional move	excellent	fair	good
Space requirement	excellent	fair	good
Handling delay	excellent	good	poor
Reliability	good	fair	excellent

5.6 Case Study II

An automated assembly line has three major operations. Part sizes vary from $0.03 \times 0.5 \times 1$ to $1 \times 2 \times 8$ cubic inches. To reduce material handling effort, parts and assemblies are placed in trays of a standard size, $3 \times 10 \times 15$ cubic inches. At each operation, parts are picked from a tray and completed assemblies are placed in the same tray. Then a completed assembly tray is sent to the next operation. Parts are kitted by suppliers. Only operations 1 and 2 consume parts, and each has one kind of kit. For convenience, parts for operations 1 and 2 are called P1 and P2, respectively. Likewise, assemblies produced by operations 1, 2 and 3 are called A1, A2 and A3, respectively. All three operations have the same size workstation, 14 feet wide. Two shifts are scheduled each day. The effective working time is 7.2 hours per shift.

At the first operation (OP1), assembly time per tray is estimated to be 24.2 minutes. There are 100 assemblies in each tray. A total of 10,000 assemblies (or 100 trays) should be produced on each shift. The trays, processed by operation 2 (OP2) and operation 3 (OP3), can hold 10 assemblies. Both OP2 and OP3 can produce 2,000 assemblies (or 200 trays) per shift. The process times per tray are 5.8 minutes for OP2 and 5 minutes for OP3.

Since 100 A1 trays should be produced in 7.2 hours and it takes 24.2 minutes to process a tray, the total number of workstations at OP1 is the least integer greater than or equal to $100 \times 24.4/(7.2 \times 60) = 5.6$. Hence six workstations are needed at OP1. Similarly, both OP2 and OP3 should have three workstations each.

Material flow can be described by the following table.

From	To	Content
Storage	OP1	P1 trays
Storage	OP2	P2 trays
OP1	OP2	A1 trays
OP2	OP3	A2 trays
OP2	storage	empty A1 trays
OP3	package	A3 trays

Since the number of operations and the number of tray types are few, a sophisticated material control system may not be necessary. To achieve maximal flexibility, the line designer wants to know if manual delivery is economical. A proposed material handling method is given in the following.

A cart with four storage units is manually handled by an operator. Each unit can hold 10 trays and has a size $11 \times 16 \times 40$ cubic inches (see Figure 5.19). The operator distributes part trays and collects completed trays, in accordance with assembly process flows. During a trip, the operator will take the following steps:

1. Load P1 and P2 trays into the cart from the storage area
2. Move the cart to OP1
3. Load completed A1 trays into the cart from workstations at OP1
4. Replenish P1 trays
5. Move the cart to OP2

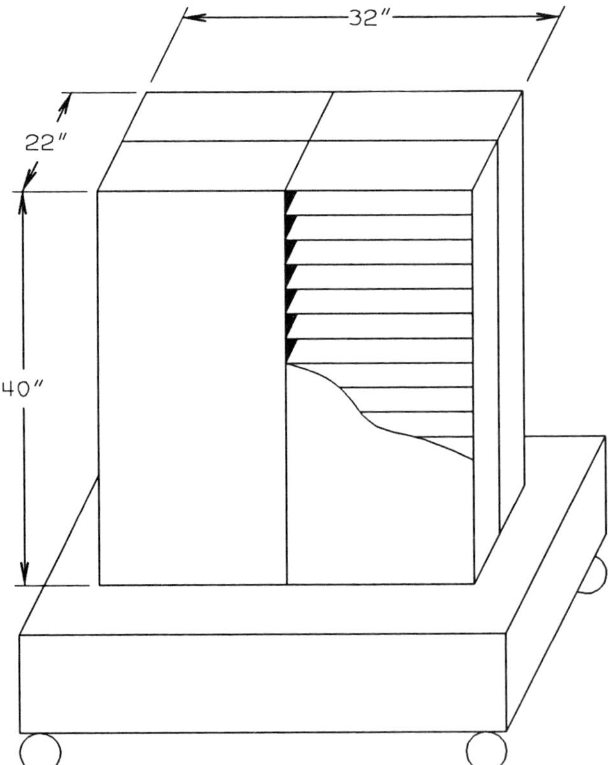

Figure 5.19 A cart with 4 storage units.

6. Load completed A2 trays and empty A1 trays into the cart from workstations at OP2
7. Replenish A1 trays and P2 trays
8. Move the cart to OP3
9. Load completed A3 trays into the cart from workstation at OP3
10. Replenish A2 trays
11. Move the cart to the packaging area
12. Unload all completed A3 trays for packaging and shipping
13. Move the cart back to the storage area and return all empty trays

Note that there are only two types of trays: (i) P1 or A1, and (ii) P2, A2 or A3. A P1 tray and an A1 tray are equivalent. The former is an input tray to OP1 while the latter is an output tray. Also, P2 trays, A2 trays and A3 trays are equivalent. Storage columns may be used at the storage area. Each column consists of a number of slots that hold part trays. After part trays have been removed from a column, the vacant slots can be used to hold empty trays returned from the line. Columns with empty trays are sent back to the supplier for replenishing.

Trays in the cart may become undeliverable, due to a full storage condition at some workstations. It is possible that some P1 trays have not been processed yet upon the arrival of the cart. In this case, only completed trays (i.e., A1 trays) will be replaced by P1 trays, and incomplete trays will not be removed. If there are undeliverable trays when the cart is returned to the storage area, the number of part trays loaded into the cart for the next trip will be reduced accordingly. This mechanism automatically adjusts the part-feeding speed. Consequently, the total amount of work-in-process in the line will not be increased indefinitely owing to any abnormal conditions, e.g., workstation breakdown. The trays in the carts will be handled in a first-in and first-out fashion.

Workstation reliability can be characterized by interfailure time and downtime. The latter is the sum of a repair time and the waiting time for the repairperson. The estimated average interfailure time is 400 hours, while the average repair time is 2.25 hours. The waiting time is dependent on the number of workstations, the number of repairpersons, interfailure time and repair time. The reliability of a workstation is given by

$$r = \frac{\text{Average interfailure time}}{\text{Average interfailure time} + \text{Average down time}}$$

For a given r, however, the number of workstation failures can be approximated by a binomial random variable for which a probability density function is given by Equation (2.11). Let n be the total number of workstations and X be the number of workstations not functioning. Then the probability of $\{X = k\}$ is defined by

$$P[X = k] = \binom{n}{k} r^{n-k}(1-r)^k$$

The approximate distributions of X for different average downtimes are given in Table 5.8. More accurate results can be obtained by employing the machine repairman model; see Examples 3.9 and 3.10.

It can be seen that the probability of more than one workstation failure is consistently less than 0.01 and the probability of more than two failures is negligible. Since all workstations are automated, synchronization of assemblies is achievable under the first-in, first-out rule.

The width of each workstation is 14 feet. Since there are 12 stations, round-trip travel distance is $D \approx 14 \times (6+3+3) = 168$ feet. Assume that an operator can push a fully loaded cart at an average speed of 2.04 feet per second. The average travel time in a round trip is $E[TT] = 168/(2.04 \times 60) = 1.37$ minutes. Let N be the number of load/unload operations performed during a round trip, and S be the time for a load (or unload) operation. The total amount of time spent for load/unload operation during a trip is given by

$$LU = \sum_{i=1}^{N} S_i \tag{5.26}$$

where S_i is the ith load (or unload) time in a trip.

Table 5.8 Distribution of the Number of Failed Workstations

	Average Downtime (Hours)							
	2.25		3.00		4.00		5.00	
X	OP1	OP2/3	OP1	OP2/3	OP1	OP2/3	OP1	OP2/3
0	.9669	.9833	.9562	.9778	.9420	.9706	.9282	.9634
1	.0326	.0166	.0430	.0220	.0565	.0291	.0696	.0361
2	.0005	.0001	.0008	.0002	.0015	.0003	.0022	.0005
3	.0000	.0000	.0000	.0000	.0000	.0000	.0000	.0000

Assume that $E[S] = 10$ seconds. The average round-trip time becomes

$$E[T] = E[TT] + E[LU] = 1.37 + E[N] \times E[S]/60 = 1.37 + E[N]/6 \text{ (min)} \tag{5.27}$$

If M round trips are completed by C operators in one shift, it must be true that

$$E[M] \times E[T] \leq C \times (7.2 \times 60) \tag{5.28}$$

900 trays will be handled in a shift. This includes 100 P1 trays, 100 A1 trays, 100 empty A1 trays, 200 P2 trays, 200 A2 trays and 200 A3 trays. Since each tray will be loaded (to the cart) once and unloaded once, the total expected number of load/unload operations during a shift is 1800. Consequently,

$$E[M] \times E[N] \approx 1800 \tag{5.29}$$

By virtue of (5.27), (5.28) and (5.29), a set of feasibility conditions can be defined for different values of C:

For $C = 1, E[N] \geq 18.68, E[M] \leq 96.35$ and $E[T] \geq 4.48$ (min)

For $C = 2, E[N] \geq 4.37, E[M] \leq 411.68$ and $E[T] \geq 2.10$ (min)

For $C = 3, E[N] \geq 2.48, E[M] \leq 727.08$ and $E[T] \geq 1.78$ (min)

These conditions should be regarded as design constraints. Since the number of load/unload operations is not dependent on the number of operators, the total operation time can be treated as a fixed cost and the total travel time as a variable cost. The above constraints show that the limit of the number of trips is an increasing function of C. Therefore, it would be better to use as few operators as possible.

Trips are scheduled periodically. The time headway or the interval between the starting times of two successive trips is a constant. If the probability that a round-trip time is greater than the planned headway is small, then the part feeding cycle at each workstation is approximately equal to the headway. Therefore the number of parts kept at a workstation during a feeding cycle should be slightly higher than the product of the headway and the parts consumption rate. For a given storage level, the buffer size at each workstation follows.

To characterize system behavior, the following assumptions are adopted:

1. The occurrences of load/unload operations are uniformly distributed over 7.2 hours. Hence the number of occurrences in a fixed interval has a Poisson distribution, given by Equation (2.12).
2. For all manual operation times, their standard deviations are about 30% of their means.
3. Travel time, load operation and unload operation times are all independent.

Under the first assumption, for a fixed headway, say h minutes, N is a Poisson random variable. Since there are 1800 load/unload operations in 7.2 hours, it follows that

$$E[N] = Var[N] = h \times (1800/7.2 \times 60) = 4.1667h \tag{5.30}$$

The average round trip time is given by Equation (5.27), while its variance can be obtained by $Var[T] = Var[TT] + Var[LU]$ (assumption 3) and the fact that

$$Var[LU] = Var[E[LU \mid N]] + E[Var[LU \mid N]] \tag{5.31}$$

To evaluate a conditional expectation, N can be regarded as a constant. Therefore, $E[LU \mid N] = N\,E[S] = N/6$ minutes. Since all load/unload operations are independent (assumption 3) and $SD[S] = 0.3E[S]$ (assumption 2), $Var[LU \mid N] = N\,Var[S] = N \times (0.3E[S])^2 = 0.0025N$ square minutes. Equation (5.31) becomes

$$Var[LU] = Var[N]/36 + 0.0025E[N] = 0.0303E[N] \tag{5.32}$$

Again, $SD[TT] = 0.3E[TT]$. Consequently, we have

$$Var[T] = (0.3 \times 1.37)^2 + 0.0303E[N] = 0.1689 + 0.0303E[N] \tag{5.33}$$

If a five-minute headway is selected, the average number of load and unload operations performed during a feeding cycle is $E[N] = 4.1667 \times 5 = 20.83$. The expected number of trips is $E[M] = 1800/E[N] = 86.4$.

Table 5.9 Operator Performance Behavior under a Manual Delivery System

h min	$E[T]$	$SD[T]$	$P[T>h]$	$E[N]$	$E[M]$	Optr. Util.
5	4.84	0.89	0.430	20.83	86.4	0.968
10	8.31	1.20	0.079	41.67	43.2	0.831
15	11.79	1.44	0.012	62.50	28.3	0.786
20	15.26	1.64	0.002	83.33	21.6	0.763
25	18.73	1.82	0.000	104.17	17.37	0.749

Furthermore, by Equations (5.27) and (5.33), $E[T] = 4.84$ and $SD[T] = \sqrt{Var[T]} = 0.89$. Operator utilization is $4.84/5 = 0.968$. Since a round trip time can be regarded as the sum of many load and unload times and the travel times between workstations, by the central limit theorem, T is approximately normally distributed. The probability that a round trip time exceeds a five-minute headway is given by

$$\begin{aligned} P[T>5] &= P[(T-4.84)/0.89>(5-4.84)/0.89] \\ &\approx P[Z>0.1798] \end{aligned}$$

where Z is a standard normal random variable.

From the normal table, it is found that $P[T>5] \approx 0.43$. For $h = 10$, 15, 20, and 25 minutes, operator performance behavior can be analyzed in the same fashion. Table 5.9 summarizes the numerical results for $C = 1$.

It can be seen that the values listed in Table 5.9 satisfy all the feasibility constraints previously discussed, and that the operator utilization is slightly less than one for $h = 5$. Therefore, $\{h \geq 5\}$ defines a feasible region. But if a five-minute headway is selected, the probability that the operator is behind the schedule is rather high. Although this situation may be improved by employing an additional operator at the cost of doubling manpower, the assembly trays can no longer be kept in a first-in, first-out fashion. If the headway is increased beyond 15 minutes, the utilization reduction becomes insignificant and the operator can almost follow the scheduled pace all the time. The value of $E[N]$ is a indicator for the level of work-in-process. The higher the value is, the more work-in-process will be needed. Hence a good choice of headway should be between 10 and 15 minutes. Each shift needs only one operator. The operator is expected to take about 80% of his or her time to make 30 to 40 trips for each shift.

5.7 Remarks

A material handling system is a major component of an assembly line. Both material handling and assembly operations demand that the right parts be placed at the right position at the right time. In this sense, material handling may be considered as an extension of the assembly process. In addition to engineering concerns, material handling delay is one of the most important design parameters. This delay, together with operation cycle time and production yield, determines the line throughput.

This chapter has discussed two basic types of material handling systems, AS/RS and conveyor. The analysis of AS/RS is based on the work in Chow (1986), where justification of using an M/G/1 queuing model for system stochastic behavior is given. This approach may be extended to other material handling systems. For example, an automatic guided vehicle system can be modeled by an M/G/1 queue. The computational procedure is almost identical to Algorithm 5.1 as defined for AS/RS. The service time of the automatic vehicle is the pick and place times plus the travel time (from the last position to the sender and to the receiver). If k vehicles are serving the same line, the material handling system can be analyzed by using an M/G/k model, discussed in Section 3.3. A manual delivery system of k operators can be treated in the same manner.

Maxwell (1982), Egbelu (1984) and Muller (1984) are useful references for automatic guided vehicles. The first two are articles that discuss system capacity planning and dispatching rules; the third one is a book dealing with many technical and economical aspects.

The traffic flow model was previously proposed for a highway traffic study by Lighthill and Whitham; see Lighthill (1955). The network flow model for conveyor capacity analysis is given by Maxwell and Wilson. Different treatments for different types of conveyors are discussed in Maxwell (1981). A different queuing analysis of conveyor systems under an assumption of exponential service times was described in Gregory (1975). Closed-loop conveyors with discrete buckets analyzed by Muth are not discussed in this book. A closed-loop conveyor is suitable for single assembly operation, i.e., each workstation performing a complete assembly. When the conveyor serves a sequential line, flow control will be difficult. The reader may consult Muth (1975) and Muth (1977) for the original works in this area. Other important results in conveyor theory may be found in a survey paper: Muth (1979).

A relatively complete work in material handling system design can be found in Apple (1972). A relatively recent review is given in Manson (1982).

REFERENCES

Apple, J. M. (1972). *Material Handling Systems Design*, John Wiley, New York.

Chow, W. (1986). An Analysis of Automated Storage and Retrieval Systems in Manufacturing Assembly Lines, *Institute of Industrial Engineers Transactions*, v. 18, pp. 204–214.

Egbelu, P. J., and J. M. A. Tanchoco (1984). Characterization of Automatic Guided Vehicle Dispatching Rules, *International Journal of Production Research*, v. 22, pp. 359–374.

Gregory, G., and C. D. Litton (1975). A Conveyor Model with Exponential Service Times, *International Journal of Production Research*, v. 13, pp. 1–7.

Lighthill, M. J., and G. B. Whitham (1955). On Kinematic Waves; II. A Theory of Traffic Flow on Long Crowded Roads, *Proceedings of Royal Society* (London), series A, v. 229, pp. 317–345.

Manson, J. O., and J. A. White (1982). Material Handling: A Review, PDRC 82–04, School of Industrial and Systems Engineering, Georgia Institute of Technology, Atlanta.

Maxwell, W. L., and R. C. Wilson (1981). Dynamic Network Flow Modelling of Fixed Path Material Handling Systems, *American Institute of Industrial Engineers Transactions*, v. 13, pp. 12–21.

Maxwell, W. L., and J. A. Muckstadt (1982). Design of Automatic Guided Vehicle Systems, *Institute of Industrial Engineers Transactions*, v. 14, pp. 114–124.

Muller, T. (1983). *Automated Guided Vehicles*, IFS (Publications) Ltd., Bedford, England.

Muth, E. J. (1975). Modelling and System Analysis of Multistation Closed-loop Conveyors, *International Journal of Production Research*, v. 13, pp. 559–566.

Muth, E. J. (1977). A Model of a Closed-Loop Conveyor with Random Material Flow, *American Institute of Industrial Engineers Transactions*, v. 9, pp. 345–351.

Muth, E. J., and J. A. White (1979). Conveyor Theory: A Survey, *American Institute of Industrial Engineers Transactions*, v. 11, pp. 270–277

6

Work-in-Process Management

Work-in-process (WIP) includes all the assembly materials in the production line. Both material storage and transfer are closely related to facility space requirements, material-handling systems, and line operation policies. Material flows may begin at a parts-receiving dock and end at a shipping dock. Therefore inventory may be held in any of the following areas:

1. *Receiving dock* Parts are delivered to this place and wait to be transferred to a staging area, a preparation area or directly to assembly operations. The size of this area depends on the delivery frequency, volumes and parts transfer speed.

2. *Staging area* This is a temporary storage area for incoming parts. Parts may be placed in a rack structure to save floor space and can be retrieved upon request. The size of the storage space is a function of delivery frequency, volume, parts-consumption rate and the planned level of protection stock.

3. *Parts-preparation area* This is the place for the preparation for assembly. In some cases, parts will be kitted before assembly. In other cases, parts must be cleaned. Sometimes, parts are placed in a special carrier to avoid handling damage. Parts preparation can be regarded as an extension of the assembly process. The size of the area is contingent on production volume, number of parts, part size, and line operation policy.

4. *Workstations* Normally there is one assembly per workstation. Additional parts may be stored at the workstation for more than one assembly.

5. *Buffers between operations* Buffers are often required to smooth production flows because of line-balancing problems, multiple products, process time variation, yield and workstation failures. Buffers are used

to hold assemblies not being processed. (Many people consider work in buffers as the definition of work-in-process.)

6. *Shipping area* After having finished all assembly tasks, good products are ready to be sent out. The area size is dependent on the shipping frequency, product size, and production rate. Sometimes, a packaging operation is established in this area to prepare finished goods for shipment.

Although WIP does not add any value to products, it is required to smooth production throughput. A certain level of inventory in the front end of the line would reduce the risk of parts shortage due to delay in delivery or excessive rejection. WIP between operations is needed to absorb variations of process time, to prevent line-stop due to workstation failures, and to support a job-rotation policy. Consider that a bottleneck operation of a line has an average process time of eight minutes but an upstream manual operation takes only one minute. To save direct labor, a line manager may have two choices: either to rebalance the line or to assign an operator to the one-minute operation for only one hour in each shift. Under the latter condition, after the first hour an operator will produce seven-hour WIP for downstream operations. But the number of operators in a line may be fewer than the number of jobs. Likewise, if an absenteeism problem occurs, a line manager may reassign operators to different operations. Each time an assignment policy is established, an unassigned operation temporarily becomes a bottleneck. Thus WIP may be accumulated in front of this operation. (Job assignment problems will be discussed in Chapter 7.)

Under an appropriate arrangement, WIP can be an indicator of line efficiency. In addition to inventory cost, excessive WIP often implies the existence of hidden problems such as low availability, poor reliability, long setup time, etc. If frequent engineering changes are anticipated, excessive WIP may lead to a large amount of material waste. Consequently, a good WIP management policy should reduce WIP to a minimal level so that line problems may be detected quickly, yet also keep the WIP level high enough to maintain a continuous line flow. Since WIP is related to many manufacturing components, a line designer must plan for the optimal WIP management policy. This chapter discusses parts-ordering and feeding methods, in-process inventory control and buffer-placement problems.

6.1 Parts Ordering and Delivery

Parts ordering quantity and delivery frequency determine the designed capacities of the receiving dock and the parts staging area. Parts are usually delivered by trucks. Typically the truck-arrival rate is time-dependent. Figure 6.1 illustrates an arrival pattern, where two peak periods are observed: from 9 to 11 a.m. and from 3 to 5 p.m. A receiving dock can be treated as

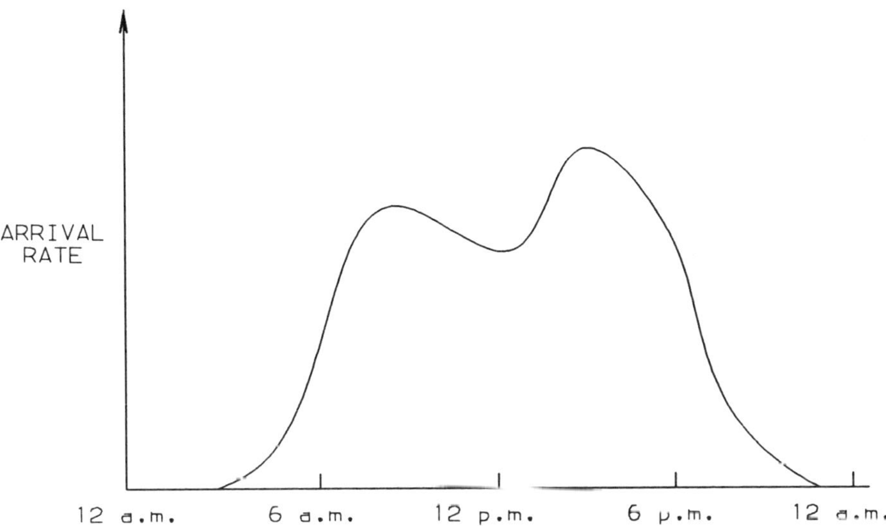

Figure 6.1 Truck arrival rate as a function of time.

a single server system. A truck is equivalent to a job whose service time is the duration of time when the truck occupies the dock. The job-arrival rate may be set equal to the arrival rate during the peak period. Then the queuing behavior may be analyzed by an M/G/1 queue. This approach assumes a homogeneous arrival pattern (i.e., the arrival rate is independent of time). Since the peak rate is used, the M/G/1 queuing model tends to overestimate the degree of congestion at the receiving dock and gives a worst case. A more accurate analysis requires advanced queuing theory, beyond the scope of this book. Another way to deal with nonhomogeneous arrival rate is simulation, which will be discussed in Chapter 9.

Material ordering is a well-known inventory problem. Under simplified assumptions, the optimal ordering quantity and the optimal ordering time can be computed by simple closed-form solutions. A common approach is to minimize the sum of inventory cost and ordering cost. Let

C = inventory holding cost per time per unit
K = ordering transaction cost per order
D = production demand rate
Q = ordering quantity

If D is a constant, the interval between two successive parts deliveries is $T = Q/D$. This interval is called the *inventory cycle*. The relation between inventory cycle and level of inventory is illustrated in Figure 6.2. It can be seen that the average inventory level is $Q/2$. Therefore the average inventory cost (per time unit) is $CQ/2$. Since each order induces an ordering cost, K, the average total cost is given by

$$F(Q) = CQ/2 + K/T = CQ/2 + KD/Q \tag{6.1}$$

Using relation (3.30), the optimal ordering quantity, Q_0 can be obtained by solving the equation

$$F'(Q) = C/2 - KD/Q^2 = 0$$

that is,

$$Q_0 = \sqrt{2KD/C} \tag{6.2}$$

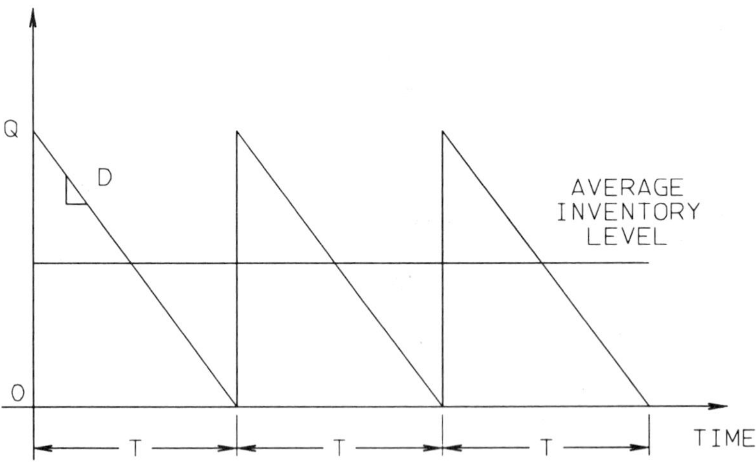

Figure 6.2 Inventory cycle and inventory level.

The above result is also known as the *economic ordering quantity* (EOQ). The optimal interval between parts deliveries is given by

$$T_0 = \frac{Q_0}{D} = \sqrt{\frac{2K}{CD}} \tag{6.3}$$

For a high ordering cost, the number of orders should be small. Therefore both T_0 and Q_0 tend to be large. On the other hand, one should place orders more frequently if the inventory cost is high. Hence, both T_0 and Q_0 are decreasing in C.

The time interval from an order release to the receipt of the order is referred to as *lead time*, and is denoted by U. To minimize inventory level, a reordering condition can be established by examining the inventory level. For a constant demand rate, D, the total amount of consumption during a lead time is $V = DU$. Thus, $R = V$ is the *reordering point*. The interval between two successive order releases is called the *reordering cycle*, denoted by S. Figure 6.3. shows the relation between reordering point and reordering cycle. If D is a constant, then $S = T$.

In many cases, however, U may be a variable and actual daily production quantity may not be fixed. Letting D_i be the production volume of day i,

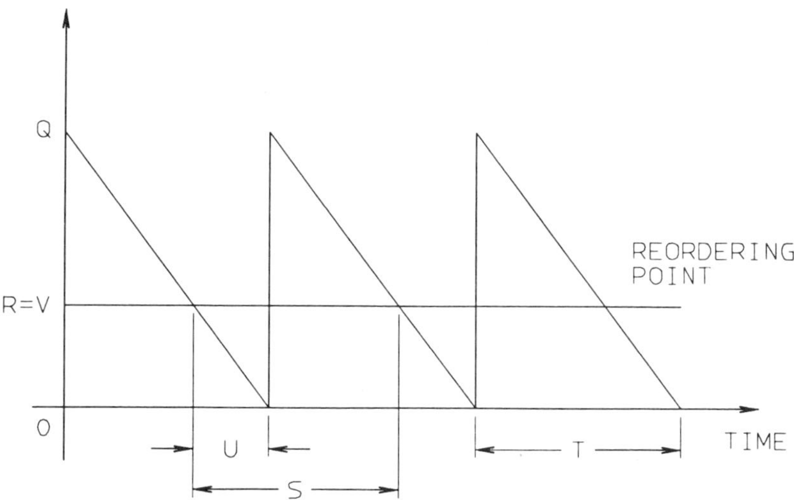

Figure 6.3 Reordering point and reordering cycle.

then the total amount of consumption during a lead time is

$$V = \sum_{i=0}^{U} D_i$$

The distribution of V can be derived from those of U and D's. Suppose that a stockout probability, α, is chosen. The least value of R such that $P[V > R] \le \alpha$ is the reordering point. For a new production line, the distribution of V may be approximated by simulation. In this case, a detailed line-simulation model should be constructed. For each value of U, the result from a model execution gives us a sample of V. Repeating this procedure, a sample distribution of V follows.

If both means and variances of D and U are known, the reordering point may be determined by using normal approximation. Assume that U and $\{D_i\}$ are independent, and D's are independently, identically distributed.

$$m = E[V] = E[U]E[D] \tag{6.4}$$

$$\sigma^2 = Var[V] = Var[U](E[D])^2 + E[U]Var[D] \tag{6.5}$$

Since V is the sum of a number of random variables, it has an asymptotic, normal distribution. Consequently,

$$\begin{aligned} P[V > R] &= P[(V-m)/\sigma > (R-m)/\sigma] \\ &\approx P[Z > (R-m)/\sigma] \end{aligned}$$

where Z is a standard normal random variable.

Example 6.1 An assembly line's average production rate is 100 pieces per day. Due to workstation failure, yield fluctuation, and absenteeism, the actual production volume varies. The estimated coefficient of variation is 0.05. Hence, $E[D]$ = 100 and $Var[D]$ = $(0.05 \times 100)^2$ = 25. A major part is supplied by a foreign vendor with a lead time uniformly distributed between four and seven days. This implies that $E[U] = (4 + 7)/2 = 5.5$ and $Var[U] = (7-4)/12 =$ 0.25. Using Equations (6.4) and (6.5), it follows that m = 550 and $\sigma = \sqrt{0.25(100)^2 + 5.5 \times 25} = 51.36$. If the stockout probability should be kept below 0.05, then from the normal distribution table, $(R -$

550)/51.36 = 1.645; the reordering point, $R = 550 + 51.36 \times 1.645 = 635$. □

It should be pointed out that if the demand volume during a lead time is a random variable, the average inventory level is no longer $Q/2$. Since V is a random variable, the amount of parts left over from lead time consumption may be positive. In this case, the inventory level after an order receipt is higher than the ordering quantity and the reordering cycle is prolonged. The optimal ordering quantity does not strictly follow Equation (6.2). On the other hand, since the amount left over is usually much smaller than the ordering quantity, Equation (6.2) will give an approximate optimal solution. Figure 6.4 presents a more realistic picture.

There are two cost items in Equation (6.2): holding cost and ordering cost. The former includes cost of capital, inventory tax, storage, obsolescence, direct and indirect labor related to inventory, insurance, etc. Some companies use opportunity cost to represent this item. The ordering cost is associated with each order placement. Typically, this includes paperwork, personnel support, purchasing, transportation and handling cost, etc. Accurate estimation of these costs is not always easy. Equation (6.2) should be used as a guide for order placement. A sensitivity analysis may help. By (6.1) and (6.2), the minimal cost at Q_0 is

$$F(Q_0) = \sqrt{2DKC}$$

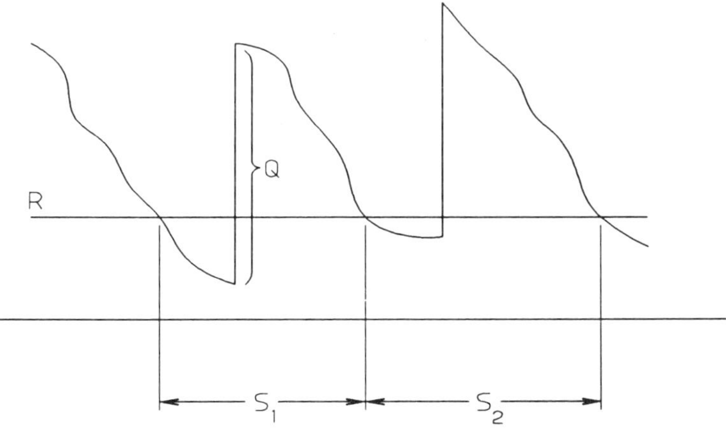

Figure 6.4 Inventory level under a variable consumption rate.

If the product of K and C is changed by a factor of $(1+\delta)$, the cost varies by $\sqrt{1+\delta}$. For example, assume that $D = 25{,}000$ pieces per year, $C = \$30$ per piece per year, and $K = \$100$ per order. The optimal ordering quantity is $Q_0 = 408$ pieces per order, and the estimated minimal cost is $F_0 = \$12{,}247.45$. If the actual holding cost is 20% higher than the estimated value, i.e., $C = \$36$, then Q_0 becomes 373 per order (8.6% change) and F_0 is \$13,416.41 (9.5% change). Therefore, if 10% error is the maximum tolerance, the cost estimation for K or C must be within ±20%.

The EOQ formula is applied to a single item or a number of items with independent demands. An assembly, however, is usually composed of many part types. For the same part type, multiple parts may be required. For a given production demand, the parts-consumption rate of each part type can be generated from the *bill of material*. As shown by the following example, the concept of EOQ under an independent demand assumption may lead to a large amount of inventory.

Example 6.2 An assembly line produces two types of products. Assembly 1 (A1) has two components and assembly 2 (A2) three components. A component can be either a subassembly consisting of a number of parts or treated as a single item procured from a vendor. The bill of material of each type is shown in Figures 6.5. Assembly 1 is composed of two type-1 components (C1) and one type-2 component (C2). C1 has three type-1 parts (P1) and two type-2 parts (P2). C2 has 1 P1 and 2 P3's. Thus, the total numbers of P1, P2, and P3 required by an A1 are 7, 4 and 2, respectively. Suppose that 25,000 type-1 assemblies will be produced each year. The parts-consumption rates of P1, P2 and P3 become 175,000, 100,000, and 50,000 per year. Parts requirements for A2 can be obtained in the same manner. The bill of material can be summarized in the following table:

Items	P1	P2	P3	P4	P5	P6	C5
A1	7	4	2	0	0	0	0
A2	2	0	4	2	2	1	4

For the given end product demand, annual demand for individual parts is computed:

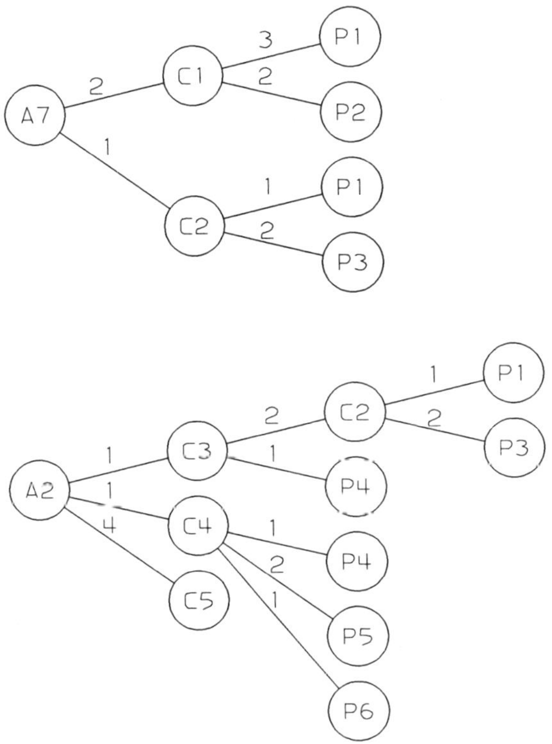

Figure 6.5 A bill of material.

Items	P1	P2	P3	P4	P5	P6	C5
25k A1	175k	100k	50k	0	0	0	0
10k A2	20k	0	40k	20k	20k	10k	40k
Total	195k	100k	90k	20k	20k	10k	40k

The last row in the table gives the total annual requirements. If the holding cost and ordering cost are available, one can invoke (6.2) to compute the EOQ for each item. The annual ordering frequency is the ratio of the annual demand to the EOQ, and the average reordering cycle (in years) is the reciprocal of the ordering frequency. The computed results are given by:

Items	P1	P2	P3	P4	P5	P6	C5
Annual demand	195k	100k	90k	20k	20k	10k	40k
Ordering cost	50	50	50	50	50	50	50
Holding cost	0.2	0.5	1.0	0.5	1.0	2.0	3.0
Ordering quantity	9874	4472	3000	2000	1414	707	1155
Ordering frequency	19.75	22.36	30.00	10.00	14.14	14.14	34.63
Reordering cycle	0.051	0.044	0.033	0.100	0.071	0.071	0.029

Assume that the consumption rate during a lead time has approximately a normal distribution with a coefficient of variation 0.3. Let m_i be the average consumed volume of item i during a lead time. If a stockout probability should be no worse than 0.05, then the reordering point is $R_i = m_i + 1.645 \times 0.3 \times m_i = 1.4935m_i$. The estimated lead times and reordering points are given as follows:

Items	P1	P2	P3	P4	P5	P6	C5
Lead time (days)	1	2	1	1	5	2	3
Daily requirement	780	400	360	80	80	40	160
Lead-time consumption	780	800	360	80	400	80	480
Reordering point	1165	1195	538	119	597	119	717
Maximum inventory	10644	5252	2256	2078	1808	785	1629

The last row in the above table shows the maximal inventory level is found at the time of order receipt. Using normal approximation, this value is equal to $Q_i + R_i - (m_i - 1.645 \times 0.3 \times m_i) = Q_i + R_i - 0.5065m_i$. □

The method used in Example 6.2 has some shortcomings. The approach assumes that the inventory of each part can be optimized independently. Under the normal assumption, stockout probability of a part is about 0.05. Since an assembly consists of a number of parts, the probability of a line-stop due to parts shortage is much higher than 0.05. For instance, A2 has six different parts or components. The probability of shortage becomes as high as 0.265 ($= 1 - (0.95)^6$). To reduce this probability, one must increase the level of parts inventory. In real life, it is not uncommon that an assembly has over 100 different part numbers. If the probability of shortage should be 0.05 or lower, then the stockout probability of each part type must be no more than 0.0005! This implies that for each part a large amount of inventory is required.

In Example 6.2 annual demand is used for the computation of EOQ. From a production-planning point of view, a fixed production demand for a whole year is not a realistic assumption. Market demand for finished products may change many times during a year. For this reason, a production plan is rarely fixed for the entire year. A planning system must reflect the dynamics of market and manufacturing changes.

Two popular techniques have been used to deal with such problems. One is called *material requirement planning* (MRP), and the other *Kanban* system. Both techniques attempt to achieve the minimal inventory level with *just-in-time* deliveries.

Material Requirement Planning

Although the concept of MRP is generic, detailed implementation may vary from one application to another. The following description is a simple outline of such a technique.

1. MRP begins with a *master schedule* for the finished product. This schedule, usually derived from market demand forecast and capacity of the production facility, determines the production quantity for each time period (e.g., a day or a week).

2. From the master schedule and the bill of material, one computes the required volume of each part in each period. The computational procedure is identical to what has been shown in Example 6.2, except that the volume is based on a time period much shorter than a year. Hence, parts requirements for each period of the entire planning horizon are obtained.

3. Given the parts requirements, the next step is to determine the ordering quantity. This can be done by using the EOQ formula, together with the schedule of parts requirements. For instance, the requirements of a part for the next five periods are 10, 10, 10, 15, and 15. Their cumulative requirements become 10, 20, 30, 45 and 60. Clearly, an ordering quantity should be one of these cumulative numbers; otherwise one must carry unnecessary inventory. Both $\{Q = 35\}$ and $\{Q = 30\}$ need two order placements, but the former will cause higher inventory than the latter. If the optimal ordering quantity based on a 10-period calculation is 35, it may be appropriate to order 30 twice for the next five periods. In other words, the optimal ordering policy may have different ordering

quantities for different periods. Since parts requirements are derived from market demand forecast, one can expect that the figures for the first five periods are more accurate and more reliable than those for the second five periods.

4. Once the ordering quantities are known, the scheduled order-receipt time can be calculated. In the previous example, if the initial part inventory is zero and $Q = 30$, then orders should be received at the beginning of periods 1 and 4. If the initial inventory is 10, the receipt times become periods 2 and 4.

5. Order receipt times, together with procurement lead times, determine when each order should be placed.

When MRP is processed by a computer, the production schedule, parts requirements, ordering quantities, and order placements can be updated in a short period of time, e.g., one day. Any deviation from the existing plan can be quickly adjusted. However, using MRP does not mean that one can overlook the relation with suppliers. Some common causes for a large amount of inventory are lack of control of procurement lead time, inconsistent parts quality, and irregular production schedules. For a variable lead time, one must increase safety stock level. If parts quality is poor, one must carry enough inventory to assure enough good parts. Finally, because of irregular demand, suppliers may have difficulties in adjusting their production quantity to match the demand.

Kanban System

This system is a multi-stage production system developed by Toyota Motor Company. Consider two work centers in tandem. The completed work from the first center is sent to the second center. Both production and delivery schedules are regulated by tickets, called *kanbans*. There are many types of kanbans; each has its own purpose. Most commonly used are two types: *production kanban* (p-kanban) and *conveyance kanban* (c-kanban). Each complete lot at work center 1 is tagged by either a p-kanban or a c-kanban. An unattached p-kanban (c-kanban) is called a free p-kanban (free c-kanban). A p-kanban is a working order for a complete lot production. A c-kanban, on the other hand, is a permit to transfer a complete lot from center 1 to center 2. Each complete lot at center 1 must have a p-kanban attached to it. At center 2, each unused lot previously drawn from center 1 must have a c-kanban. Material flow from work center 1 to center 2 is triggered by the latter, and involves the following steps:

1. If no free c-kanban is found at center 2, do nothing at center 2. If no free p-kanban is found at center 1, do nothing at center 1.

2. If there is a free c-kanban at center 2, and a complete lot at center 1, then replace the p-kanban on the lot by the free c-kanban. Thus the detached p-kanban becomes free and a lot will be produced by center 1. The lot with the new c-kanban is moved to center 2.

3. If in step 2 no complete lot is available at center 1, then the free c-kanban must wait until a lot has been produced by center 1. Since the condition of no complete lot means that all p-kanbans are free, the number of scheduled lots is the same as the number of p-kanbans. In this case, center 2 may have an inventory shortage.

4. If center 2 has begun to use material from a lot previously drawn from center 1, the attached c-kanban to the lot is detached and becomes free. This triggers another lot draw from center 1.

Since two types of kanbans are used, the process is sometimes referred to the *double kanban* system. Level of inventory between the two centers is determined by the number of kanbans and the lot size. In general, the number of c-kanbans determines the maximum inventory at work center 2, while the number of p-kanbans defines the maximum inventory at center 1.

A simplified version, known as the *single kanban* system, uses only a single type of kanban between two work centers. In this system, a kanban is a working order. The number of kanbans between the two work centers is fixed. Each free kanban in work center 1 causes a production of a predetermined lot size. After a lot has been produced, a free kanban is then attached to the lot. Center 1 will continues its production until all free kanbans are used. Upon demand, center 2 draws a complete lot from center 1. When the lot is transferred from center 1, the associated kanban is detached, i.e., becomes free. Then a production of one lot is scheduled immediately. Clearly, the maximum inventory level at center 1 is equal to the product of lot size and number of kanbans.

After having understood how a kanban system regulates the material flow and the level of inventory, it is easy to see its applications in parts ordering. Let the parts supplier be the first work center and the assembly facility the second center. A simple way to implement the kanban system is to let the lot size equal the reordering point. When the parts-staging area has one lot of inventory, an additional lot must be drawn from the supplier. This triggers the supplier to prepare another lot for the next draw. Then the maximum inventory level at the staging area equals the size of two lots.

Both the kanban system and MRP are just-in-time systems. The parts demand in the kanban system is derived from immediate production need. The parts-consumption rate in the assembly line and the supplier production rate should be synchronized in order to maintain a smooth material flow between the two parties. The kanban system emphasizes material flow control. Some Japanese factories freeze their production schedule for a significant amount of time, e.g., one month. This will simplify parts-ordering problems and ease the supplier's production plan. A good relation with the supplier is expected under a kanban system.

6.2 Parts Preparation and Feeding

Parts stored at the staging area may not be directly usable in the assembly line. For instance, in some electronics industries, assembly is performed in a controlled environment and parts must be cleaned before assembly. Parts may also be kitted to facilitate assembly. In addition to parts preparation, line feeding should be carefully studied to avoid shortages or excessive line inventory. This section discusses two major problem areas: kitting and feeding.

Parts Kitting

In the manufacturing community, people seem to have different opinions about kitting operations. One factory operates three shifts a day. The entire third shift has been dedicated to a kitting operation. Workers put necessary parts of an assembly into the same tray. The kitted trays are used by the first and the second shifts for product assembly. Another factory plans to kit major parts only. A small tray may hold multiple parts of the same kind. A kit contains multiple small trays. This will significantly reduce time for the kitting operation. The third factory does not kit parts at all. If a workstation consumes 10 different parts, then 10 different containers are sent to this workstation. Obviously, there are trade-offs between different kitting strategies. It is unlikely that everyone will concur with one general strategy.

From an assembly point of view, kitting may facilitate assembly operation. If a workstation uses 20 different part numbers, picking parts from one kit is easier than from 20 different containers. From a workstation-design point of view, 20 containers require much more space than a single kit. For a robotic assembly station, kitting operations may be necessary because of the limited working envelop. For example, an assembly robot might only cover an area with a diameter of six feet. On the other hand, kitting operations require additional working space, equipment, packing materials, and labor. Part quality is a major concern in a kitting strategy, because one single bad part in a kit will cause rejection of the entire kit. Kit design may also present a problem. What is the best kit layout to achieve the maximum packing density? How to make a kit design flexible so that the same kit can be used for different parts? If the product design is changed, is the same kit design still usable?

One way to reduce kitting effort and to keep the design flexible is the concept of *insertion*. There are two levels of packing. First parts are placed onto an *insert*. An insert holds only one type of part; then inserts with different part types can be placed in a kit. To ease the kit-design problem, the kit size can be standardized, but the insert size may follow a *binary division rule*. An insert can be the same size as the kit, or one half of the kit size, one quarter, one eighth, and so on. Hence, parts of different sizes and shapes can be fitted into different inserts. Each insert layout may be uniquely designed for a specific part type. If the design of a part is altered due to product design change, the corresponding insert may have to be modified or even replaced, but other inserts can remain intact. A conceptual kit layout is illustrated in Figure 6.6, where the kit has one 1/2, one 1/4 and two 1/8 inserts.

It is important that parts in a kit be exact for the same number of assemblies. For instance, three types of parts are placed in the same kit and their consumption rates are 1, 4 and 2, respectively. Then the part counts of these three types in the kit must be respectively n, $4n$ and $2n$ for some integer n. This requirement may be called the *constraint of compatibility*. The key problem is to pack inserts into kits to achieve the maximum density under this constraint. A higher packing density means a smaller number of kits, less part storage space, lower part-feeding frequency, and less material-handling effort.

Suppose that an operation consumes m types of parts. Hence there are m different inserts. Let s_i be the size of the insert for part i and n_i be the number of assemblies per insert i. If the kit size is 1, then s_i can only take

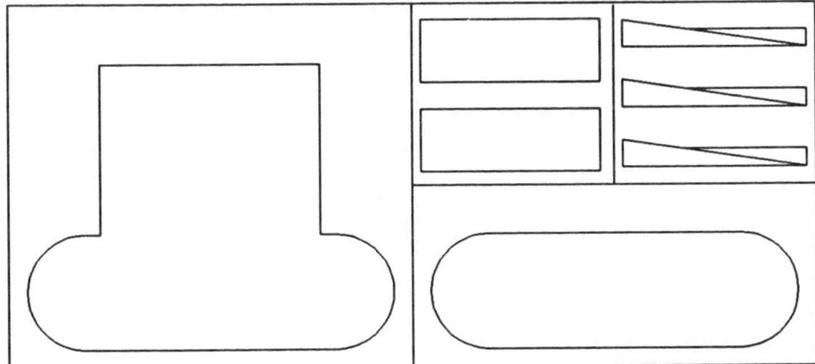

Figure 6.6 Kit layout with inserts.

a value from $\{1, 1/2, 1/4, 1/8, 1/16, \ldots\}$. If insert i is in a kit that consists of parts for k assemblies, then the space occupied by part i is $t_i = I[k/n_i]s_i$, where $I[x]$ is the least integer greater than or equal to x. An approximate solution for the minimum number of kits is given by solving the bin-packing problem (see Algorithm 3.4). However, if $t_i > 1$, then no kits that contain part i can hold enough parts for k assemblies. In this case, insert i is said to be infeasible. To resolve this problem, we need two algorithms. The first algorithm considers all feasible inserts for k assemblies, and generates (i) a list of infeasible inserts, named B, (ii) the minimal number of kits for all feasible inserts, and (iii) the kit layouts. The second algorithm deals with B.

Let A be the collection of all part types. Assume that $\{n_i, s_i \mid i \text{ is in } A\}$ and k are given. The first algorithm is defined as follows.

Algorithm 6.1

1. Let B and C be empty sets.
2. Take an element, say i, from A and do steps 3 and 4.
3. Compute $t_i = I[k/n_i]s_i$.
4. If $t_i > 1$, then put i into B. Otherwise, put i into C.
5. Go to step 2 unless all elements in A have been taken.
6. Take $\{t_i \mid i \text{ is in } C\}$ as input to a bin-packing problem. Invoke Algorithm 3.4.

At the completion of Algorithm 6.1, B contains all "unpacked" insert types, and an approximate optimal solution for insert types defined in C is obtained. The inserts in B are considered with a reduced value of k. Assume that there are p different types of inserts, and the maximum number of assemblies per kit is m. Let H be the maximum number of distinguish kits allowed. A complete solution is given by the following algorithm.

Algorithm 6.2

1. For $K = 1, 2, \ldots, m$, do steps 2 through 6.
2. Let $A = \{1, 2, \ldots, p\}$, and $k = K$.
3. Invoke Algorithm 6.1 with A and k. Record the content of each kit defined by the algorithm and the set B for all infeasible insert types with respect to k.
4. If B is not empty (i.e., infeasible insert types exist), then let $A = B$ and $k = k - 1$ (reduce k).
5. Repeat steps 3 and 4 until either $k = 0$ or B is an empty set. The former case implies that the size of an insert type is greater than the size of the kit. Hence no feasible solution exists. If the latter one is the case, a feasible solution is found. Continue with the next step.

6. Let $h(K)$ be the number of different kits defined by steps 3, 4 and 5. Inspect the occupancy of each kit. Let q be the total size of all inserts in a kit. Then the space left-over is $r = 1-q$. If $r \geq q$, then add another set of inserts for k additional assemblies. The left over space becomes $r = r-q$. Repeat this until $r < q$. Let a_j be the number of assemblies per kit $j, j = 1, 2, \ldots, h(K)$.
7. The number of kits per assembly is given by

$$f(K) = \sum_{j=1}^{h(K)} 1/a_j$$

The solution is defined by

$$z = \min_K \{f(K) \mid h(K) \leq H\}$$

In step 6, if the occupied space in a kit is less than 50%, it is more economical to increase the number of inserts in a kit. Since the optimal number of assemblies per kit is unknown, our approach is simply to try a number of candidates defined in step 1, where the maximum number is m. The value of m may be set by looking at the case of a single part per tray. If a tray contains only part type i, the number of assemblies per tray is equal to $v_i = n_i/s_i$. Then $m = \max\{v_i \mid i = 1,2,\ldots,p\}$. After having considered m cases, the best one is chosen for the final solution. A real-life example follows.

Example 6.3 A spindle assembly operation requires 9 different parts. The size of each insert, the number of assemblies per insert and the number of trays per assembly are given in Table 6.1.

The last column in Table 6.1 is obtained by dividing the insert size by the number of assemblies per insert. For example, parts in insert A can produce 12 spindles. Since the size of insert A is 1/16 (i.e., a full tray has 16 inserts), a full tray of part A can make $12 \times 16 = 192$ spindles. Thus, the number of trays per assembly is $1/192 = 0.005$ (see the first row in Table 6.1). The sum of the last column is $d = 0.369$. This value can be regarded as a reference for comparison of *packing efficiency*.

Since the maximum number of assemblies per tray is 192, the range of k under consideration is $[1, 192]$. Invocation of Algorithm 6.2 leads to the following results:

Table 6.1 Insert Data for a Spindle Assembly

Part	Insert Size	Assemblies/Insert	Trays/Assembly
A	1/16	12	0.005
B	1/4	1	0.250
C	1/16	3	0.021
D	1/16	3	0.021
E	1/16	2	0.031
F	1/16	3	0.021
G	1/16	6	0.010
H	1/16	12	0.005
I	1/16	12	0.005

1. $k = 2$, a kit composed of (A,B,B,C,D,E,F,G,H,I) making 2 spindles.
2. $k = 6$, a kit of $(A,C,C,D,D,E,E,E,F,G,H,I)$ making 6 spindles, and a kit of (B,B,B,B) making 4.
3. $k = 24$, a kit of $(A,A,F,F,F,F,F,F,F,F,H,H,I,I)$ making 24 spindles, a kit of $(C,C,C,C,D,D,D,D,E,E,E,E,E,E,G,G)$ making 12, and a kit of (B,B,B,B) making 4.

The above three cases generate respectively one, two and three kits. Their kit layouts are illustrated in Figure 6.7. The packing densities, in terms of the number of kits per assemblies, are respectively $d_1 = 1/2 = 0.5$, $d_2 = 1/6 + 1/4 = 0.417$, and $d_3 = 1/24 + 1/12 + 1/4 = 0.375$. Since inserts of the same type can always be perfectly fitted into a kit, d gives the minimum number of trays per assembly. Hence, the packing efficiency for case i can be defined by $e_i = d/d_i$. For the three cases that resulted from Algorithm 6.2,

$$d_1 = 0.500 \quad \text{and} \quad e_1 = 0.369/0.500 = 0.738$$

$$d_2 = 0.417 \quad \text{and} \quad e_2 = 0.369/0.417 = 0.884$$

$$d_3 = 0.375 \quad \text{and} \quad e_3 = 0.369/0.375 = 0.984$$

As we expect, a larger number of kits results in better packing efficiency. □

B B
B B

A A F F
F F F F
F F H H
I I

E E E E
E E C C
C C D D
D D G G

(c) THREE-KIT CASE

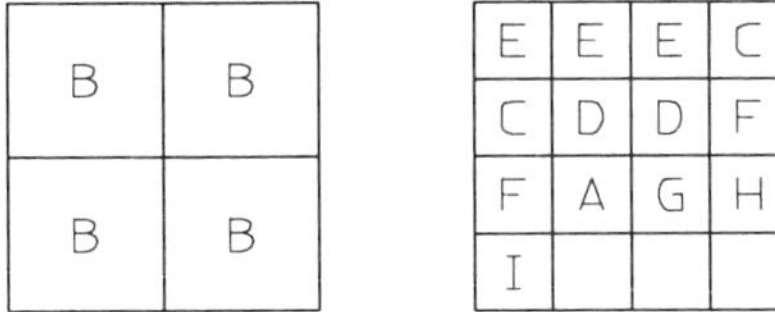

(b) TWO-KIT CASE

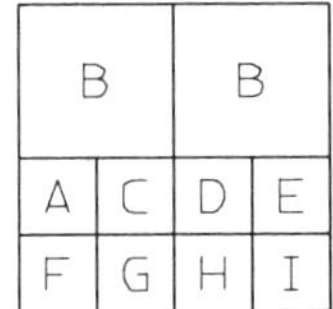

(a) ONE-KIT CASE

Figure 6.7 Kit layouts for a spindle assembly. (a) One-kit case. (b) Two-kit case. (c) Three-kit case.

Parts Feeding

Although a good packing method may help reduce the complexity of parts handling, efficiency is not achievable without solving the parts-feeding problem. It is not a rare situation that a workstation is idle simply because of shortage of a single part type. Ideally, an assembly line should have just enough parts of each kind to make the same number of assemblies. Two different approaches may be considered. One is driven by the demand at each individual operation, and is called the *individual demand* approach. The other attempts to synchronize feeding speeds of all parts through a concept called *logical kit*.

Individual Demand Approach

For each part type or each kit type, a fixed number of trays circulates in the assembly line. The tray flow begins at the parts-preparation area and involves the following steps:

1. Place parts into empty trays, if any, at the parts-preparation area.
2. Send all full trays to the workstations where parts are needed.
3. After having used up all parts from a tray at a workstation, return the empty tray to the parts-preparation area.

The maximum inventory of parts at workstations is determined by the total number of trays. The number of trays should be sufficient to keep workstations busy. On the other hand, the number of trays must not be too many so that the line does not have to carry excessive inventory. The major problem here is to find the optimal number of trays. The solution is usually dependent on individual application. A queuing network approach may be useful to find a solution quickly. Let us illustrate this approach by a simple example.

Example 6.4 Consider a parts-preparation station and an assembly station. The parts-consumption rate at the assembly station is μ trays per hour, while the preparation station can refill λ trays per hour. Let n be the number of trays. Then the flow characteristics may be analyzed by using a simple queuing network, as shown in Figure 6.8. The queue length distribution is given by Equation (3.29). Let $\rho = \lambda/\mu$ and P_i be the probability of i trays at the assembly station. Then (3.29) is reduced to

$$p_i = \frac{(1-\rho)\rho^i}{1-\rho^{n+1}}$$

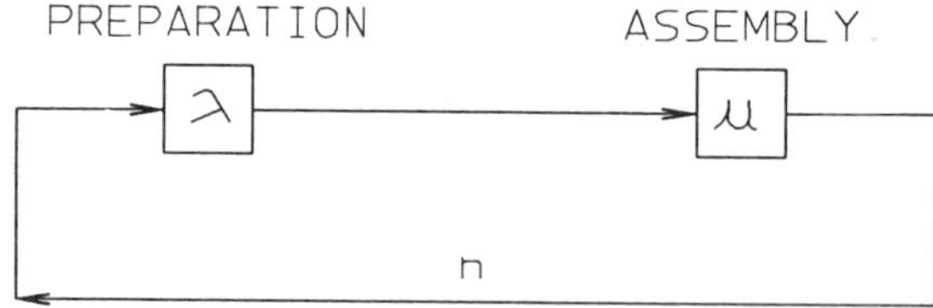

Figure 6.8 A sample queuing network model for a preparation station and an assembly station.

The probability that there is at least one tray at the assembly station is $r = 1 - p_0$. Assume that $\lambda = 30$ per hour and $\mu = 15$ per hour. Then $\rho = 2$. For n = 1, 2, 3, 4, 5, 6, and 7, r = 0.667, 0.857, 0.933, 0.968, 0.984, 0.993, and 0.996, respectively. If the desirable utilization of the assembly station is no worse than 0.99, six trays are needed. □

A real problem may be more complicated than the one illustrated above. Usually a number of assembly workstations share one or more preparation stations. An assembly workstation may use more than one type of tray. This situation may be handled by a different model which groups all empty trays as jobs to be served by the parts-preparation stations. The arrival rate of empty trays can be determined by two factors: (i) the number of trays per assembly, and (ii) the line throughput. If a line can produce 12 assemblies per hour and each assembly requires 2.5 trays of parts, then $12 \times 2.5 = 30$ empty trays will be generated in each hour. For a line with multiple workstations, the arrival stream of empty trays, as seen by the preparation stations, should be close to a Poisson process. If there are k preparation stations, their behavior can be analyzed by an M/G/k queuing model. Then the difference between the queue length and the total number of trays is the number of available part trays at workstations. If the queue length distribution is known, one can adjust the total number of trays so that the probability of parts shortage is within a predetermined limit.

Determination of the number of trays is not the only problem under the individual demand approach. Two other factors must be taken into consideration. First, material handling may be an important issue. If material-handling delay for parts delivery is significant, one may have to increase the total number of trays to compensate for the time loss. Sometimes a detailed model, e.g.a simulation model or an elaborate queuing network model, may be required. One may use a simple queuing model for a first approximation, and then conduct a more thorough analysis.

The second problem is directly related to material and production flow control. Although each operation has a fixed number of trays, the parts-flow rate may not be appropriately regulated. To see this, let us take a hypothetical case. Suppose that a workstation failure occurs in an operation. Since no more assemblies leave this operation, eventually all downstream operations stop working and no empty trays from these operations will be sent back to the parts-preparation area. But, on the other hand, all the upstream operations continue using parts to produce. Empty trays are sent back for refill. As long as parts are available, these operations will keep sending workpieces to the broken-down workstation. Hence, the level of WIP is built up. To resolve this problem, one may limit the quantity of WIP at each operation. If the WIP level at an operation reaches a predetermined threshold, then stop the immediate upstream operation. This phenomenon is called *blocking.* A blocked operation may in turn block its own immediate upstream operation. Thus, a workstation failure condition may trigger a chain reaction and eventually cause a line-stop. Determination of the blockage threshold will be discussed in Sections 6.3 and 6.4. Another method to control WIP level is to synchronize parts-feeding speed for different operations. This is the theme of the concept of logical kit.

Concept of Logical Kit

An ideal case for parts-feeding assumes a single kit that contains all necessary parts for exactly one assembly. Therefore, feeding speeds of all parts are automatically synchronized. Although parts may not be physically packed into the same kit, the principle of synchronization by way of a single "logical" kit is achievable. The question is how to link the concept of logical kit to physical material movements. This can be done by a simple scheme. Since each logical kit is equivalent to a single assembly, there is no need to distinguish a logical kit from an assembly in the following discussion. The feeding scheme can be explained by an example. If a (physical) tray has parts for three assemblies, then the tray corresponds to three logical kits. The feeding sequence of this tray is defined by a vector (1, 0, 0), where the ith element is the number of trays that should be fed to make the ith assembly. For the first assembly, a tray must be delivered. Once the tray is received, three assemblies can be produced. Therefore, no part feeds are required for the second and the third assemblies, i.e., the last two elements are zero. After the third assembly is completed, all parts are used. Thus the same feeding sequence is repeated for every three assemblies.

To implement this concept at a full scale, we need to build a feeding sequence for each tray type used by each workstation. To simplify our discussion, it is assumed, for the time being, that each operation has only one workstation. Let x be the number of assemblies per tray and z_i be the value of ith element in the feeding sequence. A complete sequence can be obtained by invoking the following algorithm.

Algorithm 6.3

1. Let $y = 0$ and $i = 0$.
2. $i = i + 1$. (Determine the value of the ith element.)
   $j = 0$. (Initial value of the number of trays.)
3. Add trays until there are enough parts for an assembly:

   a. If $y \geq 1$, go to step 4. Otherwise go to step 3b.

   b. $y = y + x$ and $j = j + 1$. Go to step 3a.

4. Define the value of z_i. Let $z_i = j$ and $y = y - 1$. (One set of assembly parts is used by the ith assembly.)
5. If $y = 0$, the situation is identical to the condition stated in step 1. Stop. Otherwise, go to step 2 to determine the value of the next element.

Two observations are of interest. First, the stop condition in step 5 will eventually occur if and only if x is a rational number. Second, if $x \geq 1$, then z_i always takes a binary value, i.e., $z_i = 0$ or 1 for all i. This implies that each element in a feeding sequence can be represented by a single bit in a computer system. Computer storage saving becomes apparent when the number of part types and the number of elements in the feeding sequence are large.

For the multiple-station (per operation) case, each station should have a z-sequence. The feeding pattern can be constructed by taking one element at a time from a z-sequence in an alternating fashion. For example, if an operation has two workstations and the z-sequence is read as (1, 0, 0), then the feeding pattern becomes (1, 1, 0, 0, 0, 0). The elements at odd positions are associated with one station, and the elements at the even positions with the other. After six feeding cycles, the pattern starts all over again.

6.3 In-Process Inventory Control

Under a "zero-inventory" policy the ideal in-process inventory (or simply called WIP) level should be driven to zero. But this is not a practical objective. There are many reasons why a positive inventory level is necessary. In an assembly, WIP is normally required due to variability of process time, resource availability, production yield, and material-handling delay. A good control policy should keep in-process inventory at a minimum level and yet maintain a desirable line throughput. Through its control process, such a policy may also be used to detect line problems. From Theorem 3.1, it is known that the average WIP level is equal to the product of line throughput and the average production lead time (i.e., the interval between when an assembly is introduced into the line and the assembly is completed at the last operation). An experienced line manager may look at these three parameters to determine whether the line operation is acceptable. Suppose that each parameter has two ratings: high and low. There are a total of eight possible combinations as listed below.

Case	WIP	Throughput	Lead Time	Comment
1	Low	High	Low	Ideal
2	Low	Low	Low	WIP too low
3	Low	Low	High	Irregular operation cycle
4	Low	High	High	Impossible case
5	High	High	Low	Excessive line capacity
6	High	Low	High	Insufficient capacity or irregular cycle
7	High	High	High	WIP too high
8	High	High	High	Impossible case

The line is running very efficiently in case 1. If all three parameters tend to be low, a line manager may raise the WIP level to increase line throughput. Case 3 indicates that variation of operation cycle at some workstations may be too high. This often happens due to availability problems such as workstation failures, absenteeism, slow parts-feeding speed, and yield problems. Since the product of two large numbers cannot be small, case 4 is an impossible case.

Case 5 has a short average lead time, but a large amount of WIP and high throughput. This often implies that the line has excessive capacity and can process workpieces very quickly. By comparison, case 6 presents a case of insufficient capacity. All three parameters are high in case 7. This could be the case of too much in-process inventory. Therefore, one may try to lower the WIP level. Finally, similar to case 4, case 8 is also impossible.

A more systematic approach is required for real-time control policies. Two policies are discussed in the following sections. One is the *pull* policy and the other the *closed-loop* policy.

Pull Policy

A pull policy is similar to a single kanban system, as introduced previously, except that kanbans are substituted by buffers. Under such a policy, buffers of finite sizes are placed between adjacent operations. Thus an operation will have both an upstream and a downstream buffers. The first (the last) operation, however, may or may not have an upstream (downstream) buffer. At the end of an operation cycle, the completed workpiece is sent to the downstream buffer and a fresh one is pulled from the upstream buffer. If the downstream buffer is full, however, the completed workpiece will stay at the workstation and the next operation cycle will be temporarily suspended until a free buffer space becomes available. The buffer-full condition is called *blocking*. On the other hand, if a buffer is empty, its downstream operation may become *starving*. If operation cycles are perfectly synchronized, no starving or blocking can possibly happen. Problems arise only if adjacent operations complete their cycles at different time epochs. This may occur under two conditions: (i) process times constant but not identical, or (ii) process times variable (but may or may not be identical). The former condition results from an unbalanced line. An operation with a short process time will be slowed down by the bottleneck, no matter how much buffer space is installed. If the bottleneck is downstream, upstream operations are blocked from time to time. If the bottleneck is located upstream, then downstream operations become starving. The line throughput is paced by the bottleneck and is independent of buffer size. The second condition is typical in reality. Because of human inconsistency in repetitive work, workstation failure, yield, material-handling delay, etc., operation process times can be regarded as random variables. Analysis of line behavior with arbitrary process times is difficult. Numerical results, however, may be obtained by simulating a series of queues, as shown in Figure 6.9, where squares represent operations at different stages of the line.

Consider a hypothetical assembly line composed of six identical workstations in tandem. Jobs must be processed at each workstation in the same sequence, i.e., $1 \rightarrow 2 \rightarrow 3 \rightarrow 4 \rightarrow 5 \rightarrow 6$. For convenience, the average process times at all workstations are assumed to be exactly one time unit, no workstation failures are possible, and the assembly yield is 100%. If all process times are constant, then operation cycles at different workstations can be synchronized so that workpieces are moved at the same time. In this case, the line throughput is equal to one. Since each workstation always holds a workpiece, the WIP size is six. Using the relation $L = \lambda W$ (see Theorem 3.1), the production lead time is $W = L/\lambda = 6/1 = 6$ time units. Each operation cycle has a zero variance, and no buffer is required.

The variability of a variable process time is characterized by its coefficient of variation. (See Sections 2.5, 2.6, and 2.7.) Let CV be the coefficient of variation and B be the buffer size between adjacent operations. Consider a six-stage assembly line, each stage having a single workstation with a mean process time of one. Table 6.2 shows simulation results for $CV = 0.2, 0.6, 1.0.$ and $B = 0, 5, 10$. It can be seen that throughput (TP) is decreasing in

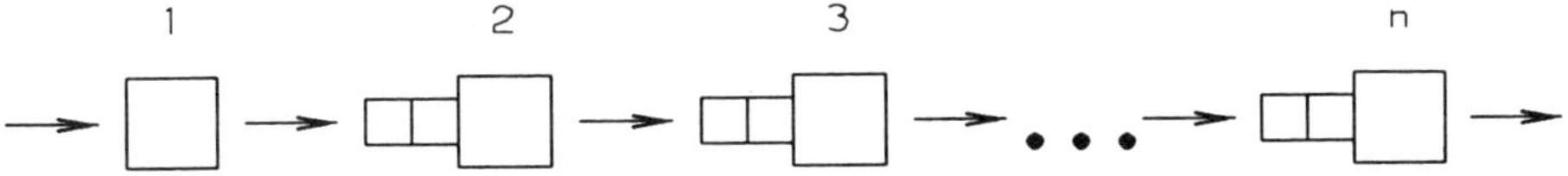

Figure 6.9 A line with finite buffers.

Table 6.2 Simulation Results of Six-Stage Lines under a Pull System

CV	0.2			0.6			1.0		
Buffer	0	5	10	0	5	10	0	5	10
TP	0.82	0.97	0.97	0.60	0.88	0.93	0.48	0.79	0.86
WIP	5.47	15.66	24.79	4.88	16.04	21.06	4.43	19.87	24.79

CV but increasing in B. When $B = 0$, the line throughput is significantly reduced. In this case, the line throughput is dropped by about 20%, 40% and 50%, for $CV = 0.2$, 0.6, and 1.0, respectively. The average WIP level is in general higher for larger buffers.

Line throughput may be determined by simulation for given buffer sizes. However, determination of buffer size is a difficult problem. A heuristic method is given in Section 6.4.

Since a workstation failure may block the incoming traffic and eventually cause a line stop, the pull policy is capable of detecting failures. However, the detection time can be a problem, when the number of operations is large. Assume that the average buffer size per operation is 4. For a line of 20 operations, the maximum amount of WIP is 100 units (20 on the workbenches and $4 \times 20 = 80$ in the buffers). If the last workstation fails, a line-stop condition will not be detected until the maximum WIP level is reached.

Closed-Loop Policy

If a closed-loop policy is implemented, one may adjust the total amount of WIP to achieve a desirable throughput. Typically a closed-loop policy can be illustrated by a cyclic queuing model as shown in Figure 6.10. This model is a special case of a closed queuing model with a fixed population size. If all workstations have the $M \Rightarrow M$ property, the model solution is given by Equation (3.29). For general process time distributions, simulation may be

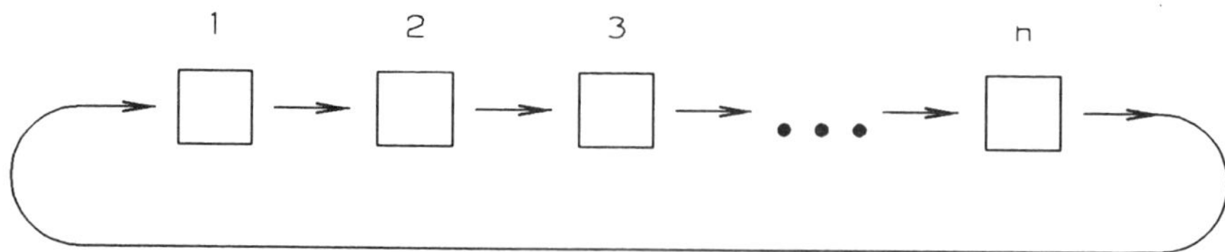

Figure 6.10 A closed loop system.

used. The population size in the model is equivalent to the WIP level. For a given line, the minimal WIP level for a predetermined throughput can be determined very quickly by a simple searching procedure. Suppose that the same assembly line, considered in Table 6.2, is operating under a closed-loop system. If the desirable throughput is no worse than 0.97, the minimum WIP levels for different CV's are shown in Table 6.3.

Comparison between Pull and Closed-Loop

Results from Tables 6.2 and 6.3 are plotted in Figure 6.11, where the average lead time is obtained by dividing the average WIP by the line throughput. It can be seen that when $CV = 0.2$, line performance under the closed-loop policy is comparable to that under the pull policy with $B = 5$. For $CV \geq 0.6$, the closed-loop policy can still keep up a line throughput of 0.97, but at the cost of higher WIP.

Under a closed-loop policy, the WIP levels are set at 15, 55, and 100 for $CV = 0.2$, 0.6, and 1.0, respectively. The WIP level, however, may be reduced by holding unused parts for the first operation. Take the case of $\{CV = 0.6\}$ as an example. Parts are fed into the line as long as the first operation is empty and the total amount of WIP is below 55. But if the first operation is busy, the line throughput will not be increased by scheduling more assembly work at the first operation. In this case, no parts should be sent to the line even if the total WIP size is less than 55. Since a closed-loop system is equivalent to a cyclic queuing system, by symmetry each operation has the same queue length distribution with a mean of 55/6. Therefore, the average WIP level in the line can be reduced to $(55 - 55/6) + 0.97 = 46.8$. The term that follows the plus sign is the utilization of the first operation.

The case of $CV = 0.6$ is close to the upper bound of CV, and can be regarded as a worse-than-average case. (See Section 2.5.) Under the closed-loop policy, 55 assemblies are needed to achieve a throughput of 0.97. The 90th-percentile of WIP distribution and the observed maximum WIP at each operation are given in Table 6.4. The average 90th-percentile is

Table 6.3 Simulation Results of Six-Stage Lines under Closed-Loop System

CV	0.2	0.6	1.0
TP	0.97	0.97	0.97
WIP	15	55	100

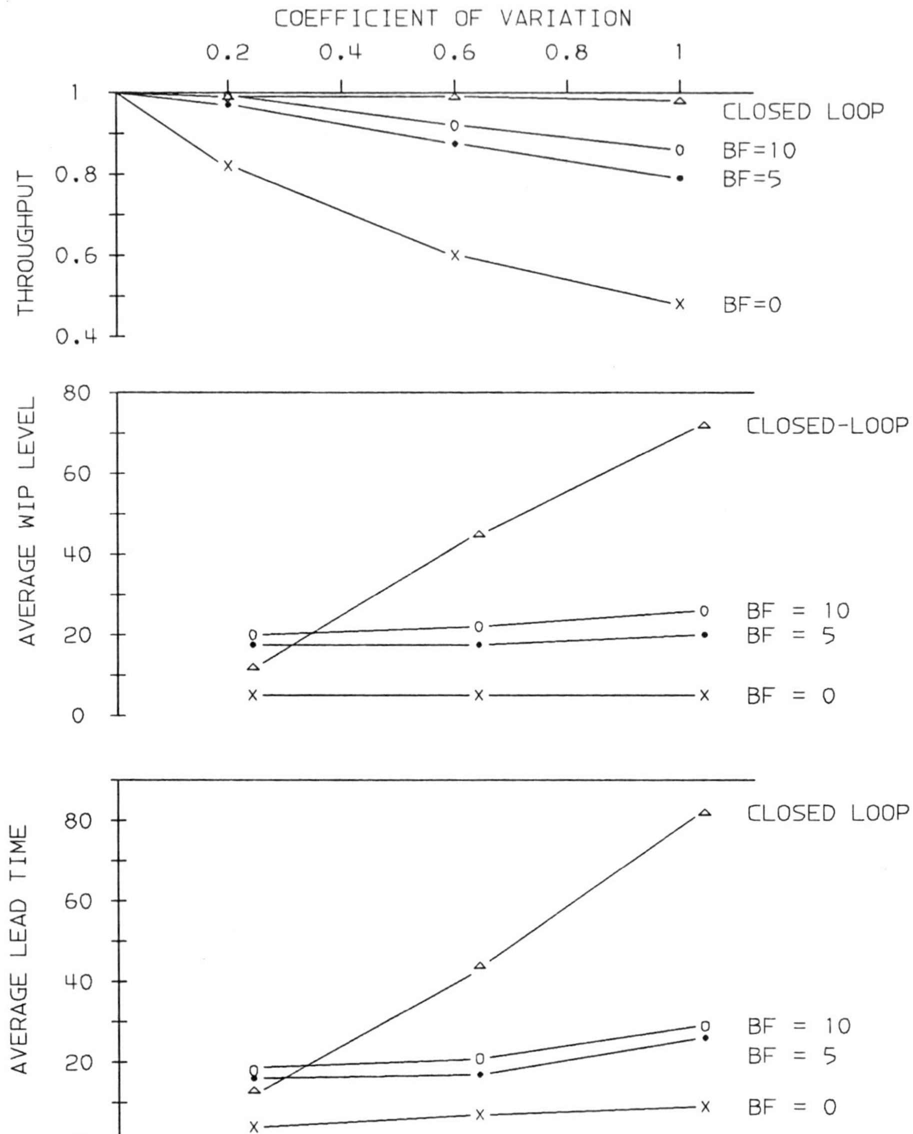

Figure 6.11 Comparison of different WIP policies in a six-stage production line.

Table 6.4 WIP Comparison between Two Different WIP Management Policies

Cases	Closed-Loop		Pull/Buffer of 20		Pull/Buffer of 15	
WIP level	90%	Max	90%	Max	90%	Max
Operation 2	19	34	16	21	16	16
Operation 3	20	27	19	21	15	16
Operation 4	17	29	19	21	14	16
Operation 5	25	44	16	21	13	16
Operation 6	20	35	16	21	13	16
Average	20.2	33.8	17.2	21.0	14.2	16.0
Average WIP	46.8		44.12		42.0	
Throughput	0.97		0.97		0.96	

20.2. This value may be considered as a first estimation of buffer size under a pull policy. Two pull policies with $B = 20$ and 15 are analyzed and also reported in Table 6.4. It is found that both their 90th-percentiles and the maximum WIP are lower than those under the closed-loop policy. However, the average WIP levels and throughputs are about the same for all three cases of Table 6.4. This implies that the WIP level at a given operation under a closed-loop policy tends to have a higher variability. On the other hand, it can be seen that a closed-loop policy requires much less total buffer space than does a pull policy—55 versus 100.

WIP Control and Parts Feeding

The two control policies for in-process inventory are compatible with the two feeding concepts discussed in the previous section. The concept of individual demand can be very easily implemented, once the number of trays has been determined. Parts-feeding speed is regulated by the consumption rate at each individual operation. If the line flow is controlled by pull logic (i.e., finite buffer between adjacent operations), consumption rates at different operations are approximately the same. Consequently, feeding paces of all parts are compatible. Furthermore, the task of parts refill is triggered by the receipt of empty trays. The parts-feeding scheme need not be computerized. The major problems are to find (i) the optimum number of trays for each part, and (ii) the optimum buffer sizes between operations. Neither of the problems has a simple answer.

Based on the concept of logical kit, a method attempts to synchronize feeding speeds of all parts. If the number of tray types is large, a computer program is required to generate successive feeding patterns. If the method is associated with closed-loop logic, the total number of assemblies is the only control variable. Therefore, searching for the optimal value is a simple task.

Our discussions so far can be summarized as follows:

1. For the same throughput, the WIP requirement under a closed-loop policy is comparable to that under a pull policy. However, at each individual operation, the maximal WIP under a closed-loop policy is higher than that under a pull policy.
2. By definition, a pull policy is a distributed storage policy. On the other hand, a closed-loop policy can be implemented with either a shared common buffer or large distributed buffers. For the same throughput requirement, the total buffer size under a pull policy is usually higher than that under a closed-loop policy.
3. If a conveyor is selected as the material-handling system, a pull policy is more suitable. Each operation may take a section of conveyor as its own buffer. If an AS/RS (automated storage and retrieval system) is used, either of the policies may be used. In this case, the WIP control logic is nearly independent of the physical storage structure.
4. The parts-feeding method should be compatible with the WIP control policy. The concept of logical kit very well matches a closed-loop policy, and the concept of individual demand may go along with a pull policy.
5. The control variable of a closed-loop policy is the total amount of WIP in the line. The optimal solution can be obtained by a few experiments on the existing line or by a simulation model. The optimal policy under a pull logic is much more difficult to derive. The number of control variables is the same as the number of operations.
6. If a pull policy is implemented, a line should have just enough buffer to absorb process time variation. Obviously, the higher the variability, the more buffer will be needed.
7. Both policies allow some WIP to smooth production flow. But they also set limits for the WIP level. These limits should be the minimum requirements for a desirable throughput under "normal" conditions so that line problems may be detected quickly (e.g., a workstation is blocked for a significant amount of time).
8. Although the two policies are different, there is no reason why they cannot be implemented simultaneously. A combination of a pull and a closed-loop policies will further reduce the WIP level and can quickly detect a line-stop condition, as compared to a pure pull or a closed-loop policy.

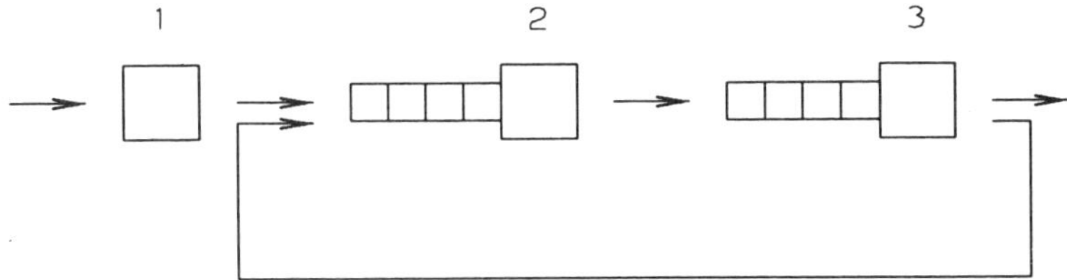

Figure 6.12 A possible deadlock situation.

It should be pointed out that a rigid pull policy may cause a deadlock when feedback flows exist. This can be illustrated by the simple example in Figure 6.12, where three operations are connected by two buffers. Between operations 2 and 3 both forward and feedback flows exist. The assembly process is given by $1 \rightarrow 2 \rightarrow 3 \rightarrow 2$. After the second pass at operation 2, the assembly work is completed. When completed work units at operations 2 and 3 simultaneously request free buffer spaces from each other, a deadlock occurs. This problem is usually resolved by introducing a "priority buffer" to operation 2. Work units in this buffer have higher priority than those in the ordinary buffer. In our example, all completed work units from operation 3 are placed in the priority buffer at operation 2 and units from operation 1 are put in the ordinary buffer. In this way, the flow would look like the one in a sequential line: $1 \rightarrow 2a \rightarrow 3 \rightarrow 2b$, where $2a$ corresponds to the ordinary buffer and $2b$ for the priority buffer. Units in the priority buffer should always be processed first. Hence, the possibility of deadlock does not exist.

6.4 Buffer-Capacity Analysis

Determination of buffer sizes under a pull policy is a very difficult problem. No general algebraic relation between line throughput and buffer sizes is known. In this section the problem under investigation is limited to sequential production lines and only a heuristic method is adopted. We shall consider two operations at a time and solve the line problem progressively. The approach can be outlined as follows:

1. Consider a line with two operations. A finite buffer is placed between the operations. Define study cases by varying the mean and the *CV* of process time, and the buffer size.
2. Simulate all cases defined in step 1. Record line performance behavior for each individual case. Information about process time and buffer size characterize the input data for the simulation program.

3. Determine the algebraic equations between line performance and input parameters by regression analysis.

Note that the regression equations are defined for two operations only. If a production line has two operations, the buffer requirement can be directly determined by the equations. If there are more operations in a line, the buffer design problem can be solved in a progressive manner. The following steps will be invoked.

4. Use the results derived at step 3 to find the proper buffer size between the first two operations.
5. Compress the first two operations as if there were a single operation. Therefore, an "equivalent" line is defined. Compared to the original line, the number of operations in the new line is one less.
6. Repeat steps 4 and 5 until all operations have been considered.

Since the above procedure includes one operation at a time, it takes a finite number of steps to find all buffer sizes. Several key problems must be addressed. From a validity point of view, the ranges of input parameters used in simulation should be broad enough to cover most cases found in the real world. The regression equations should have simple structures and simple input requirements. Since a regression analysis usually deals with empirical formulas, a complex structure is rarely useful. Furthermore, during the line design stage, information is almost never totally available. A regression model that requires too much detailed input would not be very practical. With this understanding, we shall discuss (i) experimental cases for the basic simulation model of two operations in tandem, (ii) regression models and their validation, (iii) line performance analysis with finite buffers, and (iv) buffer design optimization.

Experimental Cases

Since in-process buffers are justified by the variability of operation cycles, *CV* of process time may be selected as a parameter for our experiment. The function of a buffer is to reduce the probability of blocking or starving due to variable cycle length. However, if the process time of an operation is considerably less than those of others, then the operation will have a low utilization. In this case, blocking and starvation at this operation may not actually affect line throughput. Consequently, the mean process times should also be considered in the experiment. Without losing generality, the mean process times can be normalized so that the first operation has a mean time equal to one. The mean process time at the second operation can be interpreted as the ratio of the second mean process time to the first.

To keep the parameter set simple, an experimental case should be characterized by

1. *CV* of process times, C_1 and C_2
2. the ratio of mean process times, D
3. buffer size between two operations, B

Section 2.5 has shown that a manual process time has a *CV* between 0.1 and 0.65. Variation may also result from workstation failures at an automated or semi-automated operation. Let

p = probability that a failure occurs during an operation cycle
X = duration of an operation cycle (including downtime if any)
t = pure process time (downtime not included)
d = average downtime
c = *CV* of downtime
m = mean time between failures

From a practical point of view, it is important to keep the task at an automated assembly station as simple as possible so that good reliability can be achieved. A simple task often implies a short process time, e.g., t of one minute or less. The ratio of t/m is typically small, e.g., less than 0.001. Therefore, the possibility of two or more failures during the same operation is negligible. The failure probability is $p \approx t/m$. Since an automated operation has good repeatability, its process time can be regarded as a constant. Thus,

$$
\begin{aligned}
E[X] &\approx t(1-p) + (t+d)p = t + dp \\
E[X^2] &= E[(t+S)^2] \\
&\approx (t^2 + 2td + E[S^2])p + t^2(1-p) \\
&= 2tdp + (c^2+1)d^2p + t^2
\end{aligned}
$$

where S is the duration of downtime. It then follows that

$$
\begin{aligned}
Var[X] &= E[X^2] - (E[X])^2 \\
&\approx c^2d^2p + d^2p(1-p)
\end{aligned}
$$

The CV of an operation cycle is given by

$$\begin{aligned} CV[X] &= \frac{\sqrt{Var[X]}}{E[X]} \\ &\approx \frac{\sqrt{c^2 + p(1-p)}}{p + t/d} \\ &\approx \frac{\sqrt{(m/t)(c^2+1) - 1}}{1 + m/d} \end{aligned} \tag{6.6}$$

If t, m, d, and c are known, $CV[X]$ can be evaluated. For $m = 1000t$ and $m = 100d$, as an example, $CV[X] \approx \sqrt{0.1(c^2+1)}$. Then for $c = 1$, 1.5, and 2, $CV[X] = 0.45$, 0.57 and 0.71, respectively.

Variability of an operation cycle may result from a yield problem too. If an operation produces N bad products consecutively before a good one, then its downstream operation must wait for $N + 1$ operation cycles to see an arrival. Let X_i be the ith cycles. The interval between two successive arrivals at the downstream operation is $Y = X_1 + \cdots + X_{N+1}$. In this case, N can be interpreted as the number of failures before the first success. The probability of success is identical to the yield factor, q. Therefore, N is a geometric random variable. (See the definition by (2.9) in Chapter 2.) It can be shown that Y has a coefficient of variation

$$CV[Y] = \sqrt{q(1-q) + q(CV[X])^2}$$

To determine the appropriate range of CV for experiments, the maximum value of $CV[Y]$ must be found. The optimality condition, $dCV[Y]/dq = 0$ leads to $q_0 = [1 + (CV[X])^2]/2$. Since $0 \le q \le 1$, the possible maximum solution must be $q = q_0$, 0, and 1. At $q = 0$, $CV[Y] = 0$. Obviously this is not maximum. $CV[Y]$ is an increasing function of $CV[X]$. But for $CV[X] \ge 1$, q_0 will exceed one. Therefore $CV[Y]$ takes its maximum value at $q = 1$ if $CV[X] \ge 1$, and at $q = q_0$ if $CV[X] < 1$. Consequently

$$CV[Y] = \begin{cases} CV[X] & \text{if } CV[X] \ge 1 \\ [1 + (CV[X])^2]/2 & \text{if } CV[X] < 1 \end{cases}$$

For $CV[X] \leq 0.7$, the maximum value of $CV[Y]$ is 0.75. Therefore, in our experiment the value of CV may be confined to (0, 0.75).

The mean-process-time ratio may also affect throughput. If two operations have different speed, the throughput is dominated by the slow one. If the fast operation is temporarily blocked or becomes starving, it may not affect the throughput much. Some preliminary simulation results are shown in Table 6.5, where

TP = throughput
B = size of the buffer between the two operations
T_1 = mean process time of first operation
T_2 = mean process time of second operation
D = mean process time ratio = T_2/T_1
$C_1 = CV$ of T_1
$C_2 = CV$ of T_2

It is clearly demonstrated that

1. TP is increasing in B
2. TP is decreasing in C_1 and C_2
3. TP is most sensitive to B, C_1 and C_2 when $D = 1$, i.e.in a perfectly balanced line.
4. This sensitivity becomes less and less as D is moving away from one and very little effect can be observed if D lies outside (0.7, 1.3).

The above discussion leads to a number of experiments:

1. $B = 2$, 5, and 8
2. $D = 0.7$, 0.85, 0.925, 1, 1.075, 1.15, and 1.3

Table 6.5 Throughput as a Function of Process Time and Buffer Size

C_1	C_2	B	$D = 0.70$	$D = 0.85$	$D = 1.00$	$D = 1.15$	$D = 1.30$
0.2	0.2	2	1.000	1.002	0.985	1.003	1.003
		8	0.991	1.002	0.996	1.003	1.004
0.2	0.6	2	0.998	0.981	0.942	0.988	1.002
		8	0.998	0.997	0.984	1.002	1.002
0.6	0.2	2	0.998	0.993	0.934	0.976	0.989
		8	1.010	1.008	0.987	0.998	1.001
0.6	0.6	2	0.997	0.969	0.915	0.964	0.992
		8	1.021	1.001	0.977	1.004	1.013

3. $C_1 = 0.2$, 0.4, and 0.6
4. $C_2 = 0.2$, 0.4, and 0.6

Hence, the number of experiments is $3 \times 7 \times 3 \times 3 = 189$. A multi-stage exponential distribution is used to generate process time. The shape of the distribution looks like a gamma distribution (see Section 2.2), except that the minimal process time is assumed to be 65%, 55%, or 45% of its mean for $CV = 0.2$, 0.4, or 0.6, respectively.

The throughput and the CV of interdeparture time at the second operation are estimated by simulation. The accuracy of estimation is established by looking at the ratio of the width of a 90% confidence interval (see Section 2.3) to the estimated value. For instance, for an estimated value 0.9 and a confidence interval (0.8, 1.0), the ratio is $(1.0-0.8)/0.9 = 22.22\%$. For a good estimation, this ratio should be kept small. For a large simulation program, this ratio may be set at 10–15%. However, in our analysis, a ratio no worse than 5% is selected. Methods for construction of a confidence interval will be discussed in Chapter 9.

Regression Analysis

Two regression models are required, one for throughput and the other for CV of interdeparture time (CVD). Models are constructed in two stages. For each mean-process-time ratio, D, a pair of models are developed. Then composite models for TP and CVD are constructed for all values of D.

Since D can be regarded as a normalized process time, the two operations have mean process times 1 and D, respectively. Throughput can be written as

$$TP = \frac{1}{\max(1,D)}(1-\alpha) \tag{6.7}$$

The first term on the right-hand side of the equals sign is the throughput under the condition of an infinite buffer size. Because of a finite buffer, the throughput is reduced by a factor of α. This factor should be a decreasing function of B (buffer size) and is increasing in C_1 and C_2. After having compared several candidates, the selected form is given by

$$TP = \frac{1-(a_1C_1^2 + a_2C_2^2)e^{-\sqrt{B}}}{\max(1,D)} \tag{6.8}$$

Table 6.6 Least-Squares Results for *TP* model

D	a_1	a_2	R	SSE
0.700	−0.12	0.08	0.409	0.0010
0.850	0.05	0.20	0.923	0.0007
0.925	0.23	0.32	0.951	0.0016
1.000	0.60	0.56	0.978	0.0008
1.075	0.34	0.24	0.975	0.0012
1.150	0.25	0.04	0.922	0.0011
1.300	0.11	−0.09	0.564	0.0008

For a given D, a *CVD* model is also determined empirically

$$CVD = b_1 C_1 (1 - C_2) + b_2 C_2 + \frac{b_3 (C_2 - C_2)^+}{1 + B} \qquad (6.9)$$

where x^+ is the positive part of x, i.e., $x^+ = x$ if $x \geq 0$ and $x^+ = 0$ otherwise.

The coefficients, a's and b's, are to be determined by the least-squares method (see Section 2.3). Note that both (6.8) and (6.9) are linear equations, similar to Equation (2.17). The values of these coefficients can be computed by using relation (2.19). The results from the least-squares method are given in Tables 6.6 and 6.7, respectively. The R-value is the coefficient of correlation defined by (2.16) and SSE is the sum of squared errors, computed by (2.18).

Table 6.7 Least-Squares Results for *CVD* model

D	b_1	b_2	b_3	R	SSE
0.700	0.79	0.75	0.19	0.996	0.0071
0.850	0.52	0.85	0.29	0.999	0.0017
0.925	0.28	0.92	0.62	0.998	0.0021
1.000	0.04	0.99	0.87	0.997	0.0046
1.075	0.00	1.00	0.41	0.998	0.0031
1.150	0.00	0.99	0.23	0.999	0.0015
1.300	0.00	0.99	0.09	0.999	0.0016

The accuracy of a regression model is measured by R and SSE. Table 6.7 shows that all R-values are close to one and SSE's are small. Hence (6.9) is a good model for CVD. Results in Table 6.6 indicate that the TP model has a good fit at D near one. Though the R-value becomes poor as D moves away from one, the value of SSE is consistently small. This simply confirms what has been discussed previously—if D is deviated from one, the throughput is not sensitive to CV and B.

After having established the validity of these two models, each coefficient of a's and b's is fitted separately as a function of D. Since the values of coefficients have reverse trends when D is moving from the region $\{D < 1\}$ to $\{D > 1\}$, different treatments may be given to these two regions. Furthermore, each region has only four data points, so a quadratic function will provide an almost perfect fit. The complete models are given by

$$TP = \frac{1}{\max(1,D)}\left[1-(C_1^2,C_2^2)U\begin{bmatrix}1\\ D\\ D^2\end{bmatrix}e^{-\sqrt{B}}\right] \tag{6.10}$$

$$\text{where } U = \begin{cases}\begin{pmatrix}6.95 & -18.19 & 11.84\\ 1.90 & -5.82 & 4.48\end{pmatrix} & D \le 1\\[2ex] \begin{pmatrix}7.36 & -10.50 & 3.74\\ 16.43 & -26.64 & 10.77\end{pmatrix} & D > 1\end{cases}$$

$$CVD = \left(C_1(1-C_2), C_2, \frac{(C_1-C_2)^+}{1+B}\right)\begin{bmatrix}4.68 & -7.68 & 3.13\\ -0.50 & 2.51 & -1.05\\ & V & \end{bmatrix}\begin{bmatrix}1\\ D\\ D^2\end{bmatrix}$$

$$\text{where } V = \begin{cases}(4.76 \quad -12.72 \quad 8.84) & D \le 1\\ (20.20 \quad -32.18 \quad 12.86) & D > 1\end{cases} \tag{6.11}$$

Since both (6.10) and (6.11) are fitted in two disjoint regions ($\{D \le 1\}$ and $\{D > 1\}$) separately, discontinuities exist at $D = 1$. It is possible to smooth both TP and CVD curves. However, function smoothing will not improve the accuracy and is not necessary for practical usage.

It should be pointed out that (i) Equations (6.10) and (6.11) are derived for $0.7 \le D \le 1.3$, (ii) if $D > 1.3$, the interdeparture time is nearly identical to the second process time and $CVD \approx C_2$, and (iii) TP in (6.10) is normalized with T_1 equal to one. Consequently, the following modifications are required:

(a) If $D < 0.7$, set $D = 0.7$ then use (6.10) and (6.11)
(b) If $D > 1.3$, use (6.10) without changing D but let $CVD = C_2$
(c) TP should be converted into the actual time scale by dividing by the mean process time at the first operation, T_1

The basic relations about TP and CVD may be extended to a line of more than two stages, as discussed in the following section.

Performance Evaluation of Sequential Lines

Denote a sequential line with n operations by L_n. Let T_i and CV_i be the mean and the CV of the process time at operation i, and BS_i be the size of the buffer between operations i and $i + 1$. For given T's, CV's and B's, the line throughput can be estimated by invoking (6.10) and (6.11) repetitively. First, consider a line composed of the first two operations from L_n, and estimate TP and CVD by (6.10) and (6.11), respectively. Compress these two operations into an "equivalent" composite operation with a mean process time, $1/TP$, and a coefficient of variation, CVD. Then consider a line consisting of the first three operations from L_n. Replace the first two operations by the composite operation. Again a line of two operations is formed. Compute TP and CVD in the same manner. Then $1/TP$ and CVD become the mean and the CV of the process time at a composite operation that is equivalent to the first three operations. Repeat this process until the entire line is covered. The final value of TP is the estimated throughput of L_n. This approach is summarized in Algorithm 6.5, where L_i is a line composed of the first i operations from L_n, TP_i and CVD_i are the throughput and the CV of the interdeparture time from L_i.

Algorithm 6.5

1. $TP_1 = 1/T_1$ and $CVD_1 = CV_1$. Let $i = 1$.
2. $i = i + 1$
3. Take $C_1 = CVD_{i-1}$, $C_2 = CV_i$, $B = BS_i$, and $D = TP_{i-1}T_i$ as input; evaluate TP and CVD by Equations (6.10) and (6.11) respectively, with modifications (a), (b), and (c).
4. Let $TP_i = TP$ and $CVD_i = CVD$.
5. If $i = n$, then stop. TP_n is the estimated line throughput for L_n. Otherwise, go to step 2.

Experience shows that the error derived from Algorithm 6.5 can be expected within 5%. A numerical example follows.

Example 6.6 Assume that a sequential line has N operations with identical process times. The mean process time is exactly one time unit. Thus the theoretical maximum throughput is one. Because the process time is variable, buffers between operations are needed to keep a high throughput. The coefficient of variation of the process time is 0.35. Between adjacent operations, a buffer of size four is installed. The line can be depicted by a tandem queuing system in Figure 6.13. Line throughputs for different line sizes are estimated by Algorithm 6.5. For $N = 1$, 5, 10, 15, 20 and 25, $TP = 1$, 0.9360, 0.8974, 0.8618, 0.8427, and 0.8291, respectively. Line throughput is a decreasing function of the line size.

For the same buffer size, line throughput is also a decreasing function in CV. This can be seen from Figure 6.14, where two throughput curves are shown. The upper one is the curve for $CV = 0.25$ while the lower one for $CV = 0.35$. Both curves indicate that their decreasing rates are decreasing. This phenomenon can be explained by looking at a two-operation case. As discussed before, if the ratio of the two mean process times is moved away from one, line throughput becomes less and less sensitive to both process time variation and buffer size. (See Table 6.6.) By changing the ratio, the throughput can be arbitrarily close to the process speed at the bottleneck. When operation process times are identical, a line with $k + 1$ stages has a lower throughput than does a line with k stage. Both lines can be regarded as single composite operations. Let M_j be the mean process time of the composite operation derived from a line with j operations. Then $M_{k+1} \geq M_k$. If both lines are expanded by adding an operation at their ends, then the new lines can be approximated by two 2-stage lines, respectively. The additional operation has a mean process time of one. Thus the first line has a mean process time ratio equal to $1/M_{k+1}$ while the second one has a ratio $1/M_k$. Since the former ratio is deviated farther from one as compared to the latter, the line with $k+2$ operations is less sensitive to its last operation. This implies that the throughput reduction also becomes less, i.e., the decreasing rate is decreasing in N. □

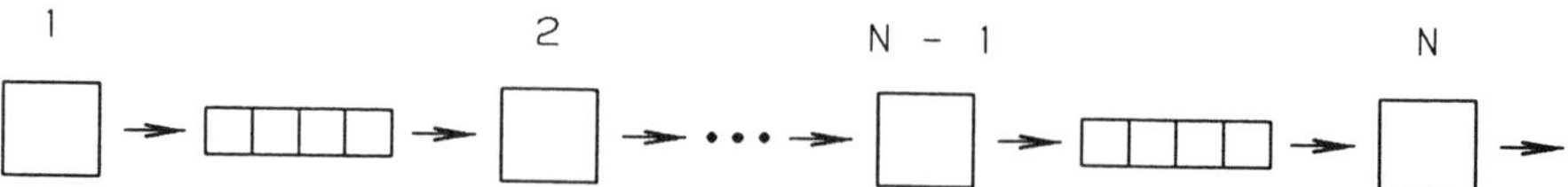

Figure 6.13 Tandem queuing model for an n-stage sequential line.

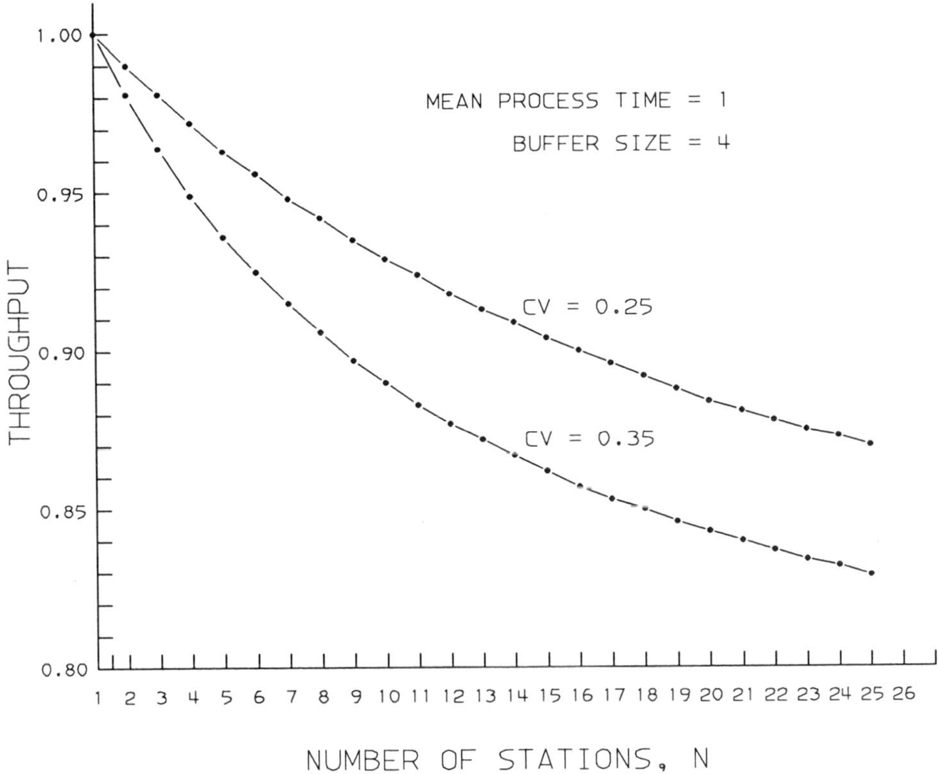

Figure 6.14 Throughput as a function of CV and n.

Since line throughput is a decreasing function of N, the number of operations in a production line should be kept small under a pull policy. If the total assembly time remains unchanged, the smaller number of operations the longer the operation process times will be. Hence, a buffer-design problem is related to the process-design problem.

From a buffer-allocation point of view, more buffer space will help increase line throughput by reducing interaction between operations. If a sufficient buffer space is placed between two operations, the line can be viewed as if there were two separate lines decoupled at the location of the buffer. Since the decreasing rate of throughput is also decreasing in N, it would be more beneficial to allocate more buffer space in the middle of the line. However, if operations have different process-time distributions, this principle may not hold. A buffer-design algorithm is discussed in the following section.

Buffer-Design Optimization

The heuristics in Algorithm 6.5 compute line throughput by progressively working with one operation at a time. From a buffer-design point of view, the optimal strategy is to keep just enough buffer so that line throughput is close to the process rate at the bottleneck. Under a normal condition, operations should be nearly decoupled by sufficient large buffers. On the other hand, the buffer size should be small so that a line problem can be quickly detected through buffer behavior.

Since Algorithm 6.5 can give us a good estimate and interactions between operations should be kept at a minimal level, a dynamic programming approach may help search for an optimal design solution. Again, the method considers one operation at a time and extends an intermediate solution to its final answer. In Equation (6.10), throughput is a function of D, C_1, C_2, and B. In a buffer-design problem, B is the controllable variable and the throughput is the objective value to be maximized. The dynamic programming algorithm must carry both *TP* and *CVD* throughout the entire computational process. Consider a line of n operations. For convenience, let

L_i = a line consisting of operations 1, 2, ..., i

$TP_i(x)$ = maximal throughput from L_i with a total buffer size of x

$CVD_i(x)$ = CV of the interdeparture time of L_i under the maximal throughput condition

The operation for maximization can be carried out by an auxiliary function,

$$G(T_1, C_1, T_2, C_2, B) = TP/T_1 \tag{6.12}$$

where *TP* is given by (6.10) with $D = T_2/T_1$, and given values of C_1, C_2, and B.

Needless to say, *TP* should be evaluated with modifications (a), (b), and (c). Sometimes, the size of a buffer between two operations must be confined to a given region because of a space or cost constraint. In this case, we let l_i and u_i be a pair of lower and upper bounds for the buffer size between operations i and $i+1$. If no constraints should be imposed, then $l_i = 0$ and $u_i = \infty$. Assume that the total buffer space for the entire line is Z. A buffer placement procedure is defined by Algorithm 6.6.

Algorithm 6.6

1. Compute $TP_2(x) = G(T_1,CV_1,T_2,CV_2,x)$, where T_i is the mean process time of operation i and CV_i is its CV. Do this for all $x \leq Z$ and $l_2 \leq x \leq u_2$.
2. Compute $CVD_2(x)$ by Equation (6.11) with modifications (a) and (b), for all x considered in step 1. Let $i = 2$.
3. $i = i + 1$. (Consider the next operation.)
4. For each $x \leq Z$ and $l_i \leq x \leq u_i$, compute

$$TP_i(x) = \max_{y \leq x} \{G(1/TP_{i-1}(x-y), CVD_{i-1}(x-y), T_i, CV_i, y)\}$$

   In the above relation, TP_{i-1} and CVD_{i-1} are obtained in the previous step. For each x, the optimal value of y is the one that maximizes the value of G. Denote this value by $y_i(x)$.
5. For every $y_i(x)$ obtained in step 4, compute $CVD_i(x)$ by (6.11) with $C_1 = CVD_{i-1}(y_{i-1}(x - y_i(x)))$, $C_2 = CV_i$, and $B = y_i(x)$.
6. If $i < n$, go to step 3. If $i = n$, the maximal line throughput is given by $TP_n(Z)$. Backtracking through the decision process, the buffer-placement policy can be obtained:
   a. Let $j = n + 1$ and $w = Z$.
   b. $j = j - 1$.
   c. Let $v_j = y_j(w)$. Then $w = w - v_j$. (v_j is the computed buffer size between operations $j - 1$ and j under the placement policy.)
   d. If $j = 2$, then stop. Otherwise go to step 6b.

Suppose that a perfectly balanced line has 10 operations. Process times of these operations are identical with a mean of one unit and a CV of 0.3. For Z = 0, 1, ..., 30, the buffer placement policies are listed in Table 6.8, where each column gives a policy determined by Algorithm 6.6. When Z = 28, for instance, the buffer between operations 4 and 5 has a size of four. It can be seen that when $Z < 9$, preference is always given to upstream operations. From the previous discussion, it is known that this policy is not optimal. However, the throughput difference between this policy and the optimal one (i.e., more buffer in the middle of the line) is normally small. When $Z = 1$, the simulated throughputs for $v_2 = 1$ and $v_6 = 1$ are 0.726 and 0.741, respectively. The relative difference is only 2.1%. The reader should be reminded that our method is not designed for a zero-buffer case. When dealing with practical problems, a nonzero buffer is always required unless $CV = 0$.

Table 6.8 Buffer-Placement Policy when $T = 1$ and $CV = 0.3$

Buffer between Operations	Total Number of Buffer Units																													
	01	02	03	04	05	06	07	08	09	10	11	12	13	14	15	16	17	18	19	20	21	22	23	24	25	26	27	28	29	30
1–2	1	1	1	1	1	1	1	1	1	1	1	1	1	1	1	1	1	2	2	2	2	2	2	2	2	2	3	3	3	3
2–3	0	1	1	1	1	1	1	1	1	1	1	1	1	1	1	2	2	2	2	2	2	2	2	2	2	2	3	3	3	3
3–4	0	0	1	1	1	1	1	1	1	1	1	2	2	2	2	2	2	2	2	2	3	3	3	3	3	3	3	3	3	3
4–5	0	0	0	1	1	1	1	1	1	2	2	2	2	2	2	2	2	2	3	3	3	3	3	3	3	3	3	4	4	4
5–6	0	0	0	0	1	1	1	1	1	1	2	2	2	2	2	2	2	2	2	3	3	3	3	3	3	3	3	3	4	4
6–7	0	0	0	0	0	1	1	1	1	1	1	1	2	2	2	2	2	2	2	2	2	3	3	3	3	3	3	3	3	4
7–8	0	0	0	0	0	0	1	1	1	1	1	1	1	2	2	2	2	2	2	2	2	2	3	3	3	3	3	3	3	3
8–9	0	0	0	0	0	0	0	1	1	1	1	1	1	1	2	2	2	2	2	2	2	2	2	3	3	3	3	3	3	3
9–10	0	0	0	0	0	0	0	0	1	1	1	1	1	1	1	1	2	2	2	2	2	2	2	2	3	3	3	3	3	3

Algorithms 6.5 and 6.6 deal with cases in which each operation has a single workstation. In many real-world problems, this may not be true. If a line is not perfectly balanced, an operation may have multiple stations. In this case, an "equivalent" mean process time should be used. This is done by dividing the mean process time by the total number of workstations at the operation. For example, an operation has two workstations with a mean process time of 10. Then the equivalent mean process time is 5. This is also the average time interval between two successive departures from the operation, when both workstations are busy.

Two real-life cases are briefly described in Examples 6.7 and 6.8. In both examples, the concept of the equivalent mean process time is adopted.

Example 6.7 A new production line has 10 operations, connected by a conveyor with an accumulation capability. Space on the conveyor between adjacent operations can be used as buffer. The operation process times, and the buffer sizes are given in the following:

Mean	0.86	0.74	0.91	0.58	0.96	0.93	1.00	0.81	0.79	0.98
CV	0.45	0.45	0.10	0.10	0.45	0.15	0.15	0.60	0.60	0.20
Buffer	—	1	2	1	2	4	4	4	2	2

Since buffer sizes are determined by the conveyor length and the sizes of workstations, increasing the buffer size means additional cost. The line designer needs to know if the existing buffers are sufficient and, if not, how much additional buffer space is needed.

The line may produce bad products. The yield is expected to be 0.98. Since the mean process time at the bottleneck is one time unit, the theoretical maximal throughput is 0.98. The objective is to achieve a 97% efficiency, that is, the throughput should be at least $0.98 \times 0.97 = 0.95$ under a finite buffer condition. Algorithm 6.6 is invoked with $l_i = 1, 2, 1, 2, 4, 4, 4, 2$, and 2, for $i = 2, 3, \ldots, 10$, respectively. If the $u_i = l_i$ for all i, then the algorithm simply gives the estimated throughput under the existing buffer arrangement. This value is found to be 0.917. The line is also simulated by using a computer program. The simulated throughput is 0.938, higher than the approximated solution by about 0.02. Since the estimated throughput is already close to 0.95, a small increment in total buffer size will achieve our objective. Furthermore, it is expected that this 0.02 difference likely remains the same after the small increment. Consequently, the target throughput may be set at 0.93.

Using the same lower limits, $\{l_i\}$, Algorithm 6.6 is invoked for different values of Z. The total buffer size under the existing policy is 22. Thus invocation of Algorithm 6.6 should be conducted with the same lower limits as described above and $Z > 22$. It is found that when $Z = 24$, line throughput is 0.93. The new policy is defined by

Buffer	—	1	2	1	4	4	4	4	2	2

That is, two additional units should be added to the buffer following operation 4. The simulated result under this new policy gives a line throughput 0.952, just slightly higher than the target. □

Example 6.8 An existing assembly line has 12 operations, including nine manual assembly operations and three automatic test operations. The test equipment represents more than 50% of the total capital investment. For this reason, the line manager has been trying to keep all test equipment as busy as possible. The previous record shows that the average production rate is about 200 pieces a day and the average total amount of WIP is nearly 400. The WIP distribution is given in the following:

Operation	1	2	3	4	5	6	7	8	9	10	11	12
Average WIP	84	31	19	11	32	22	69	24	5	6	54	20

The average production lead time is given by the ratio of WIP to throughput. In this case the lead time is approximately two days. On the other hand, the total process time (without waiting in the queue) is just a few hours. Management believes that the lead time can be shortened significantly. A pull system is installed. Buffer sizes are estimated by using Algorithm 6.6. The number of workstations per operation, the equivalent mean process times (in minutes), their CV's, and the buffer placement policy are given as follows:

Operation	1	2	3	4	5	6	7	8	9	10	11	12
Stations	2	3	2	1	2	5	1	3	1	3	25	1
Eq. Mean	5.1	5.0	3.0	4.2	4.8	2.9	4.2	4.0	4.8	4.0	4.6	4.2
CV	0.7	0.5	0.4	0.5	0.2	0.2	0.5	0.5	0.5	0.2	0.2	0.5
Buffer	—	6	4	4	6	2	2	2	2	2	4	2

After having negotiated with the line manager, the following agreement is reached:

1. Since the line operators belong to two separate departments and the partition is between operations 6 and 7, the buffer size between these two operations is increased to 23.
2. Operation 11 consists of 25 testers. The buffer size is increased to 30.
3. If a machine breaks down, the line manager has the authority to increase the buffer size ahead of the machine to minimize the risk of line stoppage.

With the above modifications, the line is running under a pull logic. The WIP level is then dropped down to the range from 140 to 200, with an average of 170. On the other hand, line throughput is actually increased slightly to about 210 a day. □

It should be pointed out that the solution method used in Algorithm 6.6 is not exact. If line throughput is considerably less than bottleneck capacity due to buffer constraints, the buffer-placement policy given by the algorithm may not be effective. In this case, no efficient method is known.

6.5 Remarks

Inventory theory has been studied and developed by many authors for over 50 years. The EOQ model is one of the simplest of all existing inventory models. Reviews and critiques in this area may be found in Aggarwal (1974) and Silver (1981). Difficulties of using an existing inventory model for practical problems come from cost structure and selection of control variables. Major cost items in a model include (1) holding cost (or carrying cost), (2) shortage cost, and (3) ordering cost (or setup cost). Even in a company with good financial records, accurate cost estimating may not be easy. The conventional way to optimize an inventory policy is to determine the best ordering quantity and reordering point. However, a significant cost reduction is achievable by shortening the lead time and reducing ordering cost. One cannot rely solely on mathematical models to control the lead time and ordering cost. To this end, a good relation with vendors is a key factor for success. Japanese experience showed that vendor relations play a key role in their inventory management program.

Conceptually, both MRP and kanban systems can be viewed as just-in-time systems, and should lead to inventory reductions. From a practical viewpoint, however, these two systems can be very different. An MRP system usually tends to accept the existing condition in a manufacturing line, and attempts to promptly adapt to changes or any variation from the line. On the other hand, a kanban system tends to force a given manufacturing line subject to a set of requirements, such as high quality, small buffer size, short setup time, etc. Comparisons of the two systems are given in Rice (1982) and Krajewski (1987). Detailed discussions about MRP can be found in Orlicky (1975) and Chalmet (1985). Discussions on kanban systems are given by Sugimori (1977), Kimura (1981), and McGinnis (1985).

A decision on kitting is difficult, and should be application-dependent. Factors that should be considered are part count and characteristics (size, shape, weight, etc.), product structure, workstation design and layout, space requirements, process design and so on. The concept of using inserts can increase flexibility in kit design. Engineering changes can only cause impact on inserts, but not on a whole kit. The concept of regulating feeding pace by recycling empty containers has been implemented in existing lines. Material flow for each individual part forms a closed-loop system. Since different parts are fed independently, synchronization must be done through a pull system, which is derived directly from the concept of kanban. The closed-loop system and the concept of logical kit were previously reported in Chow (1985).

Because of mathematical intractability, almost all existing works in the buffer- placement area deal with either a simple line (i.e., a very small number of operations) or a line with identical process times. Line structure is so far limited to sequential lines. The heuristic approach for buffer-placement strategy for sequential lines was previously presented in Chow (1987). Many authors have studied the same subject under various assumptions. To give a preferential treatment (such as higher process speed or more buffer space) to the middle operations was proposed by Hillier and Boling. They discovered that this policy is optimal (i.e., line throughput reaches its maximum) under the assumption of exponential process times. This fact is called *bowl phenomenon* in their papers: Hillier (1966) and Hillier (1967). The same observation was made in Yamashina (1983) under the assumption of normal process times. For more information on this subject, the reader may consult Anderson (1969), Buzacott (1978), Gershwin (1979), and Shanthikumar (1983).

REFERENCES

Aggarwal, S. C. (1974). A Review of Current Inventory Theory and its Application, *International Journal of Production Research*, v. 12, pp. 443–482

Anderson, D. R., and C. L. Moodie (1969). Optimal Buffer Storage Capacity in Production Line Systems, *International Journal of Production Research*, v. 7, pp. 233–240

Buzacott, J. A., and L. E. Shick (1978). Models of Automatic Transfer Lines with Inventory Banks—A Review and Comparison, *AIIE Transactions*, v. 10, pp. 197–207

Chalmet, L. G., M. De Bodt, and L. Van Wassenhove (1985). The Effect of Engineering Changes and Demand Uncertainty on MRP Lot Sizing: A Case Study, *International Journal of Production Research*, v. 23, pp. 233–251

Chow, W., E. A. MacNair, and C. H. Sauer (1985). Analysis of Manufacturing Systems by the Research Queuing Package, *IBM Journal of Research and Development*, v. 29, pp. 330–342

Chow, W. (1987). Buffer Capacity Analysis for Sequential Production Lines with Variable Process Times, *International Journal of Production research*, v. 25, pp. 1183–1196

Gershwin, S. B., and I. E. Shick (1979). Analysis of Transfer Lines Consisting of Three Unreliable Machines and Two Finite Storage Buffers, Technical Report, Laboratory for Information and Decision Systems, MIT, Cambridge, Mass.

Hillier, F. S., and R. W. Boling (1966). The Effect of Some Design Factors on the Efficiency of Production Lines with Variable Operation Time, *Journal of Industrial Engineering*, v. 17, pp. 651–658

Hillier, F. S., and R. W. Boling (1967). Finite Queues in Series with Exponential or Erlang Service Times—A Numerical Approach, *Operations Research*, v. 15, pp. 286–303

Kimura, O., and H. Terada (1981). Design and Analysis of Pull System, a Method of Multi-Stage Production Control, *International Journal of Production Research*, v. 19, pp. 241–253

Krajewski, L. J., B. E. King, L. P. Ritzman, and D. Wong (1987). Kanban, MRP and Shaping the Manufacturing Environment, *Management Science*, v. 33, pp. 39–57

McGinnis, L. F., J. Trevino, and Z. R. Toro-Ramos (1985). A Review of the Toyota Production System, MHRC-RS-85-01, Technical Report, Georgia Institute of Technology, Atlanta, Georgia

Orlicky, J. (1975). *Material Requirement Planning*, McGraw-Hill, New York

Rice, J. W., and T. Yoshikawa (1982). A Comparison of Kanban and MRP Concepts for the Control of Repetitive Manufacturing Systems, *Production and Inventory Management*, v. 23, pp. 1–15

Shanthikumar, J. G., and C. C. Tien (1983). An Algorithm Solution to Two-Stage Transfer Lines with Possible Scrapping of Units, *Management Science*, v. 29, pp. 1069–1086

Silver, E. A. (1981). Operations Research in Inventory Management: A Review and Critique, *Operations Research*, v. 29, pp. 628–645

Sugimori, Y., K. Kusunoki, F. Cho, and S. Uchikawa (1977). Toyota Production System and Kanban System, Materialization of Just-in-Time and Respect-for-Human System, *International Journal of Production Research*, v. 15, pp. 553–564

Yamashina, H., and K. Okumura (1983). Analysis of In-Process Buffers for Multi-Stage Transfer Line Systems, *International Journal of Production Research*, v. 21, pp. 183–195

7

Capacity Planning and Human Resource Management

One of the major difficulties in line design is to cope with the change of manufacturing conditions. For instance, line productivity will be different for different yield factors. If yield change is predictable, line capacity can be planned accordingly. Therefore, understanding the line improvement process is essential for effective resource planning. Unfortunately, not all manufacturing parameters are predictable. In many cases, the state of a parameter may be regarded as the sum of its *trend* and a *random fluctuation*. The algebraic expression is given by

$$X_i = m_i + Z_i \tag{7.1}$$

where i is a time index.

If X_i represents the yield during day i, m_i can be considered as the "expected" yield on that day and Z_i is the difference between the actual and the expected. For convenience, $E[Z_i]$ is assumed to be zero so that $E[X_i] = m_i$. For the purpose of resource planning, m_i should be accurately estimated. If a line can complete c assembly cycles per day, after i days of operation one can expect that the line is capable of producing cm_i assemblies per day. For a demand of D pieces per day, $I[D/(cm_i)]$ lines should be installed, where $I[x]$ is the least integer greater than or equal to x. Because of yield improvement, m_i is normally an increasing function of i. As a line design problem, resource planning may have to deal with the situation when both c and D are functions of time. In particular, c is also subject to an improvement process.

As another example, X_i can be the number of operators who report to duty on day i. Although a line usually has a fixed number of operators, X_i is not a constant because of absenteeism. In this case $m_i = m$, i.e., is independent of time. Therefore, the number of available operators is determined by Z_i. If m is the total number of operators, $-Z_i$ can be interpreted as the number of absentees and $E[Z_i] < 0$. A line manager may plan for m operators and expect $m + E[Z_i]$ of them to be available. However, there is no guarantee that a sufficient number of operators will be available every day. In this case, no trend is observable, i.e., m is a constant. One way to resolve a shortage problem is to plan for more resource. A better alternative is to adjust operation policy so that a line can be adapted to short-term fluctuations.

In this chapter, two major subjects will be discussed: (i) improvement process, and (ii) job assignment. The former is a key factor for manufacturing resource planning and line capacity planning. The latter is a good example of an adaptive operation policy that attempts to maximize line throughput by adjusting job-assignment patterns.

7.1 Improvement Process

The improvement process has been studied for over fifty years. This concept can be depicted by a curve which is referred to by different names—improvement curve, learning curve, progress curve, experience curve. The concept of an improvement process is used to set up labor standards. A trainee's performance is never comparable to that of a skilled operator. A learning period must be allowed for a new operator to gain familiarity with a task. The relation between an operation process time and its improvement process is given by Equation (2.26), that is,

$$X_n = bn^a \tag{7.2}$$

where X_n is the process time of the nth operation cycle experienced by the operator.

In Equation (7.2), the improvement speed is characterized by the parameter a. For most operations, $-0.50 \leq a \leq -0.15$. Although this relation has been widely used to characterize a long-term trend, its predictability is weakened by short-term fluctuations.

A revised model may be derived from (7.1). Consider n consecutive periods. A period can be an operation cycle, a production day, a week or a month, and so on. The cumulative values over the first n periods is given by

$$\sum_{i=1}^{n} X_i = \sum_{i=1}^{n} m_i + \sum_{i=1}^{n} Z_i \tag{7.3}$$

Without losing generality, one may assume that $E[Z_i] = 0$. For if $E[Z_i] \neq 0$, Equation (7.1) may be redefined as

$$\begin{aligned} X_i &= (m_i + E[Z_i]) + (Z_i - E[Z_i]) \\ &= m_i' + Z_i' \end{aligned}$$

Thus the new variable Z_i' has a zero mean.

Since, for $n \to \infty$, $\sum_{i=1}^{n} Z_i/n \to E[Z_i] - 0$,

$$\frac{\sum_{i=1}^{n} Z_i}{\sum_{i=1}^{n} m_i} = \frac{\sum_{i=1}^{n} Z_i/n}{\sum_{i=1}^{n} m_i/n} \longrightarrow 0 \qquad \text{for } n \longrightarrow \infty$$

The second term on the right-hand side of Equation (7.3) is negligible, at least for a large n. Consequently,

$$\sum_{i=1}^{n} X_i \approx \sum_{i=1}^{n} m_i$$

From this relation and (7.1), it is concluded that while X_i is subject to a random fluctuation, $\sum X_i$ can be approximated by a deterministic model. If m_i is a power function in the form of (7.2), $\sum m_i$ can also be characterized by a power function, that is,

$$\sum_{i=1}^{n} b i^a = d n^c + O(c-1) \tag{7.4}$$

where c and d are constant, and $O(k)$ is a polynomial with a power of k.

For simplicity, the low-power terms may be dropped. Therefore, a deterministic improvement model can be defined by

$$Y_n = \sum_{i=1}^{n} X_i = dn^c \tag{7.5}$$

Since random fluctuation has less impact on the model based on the cumulative value, model (7.5) usually has better predictability than model (7.2). On the other hand, both (7.2) and (7.5) have the same structure, and their computational requirements are comparable. Therefore, (7.5) is preferable to (7.2). In the following sections, the validity of improvement processes and their applications in resource planning will be discussed.

7.2 Line Capacity Planning

Equation (7.5) can be rewritten as a linear equation:

$$\log Y_n = \log d + c \log n \tag{7.6}$$

Thus, if $\{Y_n\}$, or equivalently $\{X_n\}$, is known, the linear regression method defined by (2.17), (2.18) and (2.19) may be employed to estimate c and d.

7.2.1 Yield Factor

Three sets of yield improvement data have been collected from existing lines over a 39-month period. These data are plotted in Figure 7.1, and named as processes 1, 2 and 3 (or P1, P2 and P3), respectively. On the horizontal axis the time scale is by month while the vertical axis is the average yield during each month. None of these three sets of data can be closely fitted by smooth curves. When Equation (7.2) is converted into a logarithmic scale, a linear equation is obtained:

$$\log X_i = \log b + a \log i \tag{7.7}$$

Regression analysis is employed to estimate coefficients in (7.7) for each of the three data sets. The three data sets and their corresponding regression lines are given in Figure 7.2. It can be seen that the regression

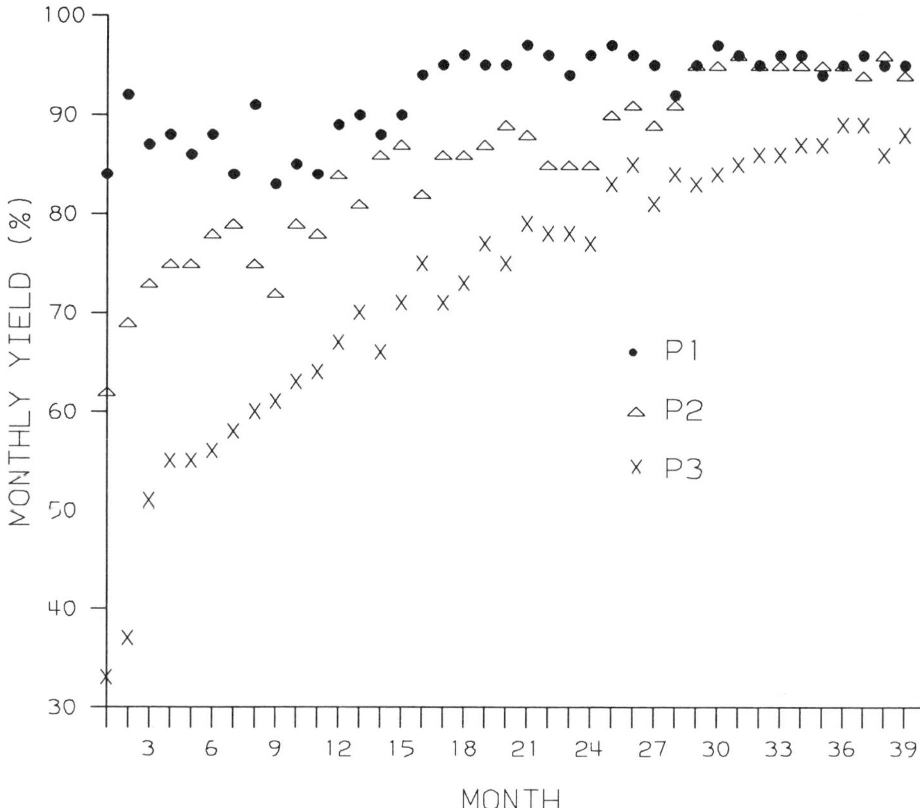

Figure 7.1 Yield improvement process.

lines can very well characterize yield improvement trends. But, on the other hand, data points are scattering around the lines. The estimated values for $\log b$ and a are listed in Table 7.1. In addition, the correlation coefficient of $\log X_n$ and $(\log b + a \log n)$ are also computed by Equation (2.16). The regression model does not have a high correlation coefficient for P1.

The same data are used for cumulative yield analysis. The regression lines and data points of $Y_n = \sum X_i$ are given in Figure 7.3. It is clearly shown that the three straight lines fit their data points very well. As previously mentioned, little fluctuation can be observed. The estimated values for the coefficients in model (7.6) are presented in Table 7.2.

From a resource planning viewpoint, model predictability is important. One way to test predictability is to compare the model parameters for

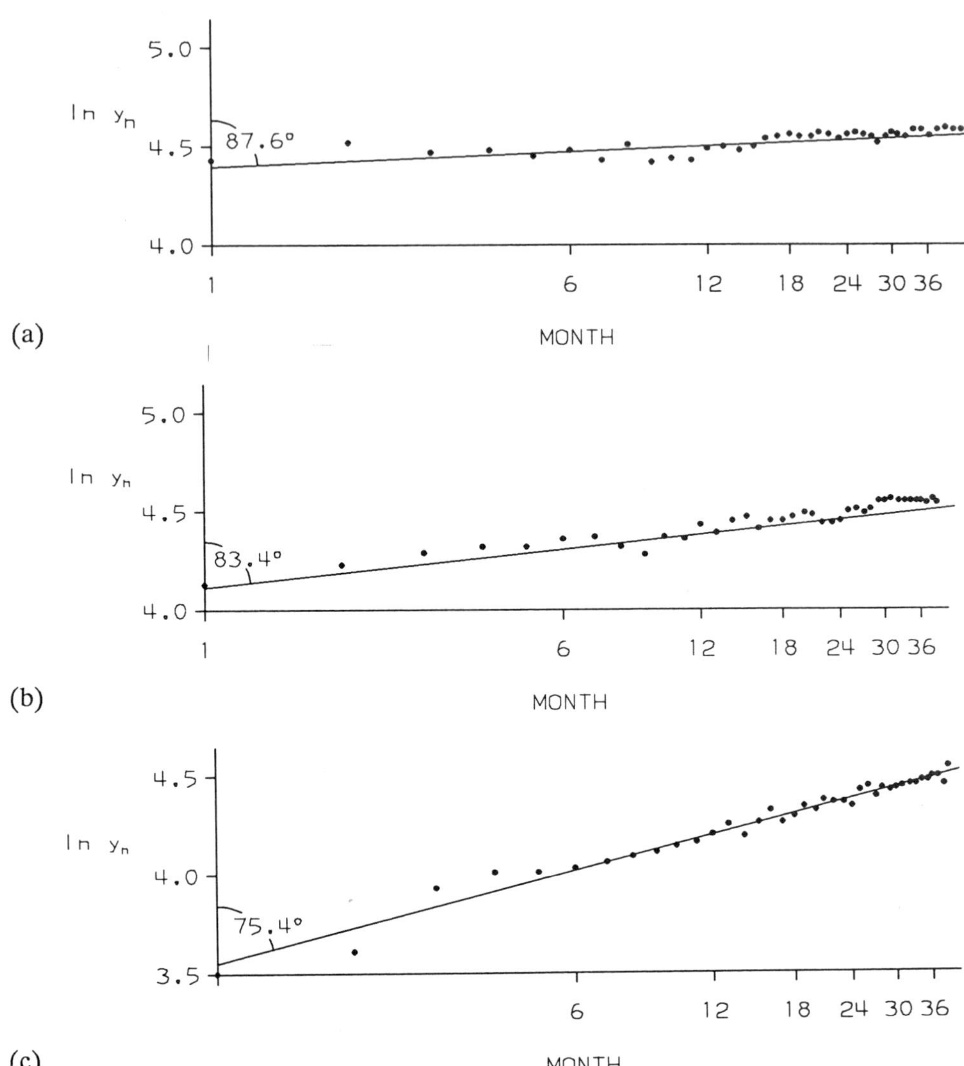

Figure 7.2 Yield in log scale and regression. (a) P1. (b) P2. (c) P3.

different values of n. For instance, model parameters can be estimated based on data observed over the first 6 months, 12 months, 18 months, and so on. The values of $\log b$, a, $\log d$, and c are given in Table 7.3 for different n. When (7.6) is invoked, six data points (i.e., $n = 6$) already characterize the improvement trend well. This is true because both $\log d$ and c do not change

Table 7.1 Regression Parameters for Model (7.7)

Process	log$b$	a	Corr. Coeff.
P1	4.4094	0.0416	0.741
P2	4.1254	0.1160	0.955
P3	3.5502	0.2605	0.987

very much as n increases from 6 to 39. On the other hand, model (7.7) is less desirable due to a relatively large change in the value of a. For process 1, the values of correlation coefficients are very poor, particularly when n is small. The correlation coefficient never exceeds 0.7 before n reaches 36.

Predictability of model (7.6) can be verified by comparing the actual yield with the predicted yield. Figure 7.4 compares the actual data observed over the 39-month period with the predicted values, based on the first six-month observations (given by smooth curves). Good predictability is observed in all three cases.

In both (7.6) and (7.7) the line slope is directly related to the rate of improvement. In (7.7) a level line or a slope of zero indicates cessation of improvement. This is the case when $a = 0$. The yield factors of all periods are identical to b. The condition of constant yield in model (7.7) corresponds to a 45-degree line or a slope of one, that is, $c = 1$. This is equivalent to the situation that $Y_n = dn$ or $X_n = d$ for all integers n. Both Figures 7.2 and 7.3 show that P1 has the highest initial yield but the slowest improvement rate. On the other hand, P3 has the lowest initial yield and the fastest improvement rate.

The rate of improvement over a given period of time usually is not a constant. A manufacturing process may experience several improvement periods over a product lifetime. A close inspection of the change of improvement rate should be done by scrutinizing local fluctuation. Data given in Figure 7.1 can be redefined by looking at defect rates, $d_i = 1 - X_i$. Denote the cumulative defect rate from period i through j by D_{ij}. Then

$$D_{ij} = \sum_{k=i}^{j} d_k = \sum_{k=i}^{j} (1 - X_k) \tag{7.8}$$

Analogous to (7.6), we have

$$\log D_{ij} = \log h_{ij} + g_{ij} \log(j - i + 1) \tag{7.9}$$

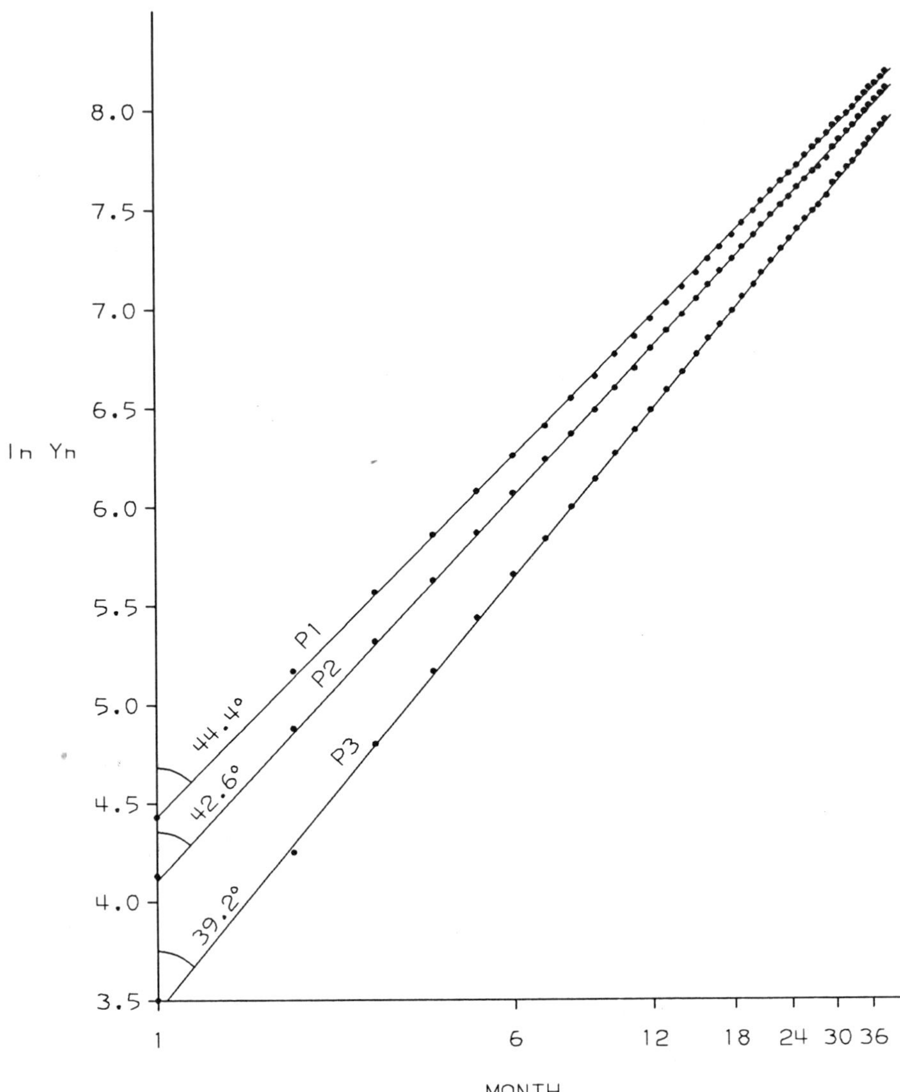

Figure 7.3 Cumulative yield in log scale.

Table 7.2 Regression Parameters for Model (7.6)

Process	log*d*	*c*	Corr. Coeff.
P1	4.4283	1.0229	1.000
P2	4.1145	1.0870	1.000
P3	3.4547	1.2254	1.000

Table 7.3 Regression Parameters for Different Models

Process	*n*	log*b*	*a*	*Corr. Coeff.*	log*d*	*c*	Corr. Coeff.
P1	6	4.4606	0.0097	0.212	4.4451	1.0195	1.000
	12	4.4718	−0.0056	−0.126	4.4563	1.0057	1.000
	18	4.4377	0.0222	0.403	4.4551	1.0065	1.000
	24	4.4152	0.0379	0.616	4.4458	1.0128	1.000
	30	4.4091	0.0417	0.689	4.4373	1.0180	1.000
	36	4.4087	0.0420	0.728	4.4308	1.0216	1.000
	39	4.4094	0.0416	0.741	4.4283	1.0229	1.000
P2	6	4.1393	0.1228	0.986	4.1264	1.0839	1.000
	12	4.1637	0.0901	0.874	4.1342	1.0749	1.000
	18	4.1504	0.1013	0.917	4.1322	1.0765	1.000
	24	4.1523	0.1004	0.930	4.1279	1.0795	1.000
	30	4.1382	0.1088	0.941	4.1238	1.0820	1.000
	36	4.1260	0.1156	0.951	4.1175	1.0854	1.000
	39	4.1354	0.1160	0.955	4.1145	1.0870	1.000
P3	6	4.1254	0.1160	0.955	3.4606	1.2235	1.000
	12	3.5247	0.2779	0.965	3.4636	1.2203	1.000
	18	3.5361	0.2685	0.974	3.4641	1.2198	1.000
	24	3.5419	0.2648	0.980	3.4620	1.2213	1.000
	30	3.5411	0.2654	0.984	3.4588	1.2232	1.000
	36	3.5449	0.2633	0.987	3.4558	1.2248	1.000
	39	3.5502	0.2605	0.987	3.4547	1.2254	1.000

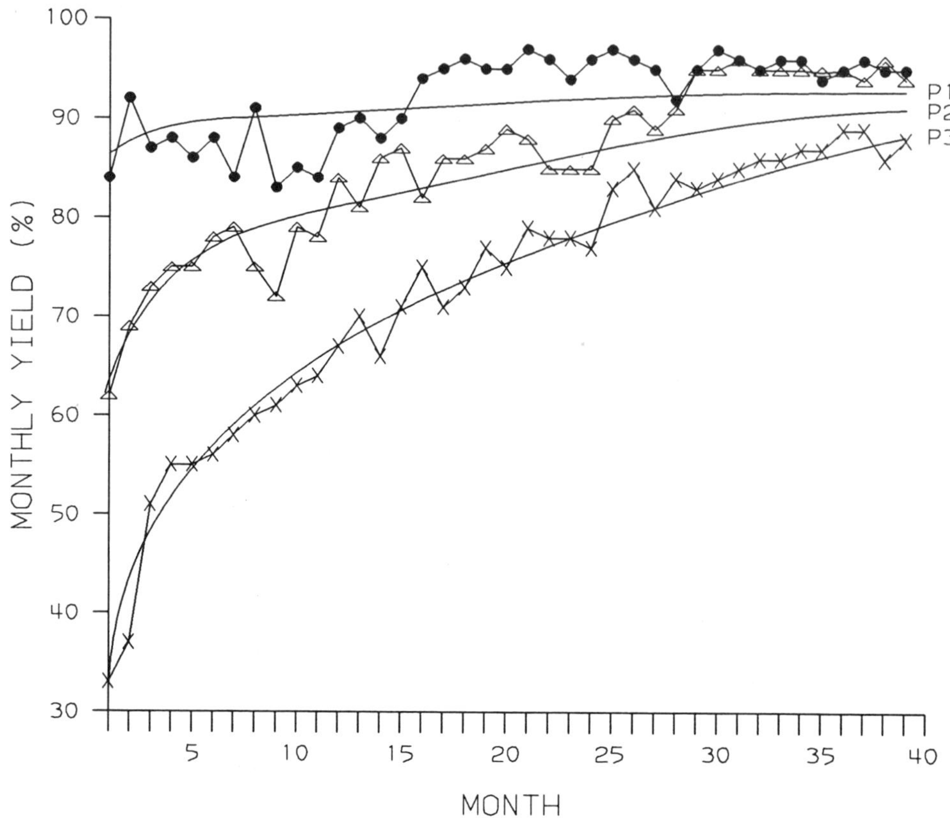

Figure 7.4 Comparison of actual and predicted yields.

Note that the coefficients g and h are dependent on i and j. By changing i and j, Equation (7.9) defines a piecewise linear regression curve. Again, a 45-degree line implies either a plateau or a stable condition. Table 7.4 presents the regression curves for P1, P2 and P3 respectively, by listing the slope of each period during which the improvement rate remains nearly constant. The curves are drawn in Figure 7.5. The first process, P1, starts with a slow improvement. After having enjoyed a period of fast improvement, yield becomes stable at about 98%. P2 has experienced two improvement periods. Between these two periods, yield did not change very much. This phenomenon can be observed from Figure 7.3, where P2 has a stable yield factor from period 15 through period 24. After 29 months, the process yield is stabilized. P3 has a relatively long improvement period. The

Table 7.4 Changes in Yield Improvement Rate

Process	Interval	Slope	Degree	Comment
P1	1—11	0.9623	43.9	small improvement
	11—17	0.7823	38.1	improvement
	17—39	0.9880	44.7	stable
P2	1—15	0.8244	39.5	first improvement
	15—24	0.9926	44.8	plateau
	24—29	0.7848	38.1	second improvement
	29—39	1.0000	45.9	stable
P3	1—13	0.8245	39.5	first improvement
	13—24	0.9221	42.7	improvement slowdown
	24—34	0.8725	41.0	more improvement
	34—39	0.9617	43.9	approaching stable

initial yield is very low (at 33%). It takes over 39 months to reach an average yield of about 88%. The process experienced different improvement rates over different periods, and has not been stabilized after 39 months.

The improvement process plays an important role in resource planning. A simple example follows.

Example 7.1 A production line has been in operation for six months. The current line capacity is 55 assemblies per day. Manufacturing records show that the past six-month yields are 0.33, 0.37, 0.51, 0.55, 0.55, and 0.56, respectively. It is expected that in two years the daily demand will be 300 pieces per day. Since the current line has already used expensive equipment at full capacity, line management wants to know how much additional equipment must be installed to meet the future demand. Equation (7.5) is employed to predict production yield in two years. Based on the six-month data, the model is given by

$$Y_n = 0.318n^{1.2235}$$

Hence, the estimated yield at $n = 30\ (= 6+24)$ becomes $Y_{30} - Y_{29} = 0.318[(30)^{1.2235} - (29)^{1.2235}] = 0.829$. For 300 pieces a day, the total number of pieces of equipment should be $300/(0.829 \times 55/0.56) \approx 4$. □

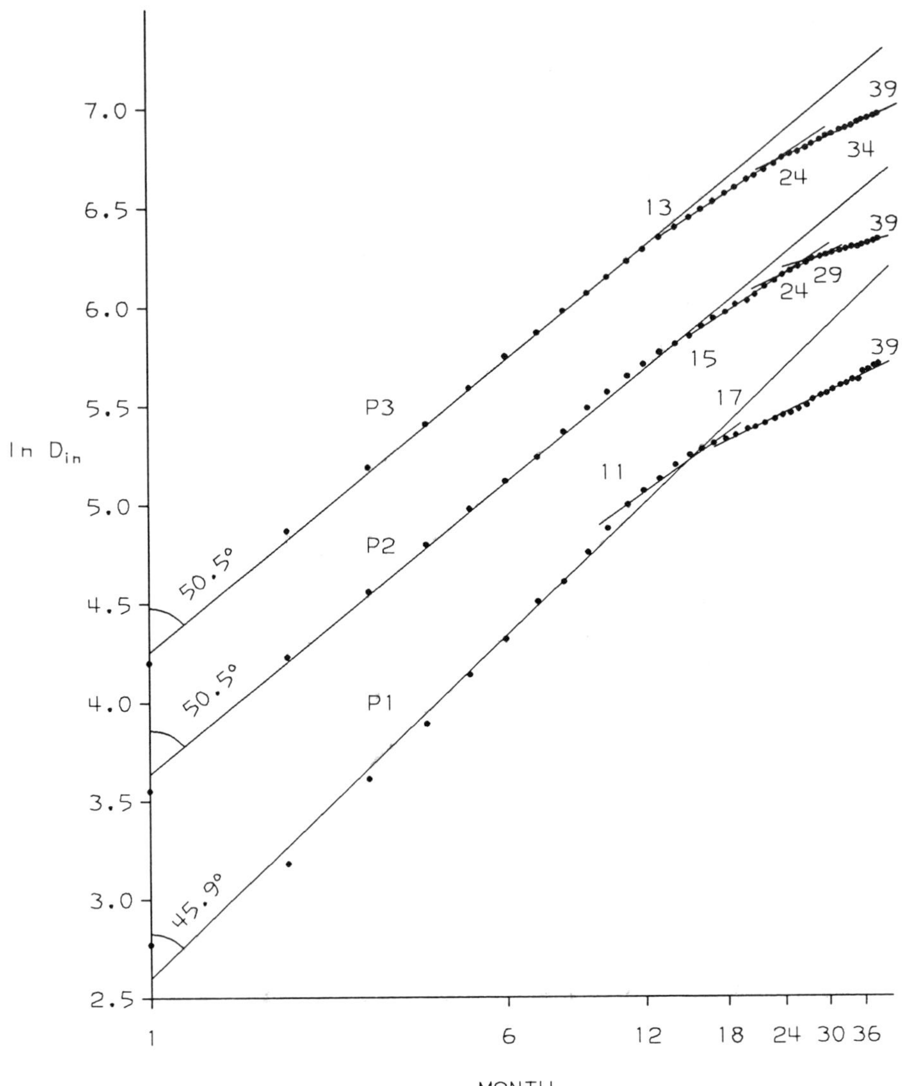

Figure 7.5 Cumulative defect rate in log scale.

7.2.2 Operator Behavior

Both models (7.2) and (7.5) may also be used to characterize the operator improvement process. An operator's improvement can be measured by either throughputs in different time periods (e.g., days or weeks) or the average operation process times of different operation cycles. One of the cases reported in Glover (1966) contains daily throughputs of a radio valve assembly task, which is reconstructed in Figure 7.6. It can be seen that the process is becoming stable after about 40 days. After 30, 40 or 50 days, the average daily throughputs are about 75.57, 75.55, and 74.35, respectively. On the other hand, their corresponding standard deviations are 9.62, 10.74 and 8.95, respectively. Daily throughput may vary from about 50 pieces to about 95 per day. Because of daily fluctuation, data points are scattered around the straight line defined by (7.7). This is illustrated in Figure 7.7, where the straight line is given by

$$\log X_n = 2.9122 + 0.3689 \log n \tag{7.10}$$

where X_n is the throughput on day n.

Note that due to the cluster property of data points after 40 days, little improvement is observed.

When model (7.6) is employed, the cumulated throughput, $Y_n = \sum X_i$, can be approximated by

$$\log Y_n = 2.5915 + 1.3869 \log n \tag{7.11}$$

Comparison of actual cumulated throughput and its estimated value is shown in Figure 7.8. Again, a slope of 45 degrees means cessation of improvement. The regression line has a slope of 54.2 degrees. After 40 days, data points begin to bend down toward a line of 45 degrees, shown by the dashed line in the figure. The mean and the standard deviation of these 27 data points (from day 41 to day 67) are 75.55 and 10.74, respectively. If the random fluctuation is subject to a normal distribution, a 90-percent confidence interval can be constructed by using a t-distribution with 26 degrees of freedom, and is given by $75.55 \pm 1.71 \times 10.74/\sqrt{27}$ or (72.02, 79.08). The estimated average throughput by (7.11) at $n = 40$ is 76.78.

Example 7.2 A line designer wants to estimate the capacity of a new assembly line over a given period. The assembly task is performed manually. Testing and inspection operations are inserted into the

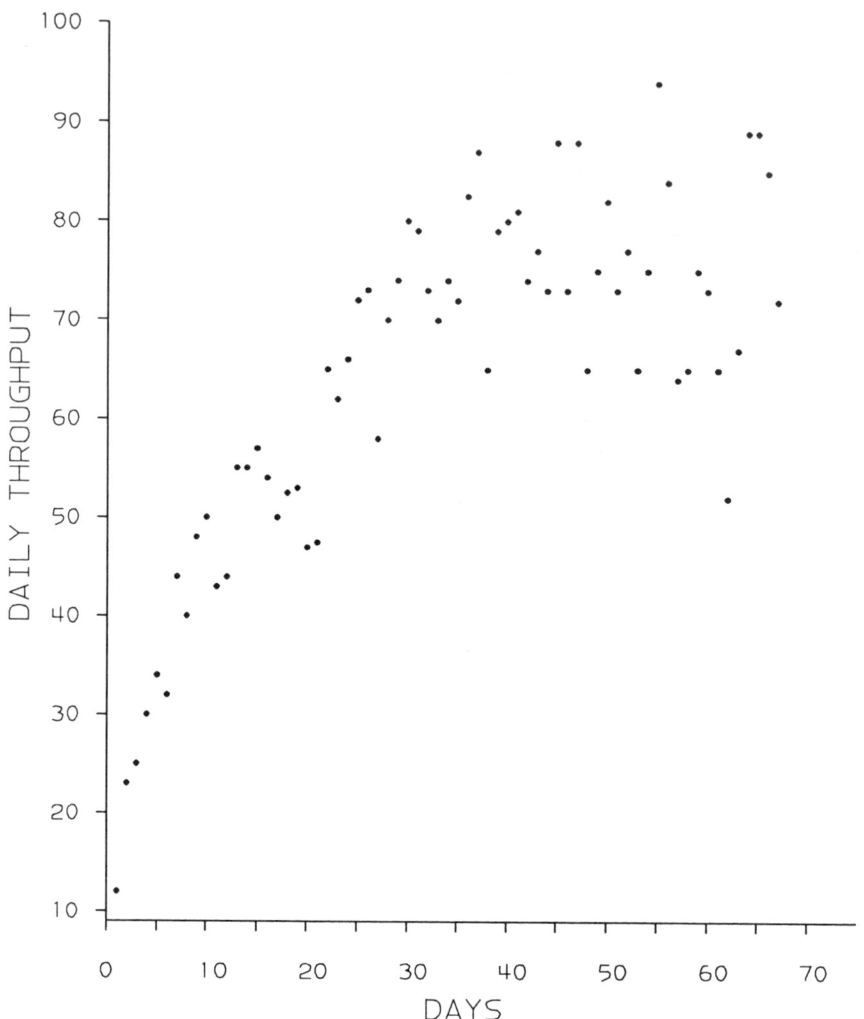

Figure 7.6 Daily performance of a trainee (a reconstruction based on Fig. 1.1 in Glover 1966).

assembly process to assure good quality. Bad products are sent to rework operations and then reintroduced to the production line later. The assembly process consists of 10 operations (P1—P10) and four rework operations (R1—R4), as illustrated in Figure 7.9. The branching probabilities from one operation to another can be derived

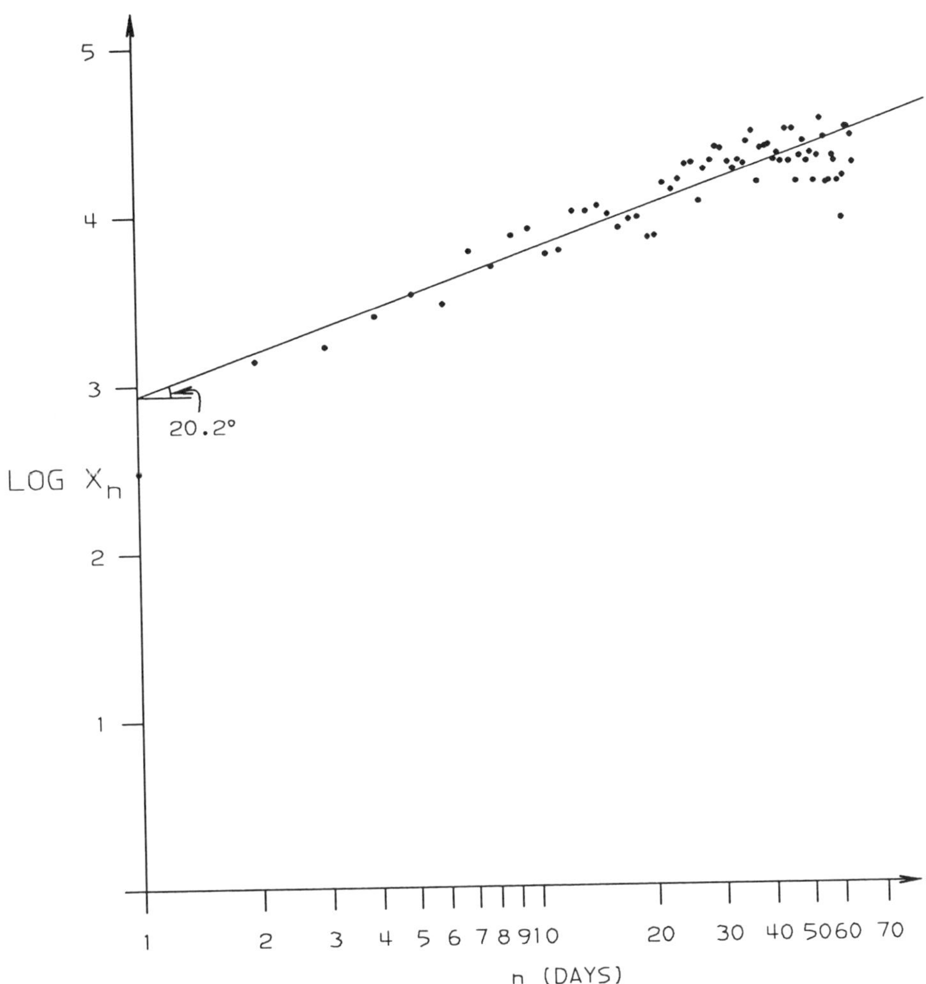

Figure 7.7 Throughput improvement in log scale.

from yield estimation. The numbers shown in Figure 7.9 are the branching probabilities, based on the ultimate yields. It is expected that 4.34% finished goods will be sent back for quality problems.

Engineers are asked to estimate (i) the initial yield, (ii) the ultimate yield, and (iii) the length of the improvement period for each operation. Therefore, an improvement curve can be constructed. For example, the initial and the ultimate yields at P8 are 0.65 and 0.98, re-

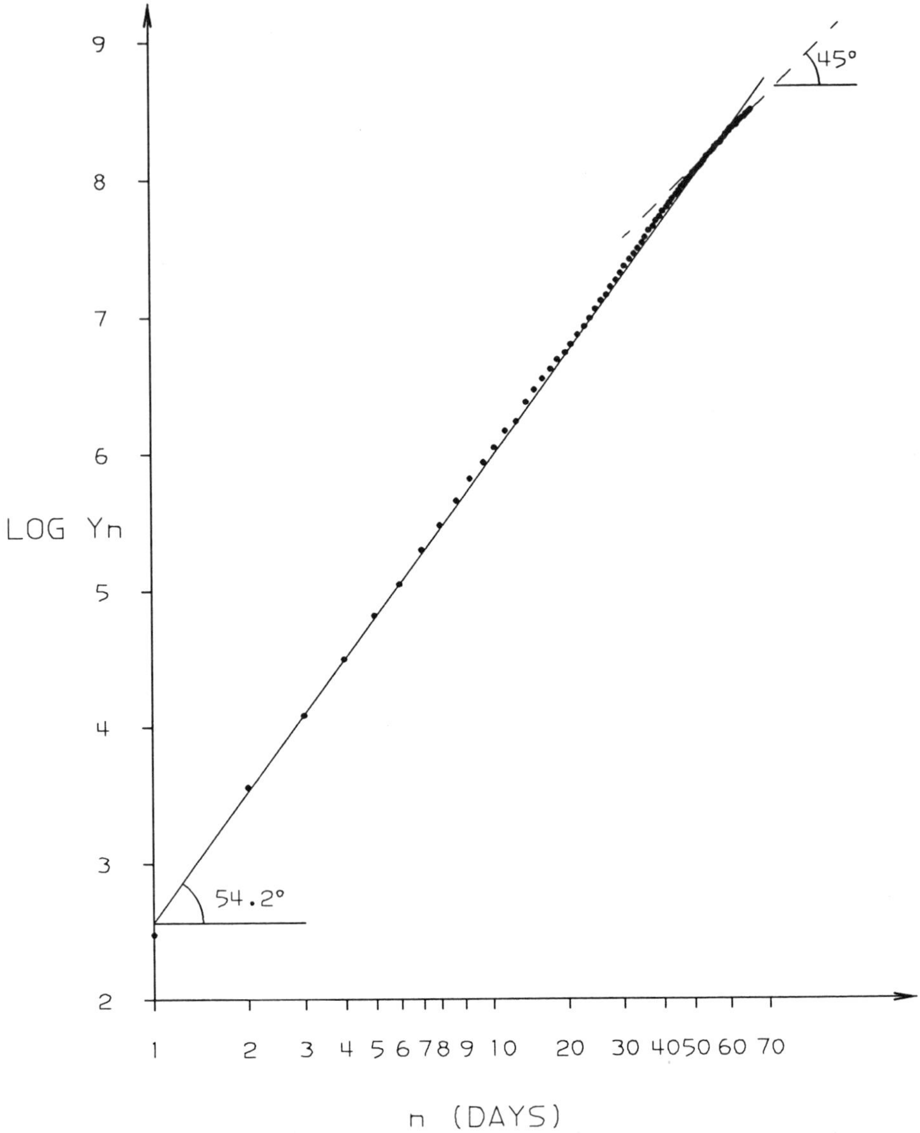

Figure 7.8 Cumulated throughput in log scale.

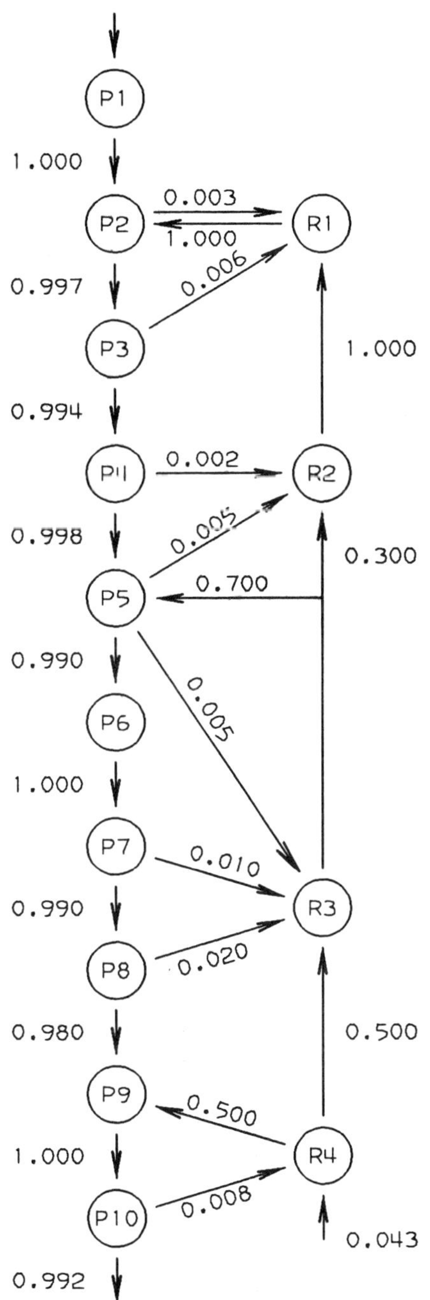

Figure 7.9 The assembly process in Example 7.2.

spectively, and its improvement period may last for three years or 12 quarters. For n = 1 and 12, model (7.5) gives us two equations with two unknowns:

$$\text{For } n = 1,\ Y_1 = 0.65 = d$$

$$\text{For } n = 12,\ Y_{12} - Y_{11} = 0.98 = d[(12)^c - (11)^c]$$

The solution of these two equations is given by d = 0.65 and c = 1.1213. The improvement curve for P8 is defined by

$$Y_n = 0.65n^{1.1213}$$

Thus, the estimated yields can be obtained and listed as follows:

Quarter	1	2	3	4	5	6	7
Cum. Yld.	0.65	1.41	2.23	3.08	3.95	4.85	5.76
Yield	0.65	0.76	0.81	0.85	0.88	0.90	0.91
Quarter	8	9	10	11	12		
Cum. Yld.	6.69	7.64	8.59	9.56	10.54		
Yield	0.93	0.95	0.96	0.97	0.98		

The same approach is used to establish an improvement process for every operation. Then for each quarter, the start factor of each operation can be estimated by solving a set of simultaneous linear equations. (See Section 4.4 for a discussion of start factor.) For instance, based on the ultimate yield, the computed start factors are given by

P1	P2	P3	P4	P5	P6	P7	P8
1.000	1.034	1.032	1.026	1.068	1.057	1.057	1.047
P9	P10	R1	R2	R3	R4		
1.052	1.052	0.036	0.026	0.063	0.052		

The operator improvement process can be constructed in a similar fashion. Instead of dealing with each individual operation, the line designer decides to characterize *group* improvement behavior. A simple approach is to assume fixed operation process times but improvement in daily productive hours. The concept of such an approach can be interpreted as follows: During the improvement process an operation process time can be, at least conceptually, divided into two parts: (i) the basic time requirement for all necessary motions, and (ii) the additional time requirement due to operator's inefficiency, such as trying to remember job steps, looking for misplaced tools or assembly parts, and so on. Since only the first part is considered to be productive time, improvement is achieved by reducing the amount of time contributed by the second part. In other words, the total amount of time that an operation spends for the basic requirement each day is increasing.

For a three-shift operation, the amount of productive time is initially 14.3 hours per day. It takes five quarters to reach 20.6 hours a day. Using Equation (7.5), the estimated productive times for five consecutive quarters are 14.3, 17.4, 18.9, 19.8 and 20.6 hours, respectively. After the fifth quarter, daily productive time is fixed at 20.6 hours.

Let

S_i = average process time at operation i
T_j = daily productive hours in the jth quarter
e_{ij} = start factor at operation i in the jth quarter

The capacity of operation i in quarter j is given by

$$c_i = T_j/(S_i e_{ij})$$

Then the line capacity in quarter j is equal to the capacity at the bottleneck, i.e.,

$$C = \min\{c_i \mid \text{all } i\}$$

The average process time of each operation in minutes is given by

P1	P2	P3	P4	P5	P6	P7	P8	P9	P10	R1	R2	R3	R4
4.2	4.5	4.1	3.6	4.5	4.7	5.0	5.0	4.6	4.7	4.6	4.5	4.1	4.2

The numerical values of line capacity are listed as follows:

Quarter	1	2	3	4	5	6	7	8	9	10	11	12
Capacity	66	102	144	174	203	216	223	227	231	233	234	234

If a demand curve is available, these numbers can be used to determine the number of production lines. □

It can be seen from the last example that operator improvement is as important as yield improvement. An accurate estimation of the degree of improvement is essential to a good resource planning program.

7.2.3 Shift Strategy

The concept of group improvement has been adopted in Example 7.2, where the degree of improvement is measured by total productive time. In addition to operator improvement, a number of other factors may affect the amount of daily productive time. From a workstation point of view, common time detractors may include scheduled (or preventive) maintenance, unscheduled maintenance (i.e.repair time), tool calibration, engineering use for experiments, setup, cleanup, breaks, lunches, and operator absence. Obviously, productive time is a decreasing function of these detractors. A trivial example is to gain productive time by improving workstation reliability or reducing the absenteeism rate. If a line is labor-intensive, sometimes it is worthwhile to investigate shift strategy. Although three-shift operation gives the maximum amount of productive time per day, poor operator utilization can be observed. The trade-off is between the labor cost and the tooling/space cost. This situation can be clearly illustrated by a numerical example.

Example 7.3 An assembly process consists of 20 operations. Since the process can be balanced at a cycle time of 5 minutes, multiple lines may be installed such that each operation has exactly one workstation and each workstation needs one operator. Assembly parts are manually delivered to each operation from the parts staging area. Two parts distributors are required to support total traffic. Operator cost, including wage, benefit, overhead due to supervision and supporting personnel, etc. is estimated to be \$50,000 per person per year. Therefore, the total labor-related cost becomes $50,000 \times 22 = 1.1$ million per shift.

The capital investment of a line includes \$2.2 million in tooling and material-handling equipment, and \$440,000 for space. This implies that labor cost of 2.4 shift-years is equivalent to a single line capital investment.

During an eight-hour shift, operators are allowed to have two 15-minute breaks. The lunch time is 40 minutes, which is not included in the eight-hour period. The workstation time detractors per shift are as follows:

Scheduled maintenance	0.1667 hours
Unscheduled maintenance	0.0333
Setup and clean up	0.0667
Calibration	0.1667
Engineering use	0.0667
Total	0.5000 hours

Operator availability is estimated to be 98%. The 2% loss is due to sickness, absenteeism, department meetings, and the like.

Under the assumption of three-shift operation, two consecutive shifts will have 40 minutes overlap due to lunch time. As an example, the first shift may start at 7:00 a.m. and is ended at 3:40 p.m. The second and the third shifts are from 3:00 p.m. to 11:40 p.m. and from 11:00 p.m. to 7:40 a.m., respectively. The expected daily productive time is $98\% \times [24 - 3 \times (0.5 + 2 \times 0.25 + 40/60)] = 18.62$ hours. The terms separated by plus signs in the parentheses are workstation detractor time, breaks, and lunch time. Since the expected line throughput is one piece for every 5 minutes, three-shift operation should be able to produce $18.62 \times 60/5 = 223$ pieces on the average. Since three shifts require 66 operators, productivity is $223/66 = 3.38$ pieces per operator.

For the two-shift case, the line can be operated in two disjointed 8-hour periods, e.g., from 6:00 a.m. to 2:40 p.m. from 6:00 p.m. to 2:40 a.m. There will be no shift overlap. Scheduled maintenance,

Table 7.5 Comparison of Different Shift Strategies

Shift	Three	Two	Difference
Number of Lines	3	4	1
Number of Shifts	9	8	−1
Number of Operators	198	176	22
Annual Labor Cost	\$9.90M	\$8.80M	\$1.10M
Capital Investment	\$7.92M	\$10.56M	\$2.64M

calibration and engineering experiments can be postponed to off-shift times. The expected daily productive time becomes $98\% \times [16 - 2 \times (0.1 + 2 \times 0.25)] = 14.50$ hours. Hence, the daily throughput is 174 and operator's productivity is $174/44 = 3.95$. Compared to the three-shift case, the productivity is increased by about 17%. By paying an eight-hour shift, the total gain in productive time is only 4.12 ($= 18.62 - 14.50$) hours.

Suppose that the total production demand is between 650 and 700 pieces a day. Comparison of different shift strategies is given in Table 7.5.

If the two-shift policy is adopted, the additional capital investment is \$2.64 million. Since the annual saving from labor cost is \$1.1 million, the payback period is $2.64/1.1 = 2.4$ years, or the return on investment is 41.66%. □

7.2.4 Capacity Adjustment through Line Rebalancing

In Chapter 4, line balancing techniques have been discussed. For a given set of work elements and their precedence relations, and a predetermined cycle time, a line balancing problem is to group work elements into operations such that (i) every operation process time is no greater than the cycle time, (ii) the sequence of operations does not violate the precedence relations, and (iii) the difference between the total process time and the product of the cycle time and the number of operations is minimal. Normally, the line capacity is inversely proportional to the cycle time. For instance, if a line has a 19-hour productive time, a five-minute cycle time implies a capacity of 228 pieces a day. Obviously, by changing the cycle time, the line capacity will be different (unless multiple workstations per operation is allowed). Hence, line capacity is adjustable by rebalancing the line.

In the case of variable production demand, rebalancing a line may be an economical solution for capacity adjustment. Let us consider the assembly

Table 7.6 Line Balancing Under Different Cycle Times

Cycle time (min.)	6	7	8	10	15	30
No. of operations	7	5	4	3	2	1
Balance efficiency	0.71	0.86	0.94	1.00	1.00	1.00
Line capacity	190	163	143	114	76	38

job depicted in Figure 4.1, where a precedence diagram of work elements is shown. By varying the cycle time, the result of line balancing will be different. Solutions obtained by Algorithm 4.2 (Largest Set Rule) are given in Table 7.6, where for each cycle time the minimal number of operations, balance efficiency, and line capacity are shown. For instance, for a seven-minute cycle time, at least five operations are needed. If one day has 19 productive hours, the planned line capacity will be $19 \times 60/7 = 163$ assemblies per day. Since the total process time is 30 minutes, line balance efficiency is $30/(7 \times 5) = 0.86$. The solutions listed in Table 7.6 are illustrated in Figures 7.10 through 7.15.

The last row of the table defines a set of feasible solutions for different demands. Assume that each operation has exactly one workstation. The maximum line capacity is limited to 190, since the largest work element takes 6 minutes. If each operation requires exactly one operator, a total of 7

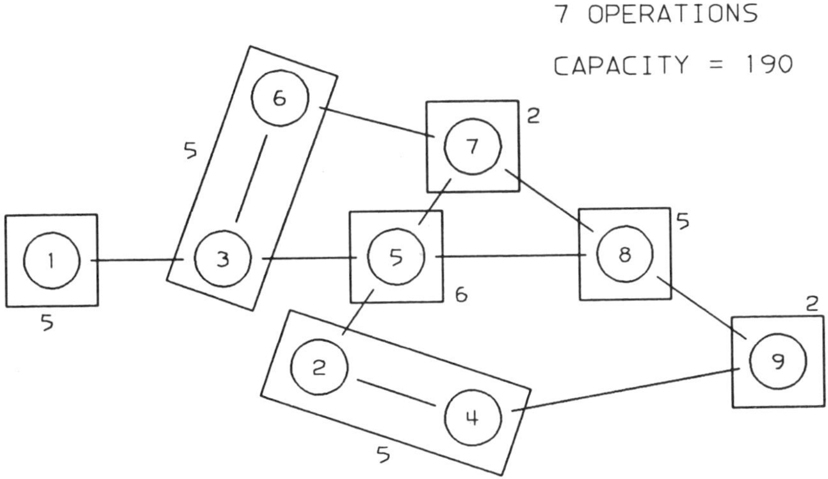

Figure 7.10 A line balance solution with a cycle time of 6 minutes.

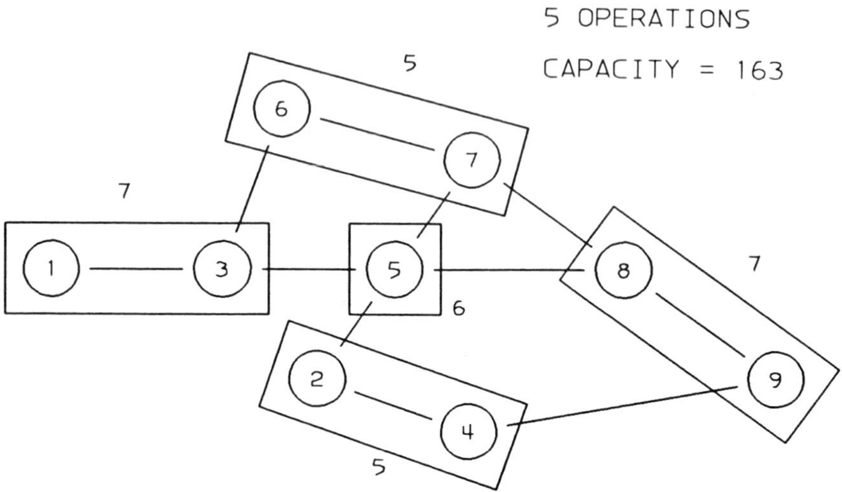

Figure 7.11 A line balance solution with a cycle time of 7 minutes.

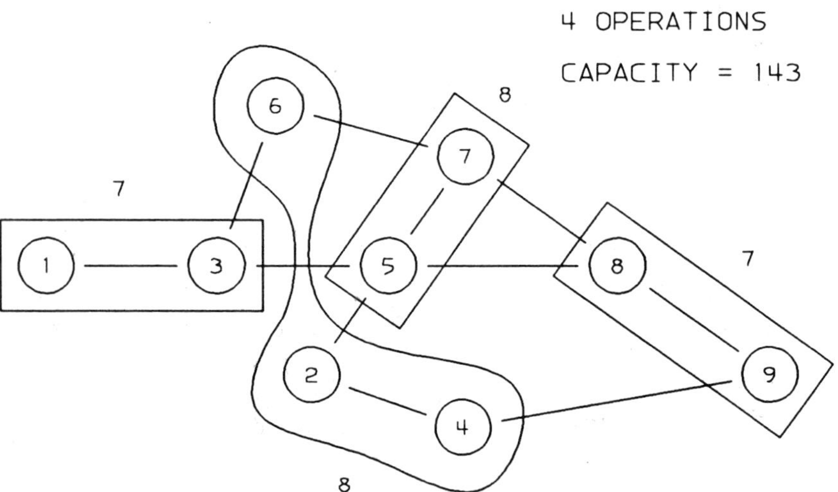

Figure 7.12 A line balance solution with a cycle time of 8 minutes.

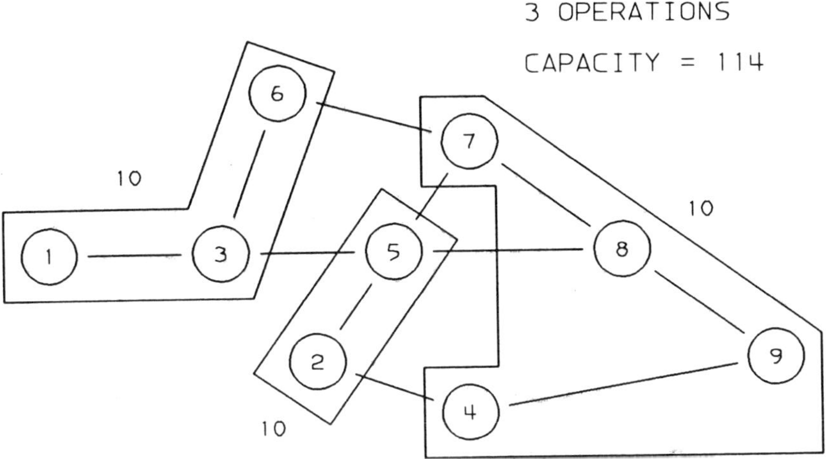

Figure 7.13 A line balance solution with a cycle time of 10 minutes.

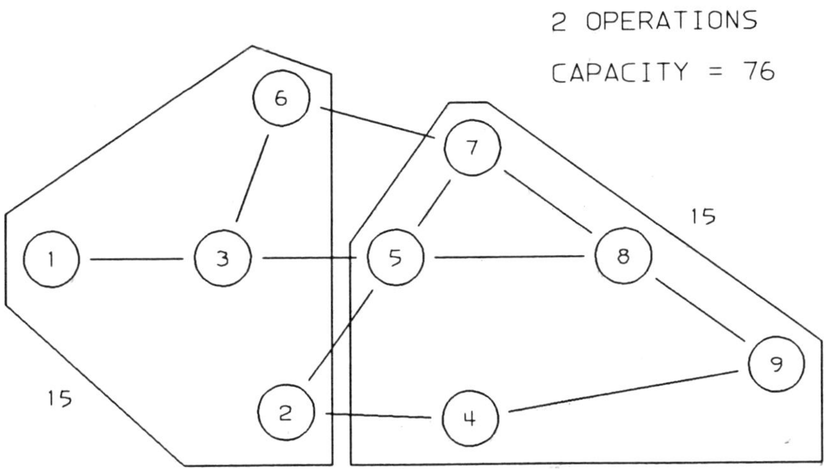

Figure 7.14 A line balance solution with a cycle time of 16 minutes.

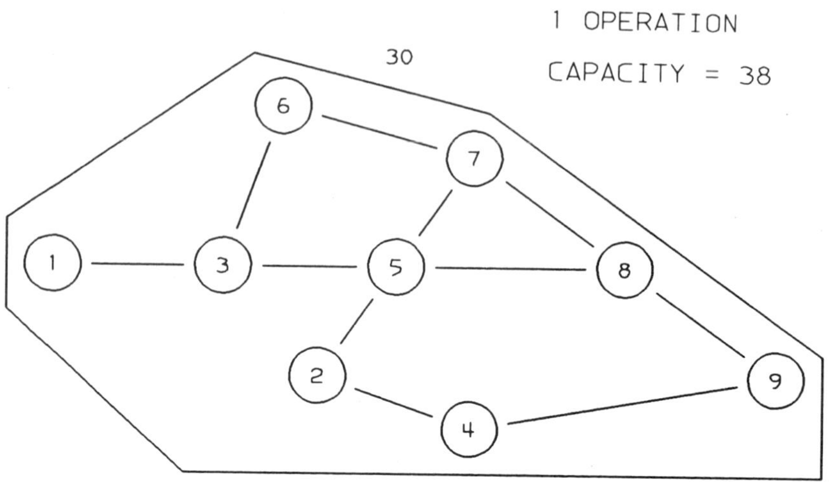

Figure 7.15 A line balance solution with a cycle time of 30 minutes.

operators is needed. If the production demand is decreased, say 100 pieces a day, the line can be rebalanced and the minimum number of operators can be reduced to 3. In this case, the line has a cycle time of 10 minutes. Line utilization is expected to be 0.88 (= 100/114).

Although the concept of line rebalancing is simple, its implementation may involve many other considerations. First, each job step in the entire assembly process must be clearly defined. This includes the work content of each step, tooling and fixture requirements, space requirements, material-handling method, parts requirements, operator skill levels, and buffer requirements. Next, tool design must be flexible enough so that once a line is rebalanced, tools can be redistributed or reassigned to different operations. Third, parts logistics should be considered. If a workstation contains too many part numbers, the operator may have difficulty accessing assembly parts. Two approaches may resolve this problem. One is to adopt the concept of inserts, introduced in Chapter 6, so that the total parts-storage area is reduced. The other relies on a local material-handling device for automatic parts indexing or parts feeding. Since work content is different after the line is rebalanced, each operator must take a cross-training program. Finally, no matter how a job is defined, ergonomic factors must be carefully considered: Is the job too long or or too short? Does the working area cover too much space for one operation? Do different job steps (in the same operation) require different working environments?

When dealing with a real-life problem, a line balancing algorithm can be computerized. Whenever production demand is drastically changed, a solution can be obtained quickly. If the line is designed for rebalancing, this approach may lead to a significant saving. In particular, a labor pool can be established for several different production lines. Once in a while, the structure of each line may be examined by rebalancing its job content, based on forecasts of short-term demand. Although the number of operators in each production line varies, the total number of operators in the pool may not be changed significantly if different product types are complementary. From an operator's viewpoint, line rebalancing simply means job reassignment. In a real-life production environment, reassignment should be done on a daily or weekly basis so that operators can refresh their skills. On the other hand, rebalancing also implies a different production schedule. From a logistics point of view, it is much easier to maintain a fixed schedule than an unstable one. Therefore, rebalancing should not be done too often. Some factories consider a schedule frozen for about a month. Hence, a possible policy is to have a new job assignment every week, but to perform line balancing every four weeks if necessary.

Finally, it should be pointed out that line rebalancing is also valuable for line expansion and contraction. When a new product is introduced, a line may have a relatively small number of (long) operations because of low demand. Later if the demand is increased, the process may be redefined by allowing a greater number of (short) operations. When the product is about phased out, the line can be rebalanced again for a smaller number of operations.

7.3 Training Program

As mentioned in Chapter 2, different operators usually perform differently. The different improvement behaviors may create two problems. The first is the problem of line balancing. For a manual operation, the process time is estimated by empirical methods. Therefore, a line can only be balanced in an average sense, i.e., the average (or standard) process time of each operation is nearly the same. Since operators are different, the actual line seldom behaves as a balanced one. This problem will be considered in Sections 7.4 and 7.5. The second problem is related to the improvement process. If a line has two or more operators, the group improvement process is dominated by the slowest learner. On the other hand, it is known that performance behavior of an operator may change for a different job assignment. If each operator's improvement behavior is predictable, it is then possible to design a training program to speed up the group improvement. This subject will be discussed next.

Cost of training can be a significant item in the total manufacturing cost. First, low productivity during a training period results in poor utilization of production facilities such as machines, equipment, and space. Next, inexperienced operators tend to produce more defective products, and consequently to increase the cost of rework and scrap. In addition, a trainee may cause more damage to fixtures, machines, and equipment than does a skilled operator. Finally, past statistics show that inexperienced operators usually have high accident rates.

The accident rate and work damage may be reduced by emphasizing discipline, but productivity can be increased by shortening the training period. An effective training program may consist of a standard for the improvement process, an incentive system based on the standard, appropriate supervision, and a good skill match for training items.

Obviously the characteristics of an improvement process depends on the job content. The improvement period of a complex job is longer than that of a simple one. For an existing job, the standard for improvement can be established, based on past observations. Note that both (7.2) and (7.5) are defined by three parameters: (i) the initial performance, (ii) the rate of improvement, and (iii) the duration of the improvement process. Suppose that a number of operators have been previously observed. For each of them, an improvement model is constructed by using the least-squares method. Then a standard improvement model can be obtained by taking the median or the average performance. For instance, if m individual improvement processes have been observed, then for each of the three parameters, a sample of size m has been obtained. Let

$\bar{d}$ = median initial performance (e.g., first-day throughput)
$\bar{c}$ = median improvement rate
S = average improvement period

The improvement standard is characterized by

$$Y_n = \bar{d} n^{\bar{c}} \qquad \text{for } n \leq S \tag{7.12}$$

If no previous records are available, the standard may be established by using a synthetic method or simply by experiment. By synthesis, it is assumed that improvement processes of motion elements are available. A job can be first broken into a number of motion elements. Let $t_i(n)$ be the average time to complete element i for the nth cycle (i.e., element i has been repeated n times). The average job process time during the nth cycle is $T(n) = \sum_i t_i(n)$. Hence, $\{T(n)\}$ defines the entire improvement process.

The length of the improvement period is the longest one among all the improvement periods of motion elements. This method requires information on $\{t_i(n)\}$. However, once the improvement standard of motion elements is established, this method can be used for any new job.

It has been discussed that model (7.6) may be used to predict improvement trends, given that a few initial measures can be made. A straightforward approach is to have a number of operators working on the same job and to take a few measures. For example, four-hour throughput may be measured for a job with an initial process time of one-half hour. For a two-day experiment, four data points are collected for each operator. Then for each operator an improvement process is defined. The medians of their initial throughputs and improvement rates define a standard improvement model. The length of the improvement period can be estimated by using the standard improvement model and the estimated ultimate job process time. The latter may be derived from a time and motion study, such as methods-time measurement (see Maynard 1948). Assume that the ultimate average process time is T and the standard improvement model is given by $Y_n = \bar{d}n^{\bar{c}}$, where Y_n is the cumulative throughput at the nth measurement cycle. The estimated improvement period, S, must satisfy the equation.

$$L/T = \bar{d}[S^{\bar{c}} - (S-1)^{\bar{c}}]$$

where L is the measurement cycle length.

Since S is the only unknown in this relation, its value is uniquely determined. Usually, a numerical method may be used.

Once the standard improvement curve is determined, an incentive program may be installed. The ultimate standard throughput is given by L/T. For the kth measurement cycle, the operator allowance is given by

$$w_k = (L/T) - \bar{d}[k^{\bar{c}} - (k-1)^{\bar{c}}]$$

The allowances for different cycles are illustrated in Figure 7.16.

If an operator produces z pieces during cycle k and $z + w_k > L/T$, a bonus proportional to the difference should be paid to the operator. When an incentive method is appropriately implemented, a significant cost saving may be expected due to a reduction in the improvement period. For example, Turban (1968) reported that "of the 23 workers checked, 15 reduced the planned learning period and 8 used the exact allowed time," and "total savings during the first year were at least twice the total expenditures."

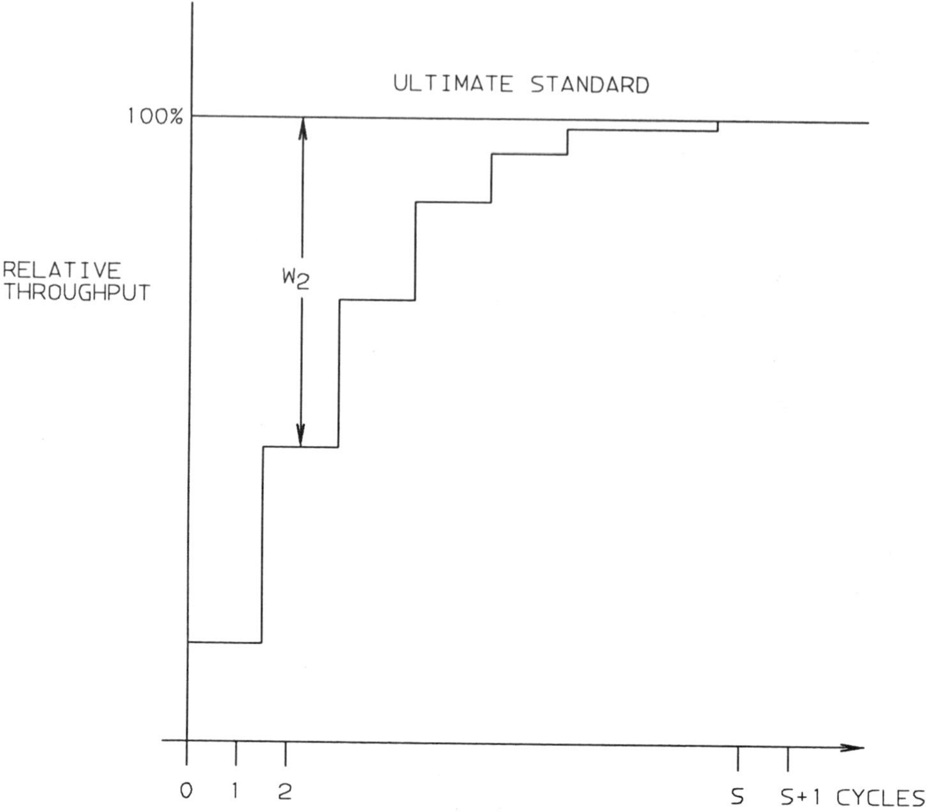

Figure 7.16 Operator learning allowance.

Supervision through the improvement process is also important. A typical improvement process has a decreasing improvement rate with multiple plateaus. Existence of plateaus may be due to job complexity, inadequate training, operator's personal problems, lack of proper supervision, or simply a random cause. Once a plateau is in the process of development, it is important to investigate its causes and to take remedial actions accordingly. Otherwise, line throughput will be kept at a relatively low level. When an inappropriate manufacturing procedure becomes habitual, the relearning process is usually much slower than the learning process. If model (7.6) is used, detection of a plateau can be done by comparing the plotted data points with a 45-degree straight line. If a local plot of three or four consecutive points are close to a 45-degree line, a potential plateau exists. This situation is illustrated in Figure 7.17, where two plateaus are shown and the second plateau leads to cessation of improvement.

Empirical data show that operator improvement processes do vary. Individual behavior is a function of age, educational background, past experience, and personality. The initial performance and the rate of improvement are the two key factors in the characterization of an improvement process. Statistically, it is then possible to select high-performance operators for a given task by investigating each individual's learning behavior. A simple approach is to let operators try the same task for a few short periods. The length of a trial period is dependent on the target task. For example, a period of two hours may be considered for a task of five minutes process time. Operator throughput for each period is observed. Using

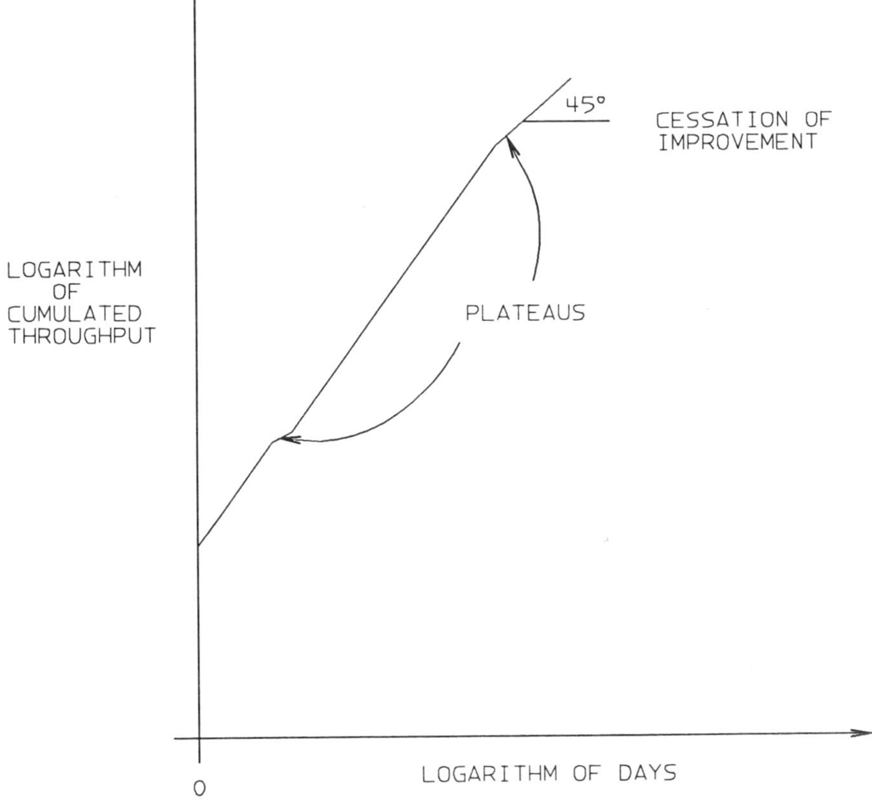

Figure 7.17 Improvement curve with plateaus.

model (7.6), operator improvement patterns can be characterized based on 4–6 observations. Each improvement pattern is compared with the standard improvement process. An operator is in the fast group if his or her improvement pattern is better than the standard one. This concept can be illustrated by the following example.

Example 7.4 A new assembly line is staffed with 12 operators. Since past experience shows that training costs can be high, the line manager is interested in training-cost reduction. All operators are asked to work on the same task for six 2-hour periods. Their throughputs are recorded and improvement patterns are developed by using model (7.6). For operator i, a pair of (c_i, d_i) are estimated by the least-squares method. The values of $\{c_i\}$ and $\{d_i\}$ are as follows:

i	1	2	3	4	5	6
d_i	5.6	6.6	5.0	5.2	5.6	4.6
c_i	1.21	1.23	1.40	1.33	1.26	1.21
i	7	8	9	10	11	12
d_i	6.4	6.4	5.2	4.4	5.4	5.2
c_i	1.28	1.33	1.30	1.40	1.30	1.26

The median of $\{d_i\}$ is 5.3 and the median of $\{c_i\}$ 1.29. The standard improvement process is defined by

$$Y_n = 5.3n^{1.29}$$

or

$$\log Y_n = 1.67 + 1.29 \log n \tag{7.13}$$

where Y_n is the cumulated 2-hour throughput.

The target training period is eight weeks, or 160 measurement periods of two hours each. Therefore, the expected number of products completed during a training period is $Y_{160} = 3694.80$, or $\log Y_{160} =$ 8.21. Operators can be classified into two groups. The fast group is composed of better-than-average operators. Figure 7.18 compares

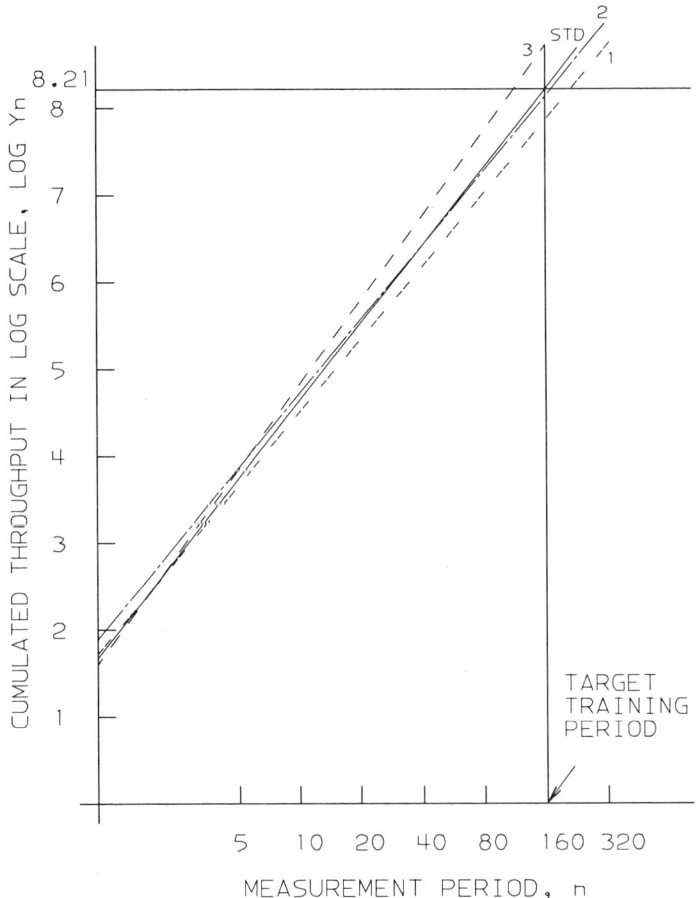

Figure 7.18 Estimated improvement patterns.

the improvement processes of the first three operators with the standard one (solid line). It can be seen that operator 1 has a poor performance, at least initially. The estimated improvement period for operator 2 is slightly longer than the target training period. Operator 3 is expected to reach the standard in 112 measurement periods and therefore belongs to the fast group. The estimated improvement period for operator *i* is defined by

$$S_i = (3694.80/d_i)^{1/c_i} \tag{7.14}$$

Numerical values of $\{S_i\}$ are as follows:

Operator i	1	2	3	4	5	6
S_i	214	171	112	139	173	252
Operator i	7	8	9	10	11	12
S_i	144	119	156	123	152	183

The average (estimated) improvement period is 161.50. If only the top six operators are allowed to continue their training program, the sum of their expected training periods is $(112+119+123+139+144+152)\times 2 = 1578$ hours. Since the other six operators have spent $6\times 6\times 2 = 72$ hours, the total expected training hours is $1578+72 = 1650$. Under a random selection policy, this value becomes $6\times 161.5\times 2 = 1938$ hours. Hence the expected saving is 288 hours. □

In the above example, the amount of saving is dependent on operator homogeneity. If operators have very much different backgrounds, a large saving can be expected. On the other hand, individual improvement is estimated by an empirical method. The observed results may be different from what has been estimated. An operator who was classified in the fast group may turn out to be a below-average performer. Learning statistics of 21 trainees in spools-preparation jobs for rug-weaving looms were reported by McGehee (1948). Their improvement periods varied from 6 to 14 months, with an average of 10.82 months and a coefficient of variation of 0.24. The improvement pattern of each individual was determined and given in Glover (1966b), based on model (7.6) with three data points. (Each data point is equivalent to one week's output.) Subsequently, nine trainees should be classified as better-than-average operators because of their short learning periods. Their actual learning periods are 6, 6, 6, 6.5, 10.5, 10.75, 10.75, 14, and 14 months, respectively. Seven out of nine cases are less than 10.82, i.e., about 80% (= 7/9) accuracy. The total saving, compared to a random selection policy, is $9\times 10.82-(6+6+\cdots+14) = 12.88$ months.

7.4 Job Assignment

It was explained in Section 7.2.4 that a production line may be rebalanced for capacity adjustment. Line balance is a key factor for cost saving. In reality, it may not be easy to achieve this objective. First, process times of work elements may not be accurately estimated, particularly for manual operations. Second, a line may not be perfectly balanced due to various constraints such as manufacturing environments, product structure, workstation design, etc. Additionally, workstation failure, operator shortage, product yield, bad parts, and difference in operators may also cause problems.

Table 7.7 presents operator performance statistics, observed from an existing assembly line with 10 operations. Operations 1 and 2 are rework operations, and the last two are testing operations. The other operations involve manual assembly tasks. For operation i, hourly throughputs of N_i operators are recorded, and their mean (M_i), coefficient of variation (CV_i), and skewness index (SK_i) are computed. The last three columns in the table are the minimum-to-mean ratio (A_i), the maximum-to-mean ratio (B_i), and the maximum-to-minimum ratio (C_i), respectively. To avoid using extreme values, the 90-percentile is used for the maximum and the 10-percentile for the minimum.

Although hourly throughput varies from operation to operation, the coefficient of variation ranges from 0.2 to 0.4. The two testing operations have relatively small CV-values and symmetric distributions, i.e., their SK-values are close to zero. This may be due to automatic testing equipment. The difference between the maximum and the minimum is rather large for all operations.

Because different operators perform differently, operator variation should be treated as a parameter for line design problems. Often engineers have to try very hard on their tool design in order to achieve line balancing. Unfortunately, benefits from their efforts can be easily destroyed due

Table 7.7 Operator's Performance Statistics

i	N_i	M_i	CV_i	SK_i	A_i	B_i	C_i
1	19	3.23	0.39	0.47	0.53	1.46	2.76
2	18	4.98	0.31	0.08	0.70	1.41	2.00
3	22	12.88	0.34	0.62	0.58	1.34	2.31
4	21	38.68	0.38	0.12	0.45	1.42	3.13
5	23	7.34	0.30	1.45	0.74	1.31	1.78
6	25	6.69	0.24	0.38	0.72	1.27	1.77
7	22	4.12	0.34	2.10	0.73	1.24	1.70
8	23	8.80	0.22	0.19	0.72	1.16	1.62
9	16	11.51	0.20	−0.07	0.76	1.22	1.59
10	15	12.49	0.21	0.01	0.77	1.30	1.69

to operator variation. Therefore, a reasonable solution may allow small differences between average operation process times during the line-design phase, and use job-assignment techniques to improve operation idle time and line throughput. According to the statistics shown in Table 7.7, if these differences are no more than 30%, the line may be reasonably balanced by invoking an effective job assignment program.

Consider a problem of assigning m operators to n (sequential) operations. Suppose that operator performance at each operation is known. The problem is to find the best job-assignment policy such that the line throughput is maximal. Assume that each operation needs exactly one operator. A complete assignment exists only if $m \geq n$. If $m < n$, a dynamic assignment policy should be considered—after a short period operators will be reassigned. Discussion on this subject will be given in Section 7.4.2. In the next section, we assume that $m \geq n$, and only "static" assignment policies are considered.

7.4.1 Job Assignment for Line Balance

Our problem can be converted into a *maximal flow problem*, and therefore Algorithm 3.2 may be used to find the optimal solution. Let c_{ij} be the throughput rate of operator i at operation j. Consider a network that consists of $m+n+2$ nodes, numbered by 0, 1, 2, ..., $m+n+1$. The network has the following structure:

1. Node 0 is a source and node $m+n+1$ a sink.
2. Nodes 1, 2, ..., m are *operator nodes*; each corresponds to an operator. From the source to each operator node, there is a link with a capacity of 1.
3. Nodes $m+1$, $m+2$, ..., $m+n$ are *operation nodes*; each correspond to an operation. From each operation node to the sink, there is a link with a capacity of 1.
4. Nodes i and $m+j$ are connected by a link if and only if $c_{ij} \geq b$, for some $b \geq 0$, $i = 1, \ldots, m$, and $j = 1, \ldots, n$. Each operator-operation link has an infinite capacity.

The network in Figure 7.19 illustrates a case of four operators for three operations. Both operators 1 and 2 can handle all three operations. Operator 3 is not allowed for operations 1 and 3, and operator 4 can only take operation 3.

Since links associated with the source or the sink have unit capacities, the maximum flow value cannot exceed min(m, n); that is, the maximum number of operators that can be assigned is limited by the minimum of the operator number and the operation number. Furthermore, each operator

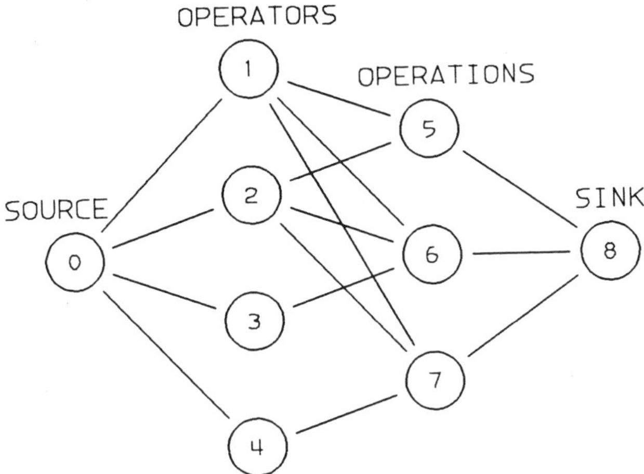

Figure 7.19 A network for job assignment problem.

can take at most one operation and each operation will have no more than one operator. Operator i is assigned to operation j if and only if the flow value between nodes i and $m+j$ is positive.

Obviously, the flow value of the network is a function of b. The larger the value of b, the fewer operator-operation links there will be. Therefore, the flow value is a decreasing function of b. On the other hand, for each operator-operation link, its corresponding throughput rate must be greater than or equal to b. If each operation has an operator, the expected line throughput should be no worse than b. Therefore, b can be regarded as a lower bound of the target throughput. By solving a sequence of maximum flow problems with different values of b, a maximum throughput assignment pattern can be obtained. Since the maximum flow is a monotone function of b, searching for the optimal value of b is straightforward.

The above discussion may be further clarified by an example.

Example 7.5 Consider an assignment problem with 5 operators for 5 operations. The throughput matrix $C = \|c_{ij}\|$ is given by

$$\begin{bmatrix} 16 & 14 & 6 & 5 & 20 \\ 15 & 5 & 18 & 15 & 13 \\ 14 & 12 & 16 & 5 & 7 \\ 6 & 10 & 16 & 13 & 17 \\ 12 & 10 & 6 & 11 & 7 \end{bmatrix} \tag{7.15}$$

where row i corresponds to operator i and column j to operation j.

Let d_i be the largest element of each row, i.e., $d_i = \max\{c_{ij} \mid j = 1,\ldots,n\}$. Then $(d_1,d_2,\ldots,d_5) = (20,18,16,17,12)$. Since $m \geq n$, a complete assignment will have n operators for n operations. The maximum throughput cannot exceed u, the nth largest value of $\{d_i \mid i = 1,\ldots,m\}$. On the other hand, if a complete assignment pattern can ever be found, line throughput should be no lower than l, the smallest element in C. Therefore, the value of b should be confined to $[l,u]$. In this example, $l = 5$ and $u = 12$. If a complete assignment pattern is found at $b = u$, then the solution gives the maximum throughput. Otherwise, another trial for a complete assignment should be made for a reduced b. If no complete assignment can be found for $b = l$, the problem does not have a feasible solution; that is, at least one operation will be idle.

To reduce total computational effort, a systematic searching procedure is required. A simple approach is to rank $\{c_{ij} \mid l \leq c_{ij} \leq u\}$. Let S be the list of c-elements ranked in descending order. Then $S = (12,11,10,7,6,5)$. If a complete assignment pattern exists, the maximum line throughput must be equal to one of the elements in S. This suggests that only the values listed in S need to be considered. Then the question is which value should be considered first. A *binary searching* method can be adopted and explained as follows.

Our computations start with the middle element, i.e., let b be the median. (In this example, b can be either 10 or 7.) If the network has a flow value of n, a complete assignment is found. Elements that are less than or equal to the median can be deleted from S. If no complete assignment is found, delete all elements that are greater than or equal to the median. Then the entire process starts over again, with only half of the elements. The process continues until S contains exactly one element. At this stage, the network-flow solution either gives the optimal assignment policy or concludes that no complete assignment exists. A decision tree is shown in Figure 7.20, where all possibilities are given.

Let us consider the initial network-flow problem with $b = 10$. The elements that are greater than or equal to 10 are given by

$$\begin{bmatrix} 16^* & 14 & — & — & 20 \\ 15 & — & 18^* & 15 & 13 \\ 14 & 12^* & 16 & — & — \\ — & 10 & 16 & 13 & 17^* \\ 12 & 10 & — & 11^* & — \end{bmatrix}$$

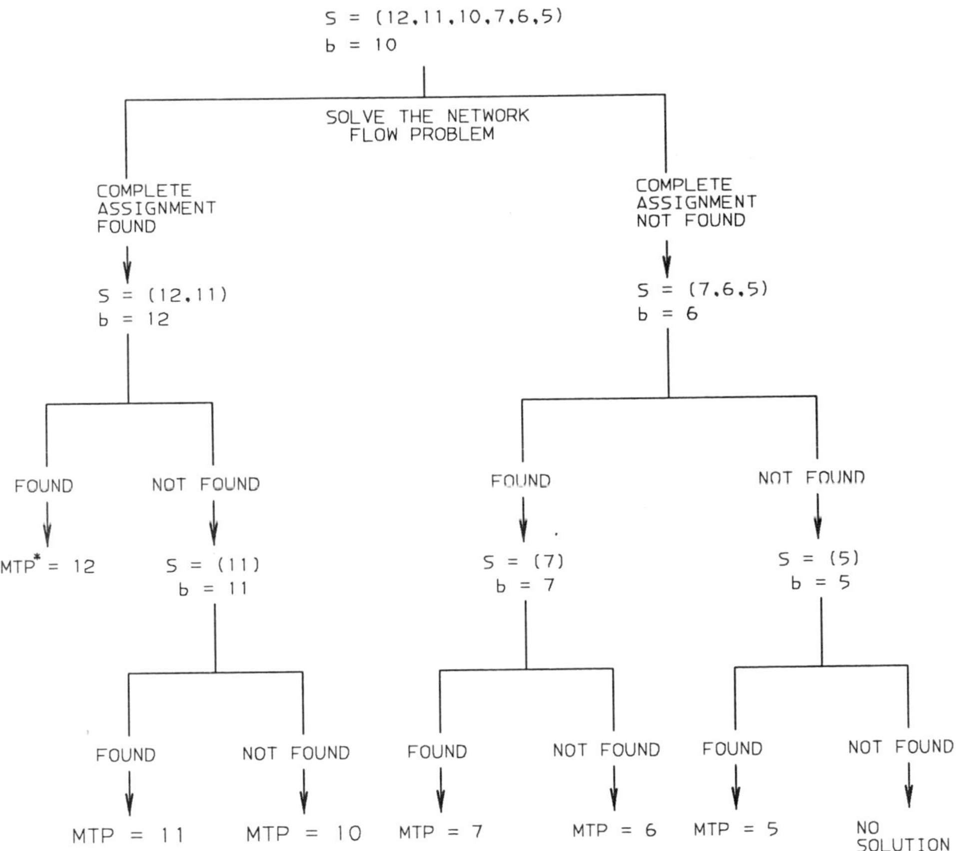

Figure 7.20 A decision tree.

The corresponding network is presented in Figure 7.21. The reader may verify that a complete assignment pattern is given by assigning a unit flow to links (1, 6), (2, 8), (3, 7), (4, 10), and (5, 9). This result is also shown by asterisks in the previous matrix. According to the decision tree in Figure 7.20, the next trial will have $b = 12$. The new

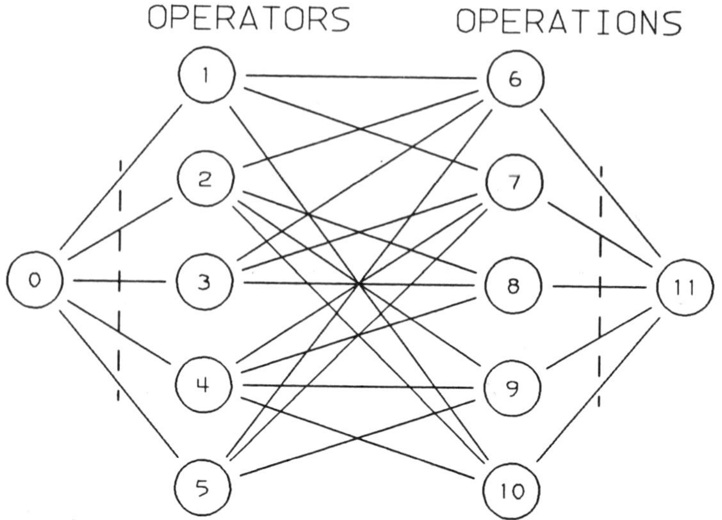

Figure 7.21 A network-flow problem defined in Example 7.5.

matrix and its solution are

$$\begin{bmatrix} 16 & 14 & - & - & 20^* \\ 15 & - & 18 & 15^* & 13 \\ 14 & 12^* & 16 & - & - \\ - & - & 16^* & 13 & 17 \\ 12^* & - & - & - & - \end{bmatrix}$$

Since a complete assignment is found, a line throughput of 12 may be achievable by letting operators 1, 2, 3, 4, and 5 have operations 5, 4, 2, 3, and 1, respectively. It can be seen that if $b > 12$, row 5 does not consist of any elements and no complete assignment is possible. □

Let a_j be the operator number assigned to operation j and B be the largest value of b that generates a complete assignment pattern. The computational procedure in Example 7.5 is summarized in the following algorithm.

Algorithm 7.1

1. Let $l = \min\{c_{ij} \mid \text{all } i,j\}$, and $u = \min_i[\max_j\{c_{ij}\}]$.

2. Let S be a descending list of all c-elements between l and u. If two or more elements have the same value, only one of them should be kept in S.
3. Let $a_j = 0$, for $j = 1, 2, \ldots, n$, and $B = 0$.
4. Let b equate to the middle element in S. (If the number of elements in S is even, then the larger value of the two middle elements is used.)
5. Construct a maximum-flow problem for the given value of b.
6. Invoke Algorithm 3.2 to find a maximum flow solution. Denote the optimal flow value by F and the flow on link (x,y) by f_{xy}.
7. If $F < n$ go to step 9. If $F = n$ (note that $n \le m$), a complete assignment pattern is found. Do the following: If $b > B$, then $B = b$. For node $m + j$, $j = 1, \ldots, n$, identify the link that carries a positive flow. Suppose that $f_{i,m+j} > 0$. Then let $a_j = i$. Go to step 8.
8. Delete from S all elements less than or equal to b. If S is empty, go to step 10. Otherwise, go to step 4.
9. Delete from S all elements greater than or equal to b. If S is empty, go to step 10. Otherwise go to step 4.
10. If $B = 0$, no complete assignment exists. Stop. If $B > 0$, the maximum throughput is B and an optimal assignment pattern is given by $\{a_j\}$. Stop.

The above algorithm will lead to a maximum throughput solution, based on operator performance. Because line throughput is determined by its slowest operation and b is a lower bound for throughput, line capacity is pushed up by raising the value of b. This approach tends to place a fast operator at the bottleneck and slow ones at less critical operations. To see this, let us consider a hypothetical case. A line has three operations with average process rates 10, 8, and 6 per hour, respectively. Assume that the first operator can perform at the average level, and the other two operators are 25% and 50% faster. The performance matrix is

$$\begin{bmatrix} 10 & 12.5 & 15 \\ 8 & 10 & 12 \\ 6 & 7.5 & 9 \end{bmatrix}$$

It is easy to see that a maximum line throughput of 9 pieces per hour is attainable by letting the fastest operator handle the slowest operation (i.e., operator 3 for operation 3) and the slowest operator take the fastest operation (i.e., operator 1 for operation 1). Under this assignment policy, the process rates are 10 : 10 : 9, and the line is nearly balanced.

7.4.2 Dynamic Job Assignment

When the number of operators is smaller than the number of operations, an operator-shortage problem occurs. Different approaches may be considered to resolve this problem. One way is to redefine the assembly process so that the number of operations matches the number of operators. However, this is not always a simple case in reality. Operator shortage usually results from absenteeism such as sickness, personal business, vacations, etc. Since shortages may happen on a daily basis, frequent process changes become inevitable. On the other hand, a new process definition may involve significant changes in work procedure, material flow, or tooling and fixtures. These changes normally take weeks or even months.

A simple but usually less effective approach is to have one operator handle multiple operations. Unless each operator has about the same amount of workload, this approach will lead to a poorly balanced line.

To resolve this dilemma, one may try a *dynamic job-assignment* policy, which dynamically determines new job-assignment patterns based on operator performance records and work-in-process (WIP) levels at each operation. If at one time an operation has accumulated enough WIP, the immediately preceding operation may be temporarily suspended. In this case, WIP can be treated as *virtual throughput.* If the number of operators is smaller than the number of operations, i.e., $m < n$, then each assignment pattern needs only satisfy m operations. The line will accumulate WIP in front of unassigned operations. Consequently, operations can be reassigned by moving operators from operations that cause downstream WIP to operations that are previously not assigned.

For convenience, $k = n - m$ dummy operators are added so that the number of operators and the number of operations are equal. At each operation, a dummy operator has a zero throughput rate. If the WIP level at operation h is $q(h)$, then for the next assignment an additional $q(h)$ (virtual) throughput should be added to its immediate upstream operation. Throughput rates of all operators at the latter operation, including dummy operators, are increased by $q(h)$. Based on the new throughput values, a new assignment pattern is determined by taking the following steps:

1. Using the same procedure defined in Section 7.4.1, construct a flow network, except that each link has a "cost" equal to its correspondent

throughput (the sum of actual and virtual throughputs). Links connected to the source or the sink have zero costs.

2. Find a flow pattern that has the maximum value and the maximum sum of link costs.
3. The links with unit flow values define the assignment pattern.

In step 2, the maximal flow value, F, is such that $m \leq F \leq n$. If $F < n$, only F operations can be assigned. In this case, the last $(n-F)$ operations can be eliminated from further consideration. The throughput matrix and its corresponding network should be reconstructed accordingly, and the new assignment problem should be solved again. This iterative process is terminated when and only when the flow value is equal to the number of operations defined in the throughput matrix. In other words, in the ultimate network every operation node has exactly one link that carries a unit flow to the sink.

Since the ultimate network flow problem always has a fixed flow value, a maximum-cost flow problem can be converted into a minimum-cost flow problem by changing link costs. Let d_{ij} be the flow cost between nodes i and j, and $D = \max\{d_{ij}\}$. Then the new cost is defined by

$$\bar{d}_{ij} = \begin{cases} D - d_{ij} & d_{ij} > 0 \\ 0 & \text{otherwise} \end{cases}$$

If the flow value is F, then the following problems are equivalent:

1. $\max(\sum_{i,j} d_{ij} f_{ij})$ such that the flow is F.
2. $\min(-\sum_{i,j} d_{ij} f_{ij})$ such that the flow value is F.
3. $\min(DF - \sum_{i,j} d_{ij} f_{ij})$ such that the flow value is F.
4. $\min(\sum_{i,j} \bar{d}_{ij} f_{ij})$ such that the flow value is F.

The first two problems are identical, because maximization of a function is equivalent to minimization of the negative value of the function. Problems 2 and 3 are equivalent due to the fact that $D \times F$ is a constant term. Finally, note that the total amount of flow carried by the links with positive d-values is F. Thus, in problem 3, $D \times F$ can be broken into F equal parts of value D, and each of them is associated with a product term such that $d_{ij} f_{ij} > 0$. This establishes the equivalence of the last two problems.

A complete algorithm is given after Example 7.6.

Example 7.6 Consider the same problem as stated in Example 7.5, except that operator 5 is not available. Element (i,j) of (7.15) is the average daily throughput of operator i at operation j, for $i = 1, \ldots, 4$, and $j = 1, \ldots, 5$. Because of a labor shortage, the line manager decides to adopt a dynamic assignment policy. The assignment pattern will be redefined every two hours. Replacing the last row of (7.15) by a null vector and dividing each element by 4, a matrix of two-hour throughput, $M = \|d_{ij}\|$, is given by

$$\begin{bmatrix} 4.00 & 3.50 & 1.50 & 1.25 & 5.00 \\ 3.75 & 1.25 & 4.50 & 3.75 & 3.25 \\ 3.50 & 3.00 & 4.00 & 1.25 & 1.75 \\ 1.50 & 2.50 & 4.00 & 3.25 & 4.25 \\ 0.00 & 0.00 & 0.00 & 0.00 & 0.00 \end{bmatrix} \qquad (7.16)$$

From (7.16), a minimum-cost flow problem can be defined. The flow network has the same structure as depicted in Figure 7.21, except that:

1. Node 5 is a dummy operator.
2. The link from node 0 to node k, or simply link $(0,k)$, has unit capacity and zero cost, for $k = 1, \ldots, 5$.
3. Link $(k,11)$ has unit capacity and zero cost, for $k = 6, \ldots, 10$.
4. Link (i,j), $i = 1, \ldots, 5$, and $j = 6, \ldots, 10$, has unit capacity if $d_{ij} > 0$ and zero capacity if $d_{ij} = 0$, and a cost $h_{ij} = 5 - d_{ij}$. (Note that $\max\{d_{ij}\} = 5$.)

Since the last row of (7.16) contains only zero entries, no complete assignment is possible, i.e., only four operations can be assigned. Therefore, the last operation is eliminated from further consideration. This implies that nodes 5 and 10 and all links connected to them should be deleted from the network. Matrix M becomes

$$\begin{bmatrix} 4.00 & 3.50^* & 1.50 & 1.25 \\ 3.75 & 1.25 & 4.50 & 3.75^* \\ 3.50^* & 3.00 & 4.00 & 1.25 \\ 1.50 & 2.50 & 4.00^* & 3.25 \end{bmatrix} \qquad (7.17)$$

The cost matrix is given by

$$\begin{bmatrix} 0.50 & 1.00^* & 3.00 & 3.25 \\ 0.75 & 3.25 & 0.00 & 0.75^* \\ 1.00^* & 1.50 & 0.50 & 3.25 \\ 3.00 & 2.00 & 0.50^* & 1.25 \end{bmatrix} \tag{7.18}$$

where element (i,j) is the cost for link $(i,4+j)$.

Solve the minimum cost flow problem by Algorithm 3.3 of Chapter 3. The minimum cost flow solution consists of four different paths as shown in Figure 7.22, where the numbers are link costs. Since each link carries a unit flow, the total cost is $1+0.75+1+0.5=3.25$. The corresponding assignment pattern is such that operators 1, 2, 3, and 4 are assigned to operations 2, 4, 1, and 3, respectively. This pattern is indicated by asterisks in (7.17) and (7.18). Since no operator is assigned to operation 5, line throughput during the first two hours is zero.

At the end of the first two-hour period, the expected WIP is given by $W = (w_1,\ldots,w_5) = (0,0,0,0,3.5)$, where w_i is the amount of WIP at operation j. In real-life applications, the actual WIP level can be observed directly from the manufacturing line. For simplicity, let us assume that actual WIP at an operation is identical to its expected WIP. Then w_j can be treated as the virtual throughput at operation

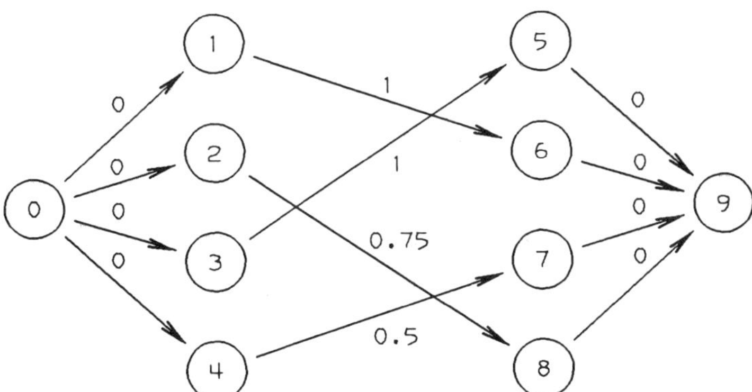

Figure 7.22 A minimum cost flow pattern for Example 7.6.

$(j-1)$. Element (i,j) of the new throughput matrix for the next two-hour period becomes $d_{ij} + w_{j+1}$. Hence,

$$M = \begin{bmatrix} 4.00 & 3.50^* & 1.50 & 4.75 & 5.00 \\ 3.75^* & 1.25 & 4.50 & 7.25 & 3.25 \\ 3.50 & 3.00 & 4.00^* & 4.75 & 1.75 \\ 1.50 & 2.50 & 4.00 & 6.75 & 4.25^* \\ 0.00 & 0.00 & 0.00 & 3.50^* & 0.00 \end{bmatrix} \tag{7.19}$$

Again the last row is associated with the dummy operator. Invoking Algorithm 3.3, a complete assignment pattern is found and indicated by asterisks in matrix (7.19). Operators 1, 2, 3, and 4 are assigned to operations 2, 1, 3, and 5, respectively. Since operation 4 has produced 3.5 pieces in the previous period, no operator should be assigned. The expected throughput for the second two-hour period is 3.5, and $W = (0,0,0,3.5,0)$.

For the third period, the expected throughput is 3.5, $W = (0,0,3.5,0,0)$ and the assignment pattern is given by

$$\begin{bmatrix} 4.00 & 3.50^* & 5.00 & 1.25 & 5.00 \\ 3.75 & 1.25 & 8.00 & 3.75^* & 3.25 \\ 3.50^* & 3.00 & 7.50 & 1.25 & 1.75 \\ 1.50 & 2.50 & 7.50 & 3.25 & 4.25^* \\ 0.00 & 0.00 & 3.50^* & 0.00 & 0.00 \end{bmatrix}$$

For the fourth period, the expected throughput is also 3.5, $W = (0,3.5,0,0,0)$ and the assignment pattern is given by

$$\begin{bmatrix} 4.00 & 7.00 & 1.50 & 1.25 & 5.00^* \\ 3.75 & 4.75 & 4.50 & 3.75^* & 3.25 \\ 3.50^* & 6.50 & 4.00 & 1.25 & 1.75 \\ 1.50 & 6.00 & 4.00^* & 3.25 & 4.25 \\ 0.00 & 3.50^* & 0.00 & 0.00 & 0.00 \end{bmatrix}$$

The expected throughput over the first eight-hour period is 3.5 + 3.5 + 3.5 = 10.5. Comparing with the result in Example 7.5, with four operators line throughput is decreased by 1.5 (= 12 – 10.5). □

It should be pointed out that although the number of operators is smaller than the number of operations in Example 7.6, the condition that $m < n$ is not necessary. If $m > n$, the minimum-cost flow algorithm will select a pattern that maximizes the sum of throughputs. The computational procedure is summarized in Algorithm 7.2.

Algorithm 7.2

1. Observe WIP at each operation. The amount of WIP at operation k that comes from operation j is denoted by w_{jk}.
2. If $m < n$, then add $(n - m)$ dummy operators. Each operator has zero (actual) throughput. On the other hand, if $m > n$, then add $(m - n)$ dummy operations. Each operator has a positive but small throughput at each dummy operation, say 10^{-6}. Now the problem has $h = \max(m,n)$ operators for h operations.
3. Let c_{ij} be the throughput of operator i at operation j. Construct throughput matrix M. Element (i,j) of M is given by

$$d_{ij} = c_{ij} + \sum_{k \text{ in } Y(j)} w_{jk}$$

   where $Y(j)$ is a set of immediate downstream operations with respect to operation j.
4. Find $D = \max\{d_{ij} \mid \text{all } i,j\}$.
5. For $i = 1, \ldots, h$ and $j = 1, \ldots, h$, $\bar{d}_{ij} = D - d_{ij}$.
6. Construct a flow network:

   a. Let node 0 be the source, and node $2h + 1$ the sink. Nodes 1, ..., h are operator nodes, and nodes $h + 1, \ldots, 2h$ are operation nodes.
   b. Place a link from the source to each operator node with a link capacity of one and a link cost of zero. Place a link from each operation node to the sink with a capacity of one and zero cost.
   c. For operator i and operation j, if $d_{ij} > 0$, then place a link from node i to node $h + j$ with a link capacity of one and a cost $\bar{d}_{ij}$. (No link between node i and node $m + j$ if $d_{ij} = 0$.)

7. Solve the minimum-cost flow problem defined in step 5. If the maximum flow value $F = h$, every operator has an assignment. Go to step 8. If $F < h$, go to step 9.
8. An optimal assignment solution is found. Links with positive flow value between operator nodes and operation nodes define the assignment pattern. If there is positive flow between node i and node $m + j$, then operator i should be assigned to operation j. An operation is unassigned if its corresponding operator node is a dummy operator. Likewise, an operator is unassigned if the corresponding operation node is a dummy. Stop the algorithm.
9. Eliminate the last $(h - F)$ operations. Let $n = F$. Go to step 2.

7.4.3 Other Considerations

A job assignment program can also be used to deal with many other problems in manufacturing lines. Depending on the nature of a problem, either a static or a dynamic assignment policy may be invoked. These problems are discussed in the following sections.

Unbalanced Line

If a line is poorly balanced, both workstations and operators will be underutilized. It is possible to reduce the number of operators by increasing WIP. To see this, let us consider a simple example. A line has three operations with mean process times 0.5, 0.5, and 1, respectively. If three operators are assigned to the line, throughput is one and operator utilization is (0.5 + 0.5 + 1)/3 = 0.66. A better policy is to let one operator take operations 1 and 2 on an alternating basis and another one be permanently assigned to operation 3. In this case, the line throughput is still 1 but operator utilization becomes 1. Since the first two operations are not processed in the same time, excessive WIP becomes inevitable. In Chow (1988), a 10-operation line is presented. Due to multiple workstations per operation and multiple workstations per operator, the line has 16 workstations and 13 jobs. Under a dynamic job-assignment policy, line throughput, utilization, and WIP for different numbers of operators are given in Table 7.8.

When the number of operators is reduced from 13 to 12, line performance remains almost the same. Hence, Algorithm 7.2 can also be used to determine the minimum number of operators.

Workstation Failures

Workstation failures will cause throughput change. Consider that an operation consists of k workstations handled by the same operator. For each

Table 7.8 Results of a Dynamic Job-Assignment Policy

Number of operators	Two-hour throughput	Operator utilization	Average total WIP
13	23.82	0.90	169.1
12	23.34	0.90	168.3
11	21.60	0.93	177.1
10	20.91	0.97	192.3
9	17.54	0.97	183.0
8	14.19	0.97	179.9

workstation failure, the operator's throughput is reduced by $1/k$. For instance, at the first operation in Example 7.6, one of the three workstations is broken. The throughput matrix (7.16), for the first assignment, becomes

$$\begin{bmatrix} 2.67 & 3.50 & 1.50 & 1.25 & 5.00 \\ 2.50 & 1.25 & 4.50 & 3.75 & 3.25 \\ 2.33 & 3.00 & 4.00 & 1.25 & 1.75 \\ 1.00 & 2.50 & 4.00 & 3.25 & 4.25 \\ 0.00 & 0.00 & 0.00 & 0.00 & 0.00 \end{bmatrix}$$

After having repaired the workstation, a new throughput matrix may be constructed for a new assignment pattern.

Job Rotation

Job rotation serves at least two purposes: job enrichment and line coverage. To many people, a short assembly operation can never be an interesting job, and may significantly affect their productivity. One way to alleviate the problem is to allow operators to change their job assignment periodically. Line coverage, on the other hand, is an effective means to deal with absenteeism problems. Operators must be trained for different operations. To keep their skills, operators should be given opportunities to practice on different operations. A job rotation pattern can be obtained by solving a sequence of assignment problems with modified throughput matrices. This approach is illustrated by Example 7.7.

Example 7.7 Assume that a throughput matrix is given by (7.15). From Example 7.5, it is known that at a throughput rate of 12 the best assignment pattern is given by

$$\begin{bmatrix} 16 & 14 & 6 & 5 & 20^* \\ 15 & 5 & 18 & 15^* & 13 \\ 14 & 12^* & 16 & 5 & 7 \\ 6 & 10 & 16^* & 13 & 17 \\ 12^* & 10 & 6 & 11 & 7 \end{bmatrix}$$

The above matrix shows that operator i should be assigned to operation j, if element (i,j) is accompanied by an asterisk. For a different job assignment, each element with an asterisk is replaced by zero; that is, a previously assigned operator must not assigned again to the same operation. Invocation of Algorithm 7.1 leads to a new assignment pattern:

$$\begin{bmatrix} 16 & 14^* & 6 & 5 & 0 \\ 15^* & 5 & 18 & 0 & 13 \\ 14 & 0 & 16^* & 5 & 7 \\ 6 & 10 & 0 & 13 & 17^* \\ 0 & 10 & 6 & 11^* & 7 \end{bmatrix}$$

The expected line throughput now becomes 11. Repeat the process, and three other patterns are given by

$$\begin{bmatrix} 16^* & 0 & 6 & 5 & 0 \\ 0 & 5 & 18^* & 0 & 13 \\ 14 & 0 & 0 & 5 & 7^* \\ 6 & 10 & 0 & 13^* & 0 \\ 0 & 10^* & 6 & 0 & 7 \end{bmatrix}$$

$$\begin{bmatrix} 0 & 0 & 6^* & 5 & 0 \\ 0 & 5 & 0 & 0 & 13 \\ 14^* & 0 & 0 & 5^* & 0 \\ 6 & 10^* & 0 & 0 & 0 \\ 0 & 0 & 6 & 0 & 7^* \end{bmatrix}$$

$$\begin{bmatrix} 0 & 0 & 0 & 5^* & 0 \\ 0 & 5^* & 0 & 0 & 13^* \\ 0 & 0 & 0 & 0 & 0 \\ 6^* & 0 & 0 & 0 & 0 \\ 0 & 0 & 6^* & 0 & 0 \end{bmatrix}$$

For the five different assignment patterns, their expected throughputs are 12, 11, 7, 5, and 5, respectively. □

Different Skill Levels

Since line throughput is dominated by the slowest operator, the problem of assigning operators to multiple lines should consider homogeneity, that is, operators with the same performance level should be assigned to the same line. This also can be done by invoking Algorithm 7.1 repetitively. Consider a problem of assigning six operators to two identical lines, each having three operations. The throughput matrix is given by

$$\begin{bmatrix} 9 & 10 & 3 \\ 8 & 9 & 12 \\ 5 & 7 & 4 \\ 3 & 8 & 7 \\ 10 & 5 & 9 \\ 7 & 4 & 6 \end{bmatrix}$$

The first invocation leads to the following solution

$$\begin{bmatrix} 9 & 10^* & 3 \\ 8 & 9 & 12^* \\ 5 & 7 & 4 \\ 3 & 8 & 7 \\ 10^* & 5 & 9 \\ 7 & 4 & 6 \end{bmatrix}$$

Thus, operators 1, 2, and 5 are assigned respectively to operations 2, 3 and 1 of line 1. The expected throughput is 10. For the second invocation, rows 1, 2, and 5 are first eliminated from the throughput matrix. Then the assignment pattern for the second line is given by

$$\begin{bmatrix} 5 & 7^* & 4 \\ 3 & 8 & 7^* \\ 7^* & 4 & 6 \end{bmatrix}$$

The total expected throughput is $10 + 7 = 17$. The reader may verify that this solution results in the maximum overall throughput.

Job Training

If a production line is also used for the purpose of job training, it is clear that the training should not be done on an individual basis. At any given time, the line should be used for a single purpose only. Trainees should not be mixed with skilled operators. If the same crew of operators occupies the line for both production and training, it is better to separate a training period from a production period. If the line is in a production mode, no training activities are allowed. On the other hand, when the line is in a training mode, all operators are assigned to their uncertified operations. Separation of training mode from production mode can be done by using the same method for different skill levels. Consider a problem of three operators for three operations, with a throughput matrix defined by

$$\begin{bmatrix} 9 & 10 & 3 \\ 10 & 5 & 9 \\ 3 & 8 & 9 \end{bmatrix}$$

Obviously, operators 1, 2, and 3 need to improve their skills at operations 3, 2, 1, respectively. The assignment pattern for the production mode can be determined from the above matrix by invoking Algorithm 7.1 twice. The two possible patterns for production are

$$\begin{bmatrix} 9 & 10^* & 3 \\ 10^* & 5 & 9 \\ 3 & 8 & 9^* \end{bmatrix}$$

$$\begin{bmatrix} 9^* & 10 & 3 \\ 10 & 5 & 9^* \\ 3 & 8^* & 9 \end{bmatrix}$$

The pattern for training mode is then given by

$$\begin{bmatrix} 9 & 10 & 3^* \\ 10 & 5^* & 9 \\ 3^* & 8 & 9 \end{bmatrix}$$

The expected line throughputs are 9, 8, and 3, respectively. The first two patterns can be used for job rotation, and the last one for job training.

7.5 Remarks

The improvement process has been studied by many authors. Model (7.2) was the first improvement model developed by Wright (1936). Since then a number of different models have been proposed. Model (7.5) was proposed by Glover (1966a). Selection of trainees for training-cost reduction was previously presented in Glover (1966b).

Training and shift strategies are often considered as interrelated subjects. A line may begin with a single-shift operation. Later, line capacity can be increased by operating the line in a two-shift or even a three-shift mode. On the other hand, adding a shift may mean more training effort. A training program can last from several weeks to several months. Since a three-shift line usually has less average productive time per shift, it may be advisable to run the line on a two-shift basis and use off-shift time for training activities.

A two-shift policy has been adopted by many companies already. Often, saving is not just from labor. Because of better maintenance and less line interruption for engineering experiments, the line has a higher hourly productivity.

Experience shows that line balancing should be a long-term activity in line design, particularly in high-technology areas. Due to frequent product or process changes, it takes a long time to stabilize an assembly process. In this chapter, line rebalancing is used to redeploy operators for production. Although the version presented in Section 7.2.4 is to cope with a demand change, this technique may also be considered to handle an operator-shortage problem. By reducing the number of operations and increasing operation process times, a line may be operated with a smaller number of operators at a lower throughput rate.

Whether operations can be effectively redefined or not is dependent on tooling design. If tooling does not have flexibility for different operations, the line balancing approach cannot be used to resolve the shortage problem.

In this case, a dynamic job assignment policy may be a good remedy. The work discussed in Section 7.4.2 was reported in Chow (1988).

Two assignment problems have been presented: static and dynamic. Both problems can be converted into network flow problems. These two problems are also known as matching problems—to match elements from two different sets. The static assignment problem is equivalent to a maximum matching problem, i.e., to maximize the total number of matched pairs. The dynamic assignment problem is similar to a maximum weighted matching problem, i.e., each matched pair has a weight and the total weights must be maximized. In the literature, matching algorithms are well developed; see for example, Edmonds (1965), Hopcroft (1973), Even (1975), and Gabow (1976).

REFERENCES

Chow, W. (1988). A Dynamic Job Assignment Policy, *International Journal of Production Research* (In press)

Edmonds, J. (1965). Path, Trees, and Flowers, *Canadian Journal of Mathematics*, v. 17, pp. 449–467

Even, S., and O. Kariv (1975). An $O(n^{2.5})$ Algorithm for Maximum Matching in General Graphs, *Proceedings of 16th Annual Symposium on Foundations of Computer Science*, IEEE, New York, pp. 100–112

Gabow, H. (1976). An Efficient Implementation of Edmonds Algorithm for Maximum Matching on Graphs, *Journal of Computing Machinery Association*, v. 23, pp. 221–234

Glover, J. H. (1966a). Manufacturing Progress Functions, I. An Alternative Model and Its Comparison with Existing Functions, *International Journal of Production Research*, v. 4, pp. 279–300

Glover, J. H. (1966b). Manufacturing Progress Functions, II. Selection of Trainees and Control of Their Progress, *International Journal of Production Research*, v. 5, pp. 43–59

Hopcroft, J., and R. M. Karp (1973). A $n^{5/2}$ Algorithm for Maximum Matchings in Bipartite Graphs, *SIAM Journal for Computing*, v. 2, pp. 225–231

Maynard, H. B., G. J. Stegemerten, and J. L. Schwab (1948). *Methods-Time Measurement*, McGraw-Hill

McGehee, W. (1948). Cutting Training Waste, *Personnel Psychology*, v. 1, pp. 331–340

Turban, E. (1968). Incentives During Learning—An Application of Learning Curve Theory and a Survey of Other Methods, *Journal of Industrial Engineering*, v. 19, pp. 600–607

Wright, T. P. (1936). Factors Affecting the Cost of Airplanes, *Journal of Aeronautical Science*, v. 3., pp. 122–128.

8

Line Integration

Line integration is a key function in the process of line design. This function coordinates all the other functions of the design team, serving as the interface between the design team and other supporting groups, and often directing the actions of the entire design team. Line integration is responsible for solving problems related to the overall line design, such as line concept, assumptions, design parameters, configuration, and cost. In the subsequent sections, cost estimating, line sizing and configuration will be discussed in detail.

8.1 Line Cost

Since the ultimate objective of line design is minimum cost, cost analysis is an essential part of the design job. This section deals with line cost; both cost methods and cost items will be considered.

8.1.1 Cost Methods

Two cost methods are frequently used. One is the *total cost method* and the other the *discounted cash flow method*. The former method computes the line cost by simply adding all expenditures together, while the latter further considers the time factor and possible tax impacts. The computational procedure for the total cost method is rather straightforward. It will be illustrated by a numerical example later. A detailed discussion on the discounted cash flow method is given in the following.

For convenience, denote

$M_i(t)$ = total amount of type-i capital investment in year t, $i = 1, \ldots, m$
$d_i(s)$ = depreciation rate of type-i investment in its s-th year, $i = 1, \ldots, m$
$N(t)$ = total amount of noncapital expenditure in t
r = discount rate
b = company's tax bracket
p_i = rate for investment credit
T = product lifetime

The total spending in year t is given by

$$A(t) = \sum_{i=1}^{m} M_i(t) + N(t) \tag{8.1}$$

The depreciation from capital investment, however, may lead to a saving from the company's income tax. If a capital investment item is made in year s, then at the end of year t, $t > s$, the depreciation rate becomes $d_i(t - s + 1)$. Since the company's tax bracket is b, the total tax saving from depreciation in year t is

$$B(t) = \sum_{i=1}^{m} \sum_{s=1}^{t} M_i(t) d_i(t - s + 1) b \tag{8.2}$$

If the tax regulation allows investment credit, the amount of saving in year t becomes

$$C(t) = \sum_{i=1}^{m} M_i(t) p_i \tag{8.3}$$

Finally, using the discount rate, all cash items can be converted into their present values. The discounted cash flow at the present value can be obtained from Equations (8.1)—(8.3) and is given by

$$f = \sum_{t=1}^{T} [B(t) + C(t) - A(t)](1 + r)^{-t} \tag{8.4}$$

This method will be used later, in Section 8.4, to compare different line configuration strategies.

If the total cost method is used, the result is given by

$$c = \sum_{t=1}^{T} [N(t) + \sum_{i=1}^{m} M_i(t)] \tag{8.5}$$

It can be seen that in this method neither the time factor nor the tax items are considered. For this reason, many financial analysts and cost engineers would prefer the discounted cash flow method to the total cost method, because the former seems more thorough. In reality, however, this may not always be the case. First, the value of a cost item is an estimate, subject to error. Next, both capital and noncapital expenditures are dependent on many line parameters, such as operation process time, yield, skill level, and so on. The actual values of these parameters can be very different from what is expected during the line design phase. For instance, it is not uncommon for the estimated value of a mean process time to be over 20% off, as compared with the actual value. Finally, it should be pointed out that the major advantage of the discounted cash flow method is its consideration of the time factor. However, this may be exactly its shortcoming, too. Since the expenditure is a function of production demand and the demand is a function of time, uncertainty is therefore introduced. If the accumulated demand remains the same, a change in demand schedule will lead to a different cash flow value. In other words, the discounted cash flow method is more sensitive to the demand assumption. Later, in Example 8.1, it will be shown that only the average demand and the peak demand are important when the total cost method is employed.

8.1.2 Cost Items

Since a line is composed of many components, line cost also includes many cost items. This section considers different cost items from a line design point of view.

1. *Workstation* A workstation may contain tooling, fixtures, parts feeders and any hardware or software needed to perform manufacturing functions in the workstation. For a type-i workstation, two kinds of costs are usually considered: (i) design and development cost, d_i, and (ii) replication cost, r_i.

If n_i type-i workstations are installed, $i = 1, ..., m$, the total workstation cost will be

$$\sum_{i=1}^{m}(d_i + n_i r_i)$$

The values of $\{n_i\}$ are determined by the peak demand and workstation yield capacities.

2. *Labor* The labor cost is simply the product of the total number of operators and the average annual cost per operator. Operators may be assigned for assembly jobs, testing, inspection, rework, packaging, parts preparation, parts distribution, or material handling. The number of operators is a function of line workload and job assignment policy.

3. *Material handling* Material-handling cost varies from application to application. The cost structure, in general, can be expressed as a linear combination of two parts, by $x + s \times y$, where x is a fixed charge, s is the size of the system (i.e., a measure of coverage of the material-handling system), and y is a unit cost. For a discrete type of system, such as an automated storage and retrieval system, x may consist of the cost of the direct access handler and all supporting software, and y the unit cost of the linear track for the handler plus the cost of the vertical rack measured by linear distance. In this case, x is normally the dominant factor. For a simple conveyor system, on the other hand, the majority of the cost comes from hardware. Consequently, y becomes more important.

Since many generic material-handling systems exist in the current market, most line-design projects would purchase the system from an outside vendor. In this case, no design and development cost is incurred.

4. *Space* Space installation cost is usually expressed as a linear function of the area. The unit space cost is dependent on the manufacturing environment. For example, a typical factory may cost \$100–150 per square foot, but a clean room may have a unit cost of \$400–500 per square foot.

Space cost may also consist of operating overhead, which may include utilities, maintenance, taxes, and insurance. Overhead may also be proportional to the total area, and charged on an annual basis. For example, in an ordinary factory environment, overhead expense is about \$3–4 per square foot per year.

5. *Parts preparation* Usually parts preparation can be regarded as an operation. For instance, an automated parts-kitting station may be equipped with a robot and a conveyor system. Consequently, the kitting station can be treated as a workstation.

6. *Engineering* Engineering cost is proportional to the number of engineer-years, including salary, benefit, and laboratory expense. Estimation of engineer-years is based mainly on experience. No scientific procedure is available.

7. *Supporting personnel* Similar to engineering cost, this cost item is also dependent on the number of person-years. Supporting personnel is composed of people from many different functions, such as quality assurance, production control, logistics, accounting, payroll, data processing, and other administration. The number of people is dependent on both the direct labor and the production volume.

8. *Quality* Quality in a manufacturing line is usually reflected by the yield factor. Yield improvement is possible through operator skill training, line discipline, reduction of handling damage, better tool precision, and early problem detection. If a cost occurs due to any of these actions, it should be distributed to the appropriate cost items. For example, cost for tooling change may become a part of workstation cost. On the other hand, a yield improvement would increase productivity—higher capacity, less rework and scrap. Thus, the quality cost may be understood by comparing the productivity gain to the cost for yield improvement.

9. *Maintenance* Maintenance cost consists of three parts: (i) maintenance tools, (ii) spare parts, and (iii) maintenance personnel. The first two items are application-dependent. The tooling cost and the price of spare parts are usually estimated by engineers. The spare parts cost is also dependent on parts inventory policy. Determination of the optimal inventory level has been discussed in Section 6.1. The cost of maintenance personnel is proportional to the number of people. For given design criteria (e.g., the mean response time to a failure, workstation reliability), the size of the maintenance crew can be computed by using a machine-repairman model (see Examples 3.9 and 3.10). Let

T = mean interfailure time
S = mean repair time
R = average delay for a repair action

Workstation reliability is given by

$$r = T/(T + S + R) \tag{8.6}$$

Clearly, this quantity is bounded by $r_m = T/(T + S)$, when $R = 0$. Hence, the efficiency of the maintenance service can be defined by taking the ratio of reliability to the upper bound:

$$e = (T + S)/(T + S + R)$$

Suppose that the design criteria are (i) $e \geq 0.99$, and (ii) $R \leq 10$ minutes. Using the repairman model, the size of the maintenance crew can be expressed as a function of the number of workstations (WS), T, and S as shown in Table 8.1, where both T and S are measured by hours.

10. *Reliability* Workstation reliability is defined by Equation (8.6). For a given efficiency requirement, workstation reliability can also be computed by using the following expression:

$$r = eT/(T + S) \tag{8.7}$$

As a cost item, reliability can be regarded as a capacity detractor, which will affect the number of workstations.

11. *Parts quality* Parts quality is an important factor for low cost production. Good quality is necessary for (i) direct line feed (i.e., no warehouse operation, and no receiving inspection), (ii) low parts inventory, (iii) high assembly yield, and (iv) automated assembly. Parts quality is measured by the number of bad parts per million (PPM). For an automated assembly line, the quality requirement is usually set at 100 PPM or better. Parts quality cost may be treated in the same way as product quality cost (see item 8).

Table 8.1 The Size of the Maintenance Crew

T	S	WS = 50	WS = 40	WS = 30	WS = 20	WS = 10
100	0.5	1	1	1	1	1
	1.0	2	2	2	2	1
	2.0	3	3	2	2	2
250	0.5	1	1	1	1	1
	1.0	2	1	1	1	1
	2.0	2	2	2	2	1
500	0.5	1	1	1	1	1
	1.0	1	1	1	1	1
	2.0	2	2	2	1	1

12. *Work-in-process* Work-in-process (WIP) can be treated as an interest-loss. Let C be the product cost, p the interest rate, and Q the average amount of WIP in line. The expected annual WIP cost is approximately equal to $(C/2)\times Q\times p$. The average product cost can be estimated by dividing the total spending by the total production volume. The value of Q may be estimated by a simulation model (discussed in Chapter 9) or a queuing model (see Chapter 3). If a pull system is installed (see Section 6.3), the maximum amount of WIP is equal to the sum of total buffer spaces and the number of workstations. The interest rate may also be interpreted as an opportunity cost. Therefore, p can be the prevailing interest rate in the market or the expected annual return on investment projected by the company.

8.1.3 An Approximation of Line Cost

Due to (i) lack of information, (ii) existence of uncertainty, and (iii) multidisciplinary activities, it is not simple to estimate line cost during the design phase. On the other hand, many important decisions must be based on cost. To resolve this dilemma, a practical approach may be useful. The line designer may conduct a simple cost analysis for decision-making. After better cost figures have been obtained, the line cost may be revised. This section outlines such a procedure.

The cost items discussed in Section 8.1.2 can be classified into three categories. The first category is determined by the peak demand and ultimate line performance; this includes most capital investment items such as workstations. The second is dependent on the average demand and average performance. In this category, cost items such as labor are often treated as variables. The last category is independent of production volume, such as engineering cost. A cost item of this type is usually a fixed charge.

From a capacity-planning point of view, line performance is determined by the improvement process (see Sections 7.1 and 7.2). In a high-volume assembly line, the operator learning period can be very short as compared to the product lifetime, say 2 months versus 5 years. It is the yield factor that is more important. Consider a *continuous* improvement model, where the initial production period is at $t = 1$. Then

$$y(t) = y(1)t^{a-1} \qquad t \geq 0$$

where $y(t)$ is the yield at time t and a is a model parameter.

Let x be the improvement period. The ultimate yield is $y(x)$. Then, the parameter

$$a = 1 + \frac{\log[y(x)/y(1)]}{\log x}$$

The average yield over the entire product lifetime, z, is given by

$$\begin{aligned} \bar{y} &= \frac{1}{z}\left\{\int_1^x y(1)t^{a-1}dt + y(x)(z - x + 1)\right\} \\ &= \frac{1}{z}\{y(1)(x^a - 1)/a + y(x)(z - x + 1)\} \end{aligned} \tag{8.8}$$

This means that the average yield is a function of the initial yield, the ultimate yield, the improvement period, and the product lifetime.

If a line size is chosen to be c pieces per day and the demand is d, the number of workstations of operation i is given by

$$n_i = I[d/c]I[(cs_i)/(Ty_ir_i)] \tag{8.9}$$

where s_i, y_i, r_i are respectively the mean process time, the yield and reliability of type-i workstation, T is the total productive time in a day, and $I[x]$ is the least integer greater than or equal to x.

Note that yield factors, instead of start factors, have been used in Equation (8.9). (See Example 4.4.) To estimate the start factor, one must know the yield factor and a completed assembly flow process, including rework paths. In an early design stage, a well-defined flow process may not be available. Therefore, it is relatively easy to deal with yield. If the yield factor is close to 1.00, the difference may be negligible. It should also be pointed out that $I[x]$ can be sensitive to the value of x. If the capacity is only a few percent below the demand, it may be more economical not to add a line. In this case, $I[x]$ is the least integer greater than or equal to $(1-\alpha)x$, for a small but positive α, say 0.05.

If d is the average demand over the product lifetime and y_i is the average yield, as computed by Equation (8.9), then n_i is the average number of workstations of type i. The maximum number of workstations of type i can be obtained by using the peak demand and the ultimate yield.

Results from Equation (8.9) form the basis for cost estimation. This concept is illustrated in Example 8.1.

Example 8.1—Impacts of the Line Cycle Time A product has an assembly time of about 45 minutes. The entire assembly process consists of 13 operations as shown in Figure 8.1. Both the tooling cost and the operation process times have been estimated by manufacturing and industrial engineers. Since most mean process times are close to 3.5 minutes, the planned capacity for a 3-shift operation (18.6 hours) is approximately $18.6 \times 60/3.5 \approx 300$ pieces per day. The line designer believes, however, that a cycle time of 3.5 minutes might be too short for manual operations. Therefore, a study of job enrichment is conducted. One possibility is to combine adjacent operations. The tooling costs in most cases are additive. Since every workbench has a depth of 3 feet, it suffices to measure the workstation size by its width. In the 3.5-minute case, all workstations have the same width of 5 feet. The case of combined operations will have different workstation widths. The mean process times, ultimate yields, tooling costs and the space requirements per tool are given in Table 8.2.

The space requirement for a line can be approximated by looking at the workstation size, the space needed for maintenance and material handling, and the aisle space. Let w be the width of a workstation and d the depth. Each workstation has 3 feet for the material-handling system, 3 feet as maintenance-accessible area, 3 feet for the operator, and 3 feet for the pedestrian aisle. Hence, the total space requirement for such a workstation will be $w \times (d + 4 \times 3)$ square feet.

The average yield is computed by using Equation (8.10), assuming that (i) the initial yield is 85% of its ultimate, (ii) the improvement period is 3 years, and (iii) the product lifetime is 5 years.

The following cost items that are affected by cycle time change:

1. *Tooling*: The design and debug costs remain the same. The total replication cost is determined by the maximum number of workstations.
2. *Space*: Installation cost is \$140 per square foot and there is overhead expense of \$4 per square foot per year. The

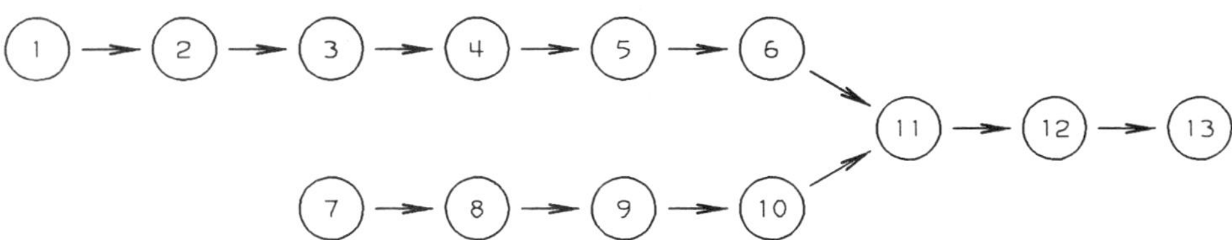

Figure 8.1 Precedence relations of an assembly process.

Table 8.2 Workstation Requirements under Different Cycle Times

Operation number	Process time	Ultimate yield	Tooling cost($k)	Operation number	Process time	Ultimate yield	Tooling cost($k)	Width (feet)
001	2.9	0.998	128					
002	3.6	0.998	11	001/002	6.5	0.996	139	8
003	2.2	0.988	56					
004	3.8	0.988	101	003/004	6.0	0.991	157	5
005	2.5	0.995	75					
006	3.5	0.995	32	005/006	6.5	0.990	107	8
007	3.5	0.980	85	007	3.5	0.980	85	
008	2.3	0.977	110					5
009	3.1	0.974	70	008/009	5.6	0.971	180	7
010	5.4	0.978	125	010	5.4	0.978	125	5
011	3.5	0.990	125	011	3.5	0.990	125	5
012	3.0	0.996	4					
013	3.5	0.996	50	012/013	6.5	0.992	54	7

total space requirement is based on the maximum number of workstations.

3. *Labor*: Each operation has one operator, except that operation 007 is an automated station. Operator cost is $35,000 per person per year. The crew size is based on the average number of workstations.
4. *Material handling*: This costs $400 per linear foot. The total length is equal to the sum of workstation widths. The maximum number of workstations is used.
5. *Maintenance*: This costs $85,000 per person per year. The size of maintenance crew is determined by using the results in Table 8.1.
6. *WIP*: The product unit cost is $1600 and the interest rate is 30%. Each workstation has a queue length of 5 (1 on the workbench and 4 in the buffer).

The WIP level is a rough estimate, based on an assumed of buffer size of 4. This estimate presents the worst case. More accurate results may be obtained by a modeling approach, which requires much more detailed information and can be very time-consuming. Although WIP reduction is an important subject in line management, WIP only contributes a small portion of the total cost in this example.

Both cases have the same tooling concept, but with different workstation layouts. Since workstation rearrangement does not involve major effort, no significant difference in engineering cost is expected. Supporting personnel, part-preparation cost, and part-quality cost are mainly dependent on the total production volume. These cost items are not considered in this example.

Table 8.3 Cost Comparison of Two Different Cycle Times

Cost item	Short cycle ($k)	Long cycle ($k)
Tooling/fixtures	4388	6804
Space	656	815
Direct labor	25200	22386
Material handling	168	210
Maintenance	3825	3825
Work-in-process	259	237
Total	34496	34277

The line sizes are 300 and 170 pieces per day for the short and the long cycles, respectively. A cost comparison is given in Table 8.3.

The major cost differences between the two cases are the tooling and the labor. The long cycle time requires more workstations but fewer operators. The cost difference between the two cases is 0.6% (= 1 − 44277/34496). This is a very small value, as compared to the possible errors from parameter estimation, such as operation process times, yield, and hardware/software costs. Since the cost difference is negligible and job enrichment would help productivity, the long cycle time is recommended. □

From Equation (8.9), it can be seen that the line size is also an important design parameter. Line size is usually dependent on line cycle time. In Example 8.1, a small line size (or equivalently, a long cycle time) is preferred. However, the structure of line cost may vary from one application to another. In the case of automated assembly, as an example, a short cycle time (or a large line size) is desired for simplicity. Line size is also a function of line organization. If multiple workstations per operation are considered in a line, then the line organization is a more important design factor than the line cycle time. This subject will be discussed in the next section.

8.2 Line Size

Determination of line size is a complex problem. Although it is convenient to assume an identical line size for multiple-line facilities, this assumption may not lead to the most economical solution. For a given demand, it is possible to find the optimal combination of line sizes. In this section, it is assumed that the peak demand is given, at least from a planning point of view. Methods dealing with demand change are presented in Section 8.3, where line configuration will be discussed.

Selection of a proper line size may be based on a cost analysis or facility requirements such as physical space constraints. Simple criteria are often used by line designers. One of these criteria is based on the ratio of cost to capacity. A line size with a small ratio will be chosen. Another criterion can be established by looking at the *excess capacity index* (ECI). This index is a measure for capital utility and is defined by

$$\mathrm{ECI} = 1 - \sum_{i=1}^{m} \frac{n_i u_i c_i}{TC} \tag{8.10}$$

where

m = number of workstation types

n_i = number of type-i workstations
u_i = utilization of a type-i workstation
c_i = cost of a type-i workstation
$TC = \sum n_i c_i$

For a given line size, the number of workstations of each type can be determined by Equation (8.9) with $d = c$ (i.e., demand equal to the line size). Cost items such as space, material-handling system, and labor may be expressed as functions of workstation requirements. Therefore, they can also be included in Equation (8.10) as workstation cost. After having computed ECI, line sizes with relatively small ECI can be chosen. A numerical example, based on a real case, follows.

Example 8.2 Consider an assembly process of 13 operations, as shown in Figure 8.2. The peak demand is 1150 pieces per day. The line capacity is measured by its *daily going rate* (DGR). The cost of a complete tool set plus space installation cost is $6,247,000 and is capable of producing 16 pieces per day, i.e., 16 DGR. The cost-to-capacity ratios and their corresponding ECI's are computed and listed in Table 8.4.

The values in the last two columns are decreasing in line size. Line sizes not listed in Table 8.3 and less than 576 have relatively higher ratios and larger values of ECI. For example, any size between 525 and 576 still costs $47.35 million due to tool breakage and, therefore, its cost-to-capacity ratio is greater than or equal to 82.35.

The cost-to-capacity ratios and ECI's are plotted in Figure 8.3. It can be seen that the ECI curve becomes flat after the line size exceeds 200 DGR. Based on this observation, a line size of 230 DGR. may be selected so that 5 lines are needed for a demand of 1150 DGR. □

When multiple lines of different sizes are installed, the line sizing problem can be treated as a mathematical programming problem. Assume that n possible line sizes are under consideration. For $k = 1, 2, \ldots, n$, define

s_k = size of line k
c_k = cost of line k
x_k = number of lines of size s_k
d = demand

The line sizing problem may be formulated as follows:

$$\text{minimize} \sum_k c_k x_k$$

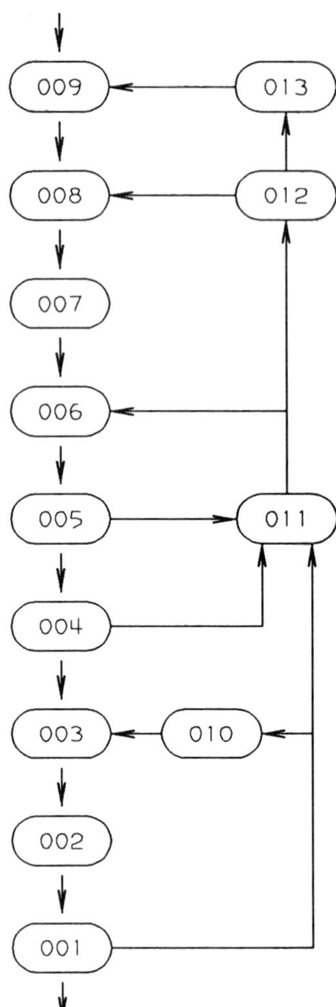

Figure 8.2 An assembly flow process.

Table 8.4 Tooling and Space Installation Cost per Line

Number	Line size	Cost ($m)	Ratio ($k)	ECI
1	16	6.25	390.44	0.81
2	33	6.54	198.14	0.63
3	48	6.83	142.31	0.48
4	67	9.05	135.07	0.45
5	84	9.34	111.21	0.34
6	90	9.63	107.04	0.31
7	135	13.54	100.26	0.26
8	139	13.83	99.47	0.26
9	145	14.30	98.59	0.25
10	169	16.52	97.72	0.24
11	180	16.81	93.37	0.21
12	193	17.94	92.97	0.21
13	231	20.77	89.90	0.18
14	237	21.00	88.61	0.17
15	270	23.69	87.75	0.16
16	288	25.12	87.21	0.15
17	321	27.94	87.04	0.15
18	338	28.70	84.91	0.13
19	436	36.53	83.78	0.12
20	484	40.23	83.49	0.12
21	525	43.23	82.35	0.10
22	576	47.35	82.20	0.10
—	1152	92.05	79.90	0.09

$$\text{Subject to } \sum_k s_k x_k \geq d$$

$$x_k \geq 0, \text{ integers } k = 1,\ldots,n \tag{8.11}$$

This problem is also known as the *knapsack* problem, and may be solved by a dynamic programming technique. The method described in the following requires two auxiliary functions:

1. $f_k(y)$ is the minimum cost value for a demand of y, after having considered lines $1, \ldots, k$

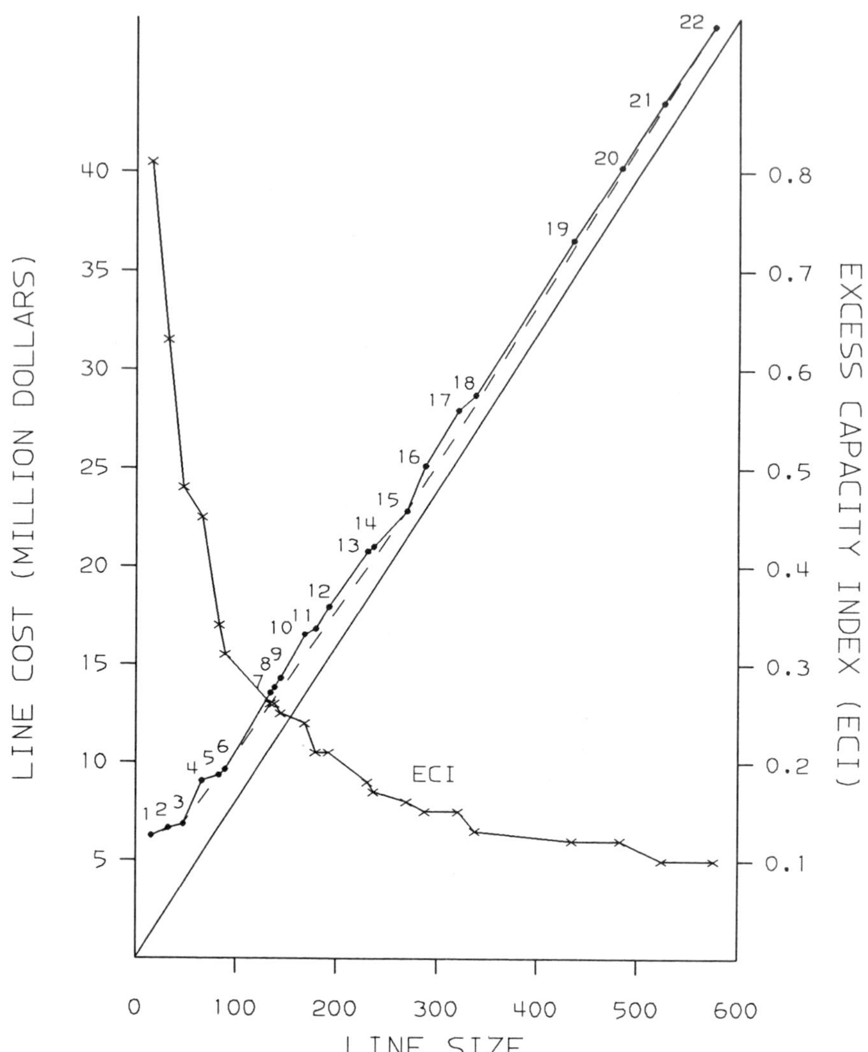

Figure 8.3 Excess capacity index and line cost.

2. $\{h_k(y) = i\}$ means that a line of size s_i should be installed to support a total demand of y, after having considered lines 1, ..., k

Then a recurrent relation can be defined:

$$f_k(y) = \min\{f_{k-1}(y), f_k(y - s_k) + c_k\} \tag{8.12}$$

If $f_k(y - s_k) + c_k$ is smaller, it will be more economical to place a line of size s_k and, therefore, $h_k(y) = k$. Otherwise, line k will not be considered and $h_k(y) = h_{k-1}(y)$. A computational procedure is summarized in Algorithm 8.1.

Algorithm 8.1

1. Select line sizes $\{s_k\}$.
2. Determine line costs $\{c_k\}$.
3. Let $f_0(0) = h_0(0) = 0$ and $k = 0$.
4. $k = k + 1$.
5. Invoke the recurrent relation (8.12) for all y between 0 and d (the demand), and assign the value of $h_k(y)$ accordingly.
6. If $k = n$ (the maximum line size under consideration), stop. Otherwise, go to step 4.

Example 8.3 Consider three line sizes: $s_1 = 100$, $s_2 = 200$, and $s_3 = 400$, with line costs $c_1 = 20$, $c_2 = 30$, and $c_3 = 50$, respectively. The computational results can be summarized into two tables. Table 8.5 gives the values of $\{f_k(y)\}$ and Table 8.6, the values of $\{h_{k(y)}\}$ for $y = 100, 200, ..., 900$.

To see the computational logic, let us consider $k = 3$ and $y = 500$. Two possibilities must be investigated: (i) $f_2(500) = 80$ and (ii) $f_3(500 - s_3) + c_3 = f_3(100) + 50 = 20 + 50 = 70$. Since the second case has a smaller value, $f_3(500) = 70$ and $h_3(500) = 3$. This means that a line of size 400 should be installed. To support a demand of 500, another line

Table 8.5 Cost values of $f_k(y)$

Demand	100	200	300	400	500	600	700	800	900
$k = 1$	20	40	60	80	100	120	140	160	180
$k = 2$	20	30	50	60	80	90	110	120	140
$k = 3$	20	30	50	50	70	80	100	100	120

Table 8.6 Index values of $h_k(y)$

Demand	100	200	300	400	500	600	700	800	900
$k=1$	1	1	1	1	1	1	1	1	1
$k=2$	1	2	2	2	2	2	2	2	2
$k=3$	1	2	2	3	3	3	3	3	3

of size 100 must be placed. The line costs are 50 and 20, respectively, and the total is 70.

The function h is used to backtrack the solution. Since $h_3(900)=3$, a line of size $s_3=400$ is needed. To support the remaining demand, the value of $h_3(900-400)$ is inspected. For $h_3(500)=3$, the second line of size 400 should be placed. From the fact that $h_3(100)=1$, it is concluded that the optimal solution is composed of two 400-lines and one 100-line. The minimum cost value is $2\times c_3+c_1=120$. □

Algorithm 8.1 may also be used for constrained problems. For instance, an upper bound for line size may be imposed due to space limitations or management concerns. This concept may be illustrated in Example 8.4.

Example 8.4: Line Sizing under Constraints The same data presented in Example 8.2 are considered here. Again, to support a demand of 1150 pieces per day, multiple lines may be installed. However, all the individual lines may have different sizes and are bounded from above. By relaxing the upper bound, a sequence of problems is defined. Denote a problem with an upper bound b by $P(b)$. The problem formulation is the same as (8.14), except that $c_k \le b$ for all k. Algorithm 8.1 is still valid.

For $b_1 > b_2$, the solution space of $P(b_2)$ is a subset of that of $P(b_1)$. Therefore, the latter has a smaller cost value than the former. This is clearly demonstrated by Figure 8.4, where the horizontal axis is the maximum line size, b.

Solutions and their cost values are also given. For instance, the highest solution point shows that 13 lines are needed; 10 of them have a capacity of 90 and 3 have a capacity of 84. The total cost value is \$124.37 million.

It has been mentioned in Example 8.2 that a line size of 230 may be considered. However, the curve in Figure 8.4 suggests that significant cost savings are possible with larger line sizes. □

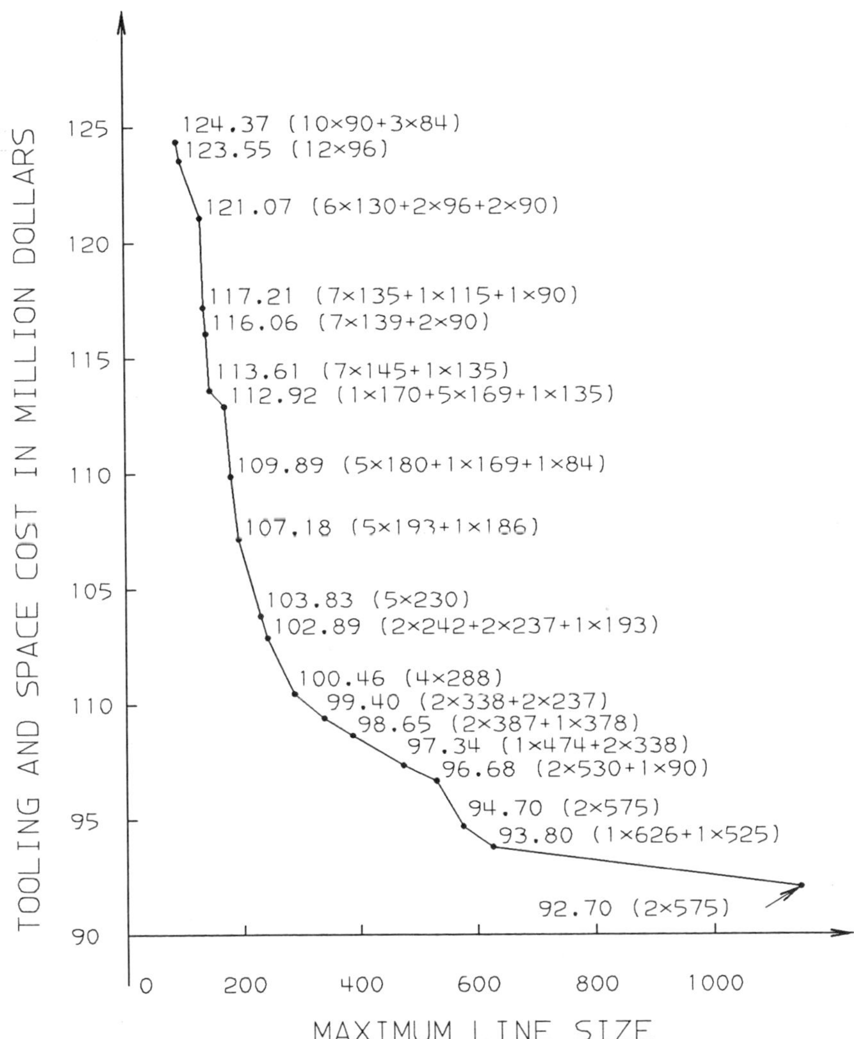

Figure 8.4 Minimal cost solutions under different line size constraints.

8.3 Line Configuration

Another one of difficult design problems is to optimize the line layout. The complexity of the problem is due to multiple design parameters, such as line size, buffer size, workstation space requirement, material flow pattern, material-handling system, and building space. Most layouts have been done by trial-and-error methods. Flow analysis has been a suggested as a tool that may lead to an optimal layout. Sometimes a computerized program, such as CRAFT (see Buffa 1964), can be used to lay out a line. But these methods usually have very limited design objectives. Demand uncertainty is also major problem. Ideally, the line capacity should closely follow the production demand. Unfortunately, the latter is derived from market forecast, and is subject to change. Engineering changes can be another source of uncertainty. When such a change occurs, line capacity may be different. Sometimes, process flow is altered so that a line must be rearranged for a better flow pattern. On the other hand, because of learning effect, both operation process time and product yield may be improved as the accumulated production quantity increases. This means that line capacity is increased. If the product has a new model, the line capacity can be temporarily dropped due to the initial yield problem. All these indicate that a line layout problem in reality is much more than flow analysis. A different approach should be adopted.

A layout problem may be solved in two steps. First, only the line configuration problem is under consideration. Then a configured line is placed in a given building space. The second step is relatively straightforward, if an effective solution can be found in the configuration stage. The following discussion concentrates on the subject of line configuration.

It is known that for the same amount of total space requirement, small objects can usually fit a given space better than large ones. Based on this observation, two configuration approaches are considered. One is called *line replication* and the other is *line modularization*. The latter approach is to organize the line in a modular fashion so that line capacity can be adjusted by adding or dropping line modules. The concept of line replication is rather simple and is discussed first.

Line Replication

Under this approach, a line of a small capacity is installed. Whenever the production demand is increased, an identical line is added. If the demand is reduced, a line may be shut down. Under this approach, line granularity and balance efficiency are two major factors for design consideration. Since market demand is never controlled by the manufacturer, a line should be flexible enough for demand changes. When the production capacity is adjusted by adding or deleting lines, a small line size means a small capacity increment and, therefore, good flexibility.

On the other hand, the line must be well balanced. For a poorly balanced line, an economical way of capacity expansion is to add extra workstations to the bottleneck operations. This addition will also help to improve balance efficiency. However, the line-replication approach always adds a line, instead of workstations. A small capacity increment becomes very costly. Consequently, manufacturing resources are wasted. To see this, consider a line of two operations; each has a single workstation. Their mean process times are 3 and 5 minutes, respectively. The line is capable of producing 12 pieces in an hour. If the line is operating at its capacity, workstation utilization should be close to 0.6 and 1.0, respectively. If a higher capacity is needed, a workstation may be added to the second operation. Workstation utilization becomes 1.0 and 0.83 (= $20/2 \times 12$), respectively. If the line-replication approach is employed, there will be two identical lines. The workstation utilization remains unchanged. The first operation is always underutilized. Therefore, this approach may not be suitable for a capital-intensive production.

If a line is well balanced or capital investment is a relatively small portion of the total line cost, line replication can be a very effective solution to line layout. Since the entire production facility consists of multiple small lines, all the advantages of the focused-line concept discussed in Section 4.3 apply. Usually each line is composed of single-workstation operations. Material flow has a simple pattern. Assemblies always go from one station to its adjacent neighbor. When dealing with multiple product types, it is likely to schedule one type at a line, i.e., no multiple products share the same line. Consequently, parts logistics is simple. Because of simple material flow patterns and logistics, the material-handling system may be simplified. Furthermore, engineering changes may be done one line at a time. Production capacity will not be totally lost during these changes.

Line Modularization

The basic concept of line modularization is to partition a large line into a number of small modules. Capacity is increased by adding one or more workstations at a time. The newly added stations may be placed in existing modules or new modules. This approach aims at capital saving and space utilization. Since a perfectly balanced line is practically nonexistent, adjusting capacity by looking at the workstation level improves utilization and balancing efficiency. Although the line is composed of modules, an individual module may not contain all the operations that are necessary for a complete assembly process. The concept of modularization usually leads to a single line of a large size. Multiple products, therefore, must share the same line. For this reason, material flow control becomes a complicated problem and efficient line operation must depend on an intelligent control system.

Table 8.7 Comparison of Two Different Line Configuration Concepts

Factor	Replication	Modularization
Capacity adjustment	Line level	Workstation level
Cost structure	Not capital intensive	Capital intensive
Balance efficiency	Should be high	Less important
Material handling	Simple	Need sophisticated system
Logistics	Simple	Complex for multiple models
WIP management	Relatively easy	More complex
Flow control	Simple	Complex
Capacity increment	Large granularity	Small
Space utilization	Relatively poor	Good

Since line expansion is treated at the workstation level, the traffic within the line often presents a problem. Adding workstations may disrupt line operation, or may create a traffic bottleneck that seriously degrades line performance. On the other hand, if the capacity adjustment is to decrease production quantity, removing workstations may also disrupt the line and waste space. To resolve this problem, a flexible material-handling system such as AS/RS (see Chapter 5) may be considered and a systematic line-configuration procedure is required; a detailed discussion is given in Section 8.4. A comparison of the line-replication and line-modularization approaches is summarized in Table 8.7.

8.4 Design for Line Flexibility

Determination of line capacity is a major problem for the line designer. Despite short-term fluctuation, production demand is never a constant over the product lifetime. A typical demand curve, based on market research on a future product, is shown in Figure 8.5. If the line is designed for the peak demand, an underutilization condition occurs at both early and late stages. If the design is based on the average demand, the line would fail to support the peak period. It is clear that line capacity should be dynamically adjusted to meet production demand at minimum cost. In this section a systematic procedure, based on the line modularization concept, is introduced. A numerical example will be presented to compare cost differences between the modularization and replication approaches in Section 8.5.

To facilitate the traffic flow, an AS/RS is adopted as the material-handling system. Since this system has direct-access capability, line layout or configuration does not have to follow assembly flow process closely. As discussed

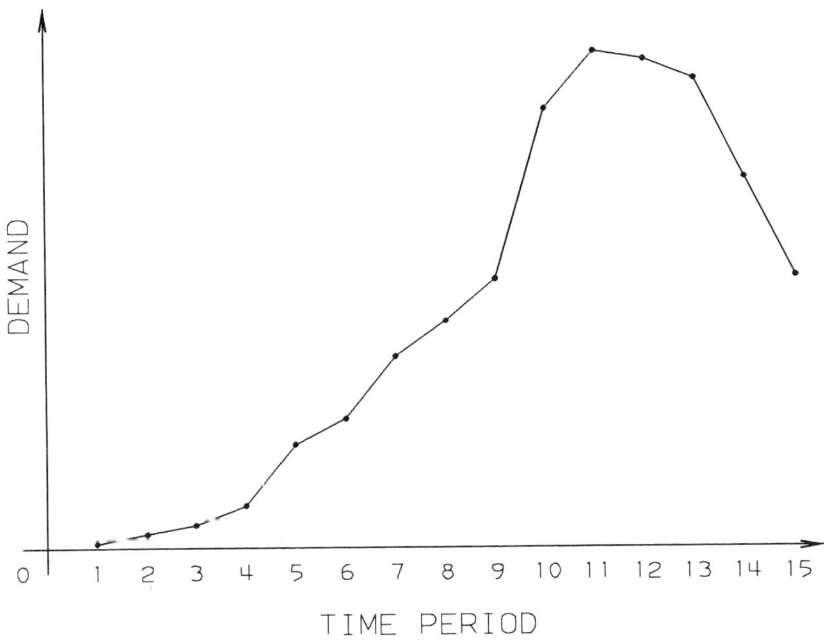

Figure 8.5 A typical demand curve.

in Section 5.2, an AS/RS may contain a number of direct-access handlers (DAH). A DAH may directly interface with a number of workstations. Vertical storage racks may be used as buffers for WIP.

If a line is partitioned into a number of modules, workstations in the same module are served by one and only one DAH. The configuration procedure consists of three major steps: (i) for a production demand pattern, determine the number of workstations for each operation, (ii) compute traffic volume between workstations, and (iii) distribute workstations into modules, according to the characteristics of a given AS/RS. The first two steps can be accomplished in a straightforward manner, it is the last one that needs our attention. From the discussion in Section 5.2.2, it is known that a DAH has a stable performance region. This region can be characterized by looking at the relation between the average material handling delay and DAH utilization. Under the existing technology, this region is defined by limiting the traffic volume, as seen by a DAH, such that the average material handling delay should not be greater than 60 seconds. (See Figure 5.6). For different hardware parameters, the stable performance region may be different, and can be redefined by employing the procedure discussed in Section 5.2.2.

In addition, the impact of material-handling delay must also be assessed from a workstation point of view. Once the delay becomes a part of the operation cycle at a workstation, the line throughput may be significantly reduced. For instance, if the mean process time at a workstation is 5 minutes and each operation cycle experiences a 60-second delay on the average, the workstation capacity is then reduced by 20%. If this workstation happens to be the bottleneck, the line capacity is decreased by 20%. A simple solution to this problem is to create a buffer of size 1 at each workstation. Removal of a work unit from the buffer to the workbench triggers a service request for a new unit. If the operation cycle time is longer than the material handling delay, then the latter becomes transparent to the workstation, i.e., the production cycle is independent of the delay. For practical problems, one may keep the 90th percentile of the material-handling delay below the average process time. The results from Section 5.2.2 show that the 90th percentile is approximately equal to the mean plus 1.5 standard deviation. The computational procedure for the mean and the standard deviation is sketched as follows. (A detailed discussion is given in Section 5.2.1.)

Let

tt_{ij} = travel time from locations i to j
r_{ij} = arm rotation time from locations i to j
g = pick time
h = place time

The travel time can be derived from the travel distance, top speed and acceleration/deceleration. The arm rotation time, on the other hand, is a ratio of the degree of rotation to rotation speed. If a service involves moving an object from locations j to k while the DAH is currently at location i, then the service request is called an (i,j,k) request and its service duration is given by

$$s_{ijk} = \max(tt_{ij}, r_{ij}) + g + \max(tt_{jk}, r_{jk}) + h \tag{8.13}$$

Given that a service request occurs, the probability of an (i,j,k) request is proportional to the product of the receiving rate at location i and the request rate from j to k. Hence, the service time distribution can be defined by the characteristics of the DAH, the line layout, the flow pattern, and the traffic volume. Using an M/G/1 model, the average material handling delay is given by

$$E[W] = E[S] + \frac{E[S]\rho}{2(1-\rho)}(1 + C^2) \tag{8.14}$$

where

S = service time
W = material handling delay
ρ = DAH utilization or product of arrival rate and $E[S]$
C = coefficient of variation of S

The variance of the material handling delay can also be obtained by using the M/G/1 queuing model. The following relation is copied from Equation (5.11):

$$Var[W] = Var[S] + \frac{\rho}{1-\rho}\frac{E[S^3]}{3E[S]} + \left(\frac{\rho}{1-\rho}\frac{E[S^2]}{2E[S]}\right)^2 \tag{8.15}$$

The above discussion may be summarized into two design criteria:

1. The average material handling delay within a module should be kept below 60 seconds.
2. The 90th percentile of the material handling delay within a module should be less than the minimum mean process time among all workstations in the module.

The configuration procedure consists of two computational algorithms. Let

m = number of operations
n_i = number of type-i workstations
t_i = average process time at operation i
a_i = availability of a type-i workstation
v_i = start factor of operation i
D = production demand
T = effective working time during a day

As shown by Example 4.4, the start factors can be obtained by solving a set of simultaneous linear equations that involve product yield at different operations. Since the total available time of a type-i workstation during a day is a_iT, the capacity of operation i is given, in terms of number of products per day, by

$$q_i = \frac{n_i a_i T}{t_i v_i} \tag{8.16}$$

The number of workstations per operation at different tool breakage points is given by the algorithm that follows.

Algorithm 8.2

1. Let $n_i = 1$ for $i = 1, 2, \ldots, m$ and $r = 1$.
2. Compute q_i, $i = 1, 2, \ldots, m$, by using Equation (8.16).
3. Find $Q_r = \min\{q_i \mid i = 1,2,\ldots,m\}$
4. Record Q_r and $S_r = \{k \mid q_k = Q_r\}$.
5. If Q_r exceeds the peak demand, stop. Otherwise, replace the value of n_k by $n_k + 1$ for all k in S_r, increase r by 1, and go to step 2.

At the termination of Algorithm 8.2, a sequential process for line expansion is defined by $\{Q_r\}$ and $\{S_r\}$. Initially, the line capacity is Q_1 and each operation has exactly one workstation. If one workstation is added to operation k for all k in S_i, the line capacity is increased to Q_2. The difference, $Q_2 - Q_1$, is the smallest possible increment. If all the operations recorded in S_2 have one additional workstation, the line is capable of producing Q_3, with a capacity increment $Q_3 - Q_2$. On the other hand, one may also choose to increase the line capacity from Q_1 to Q_3 by adding one workstation to each operation in S_1 and then to each in S_3. Consequently, a line expansion and contraction schedule can be determined by taking the following steps:

1. Invoke Algorithm 8.2 to obtain $\{Q_i\}$ and $\{S_i\}$.
2. Determine capacity increments or decrements, based on $\{Q_i\}$.
3. For each operation, find the number of workstations to be added according to step 2 and $\{S_i\}$.
4. Place workstations into modules such that the average material-handling delay of the DAH never exceeds 60 seconds and the 90th percentile is no greater than the minimum mean process time in the module.

The first three steps are straightforward. The fourth one needs more discussion. For simplicity, let us consider the problem of line expansion only. The line contraction procedure is just opposite to the one used for line expansion. Furthermore, it is assumed that (i) for each expansion only the smallest capacity increment will be chosen and (ii) exactly one workstation is added for each expansion.

The first assumption is made for simplifying our discussion and will be eliminated later. Unless workstations of different types have a common multiple, the second assumption follows automatically under the first one. Thus, the cardinality (i.e., the number of elements) of S_i is always 1. Let Q_p be the smallest capacity in the expansion schedule that is greater than or equal to the peak demand, D, and k_i be the element in S_i. Algorithm 8.3 defines a configuration procedure for a flexible line.

Algorithm 8.3

1. Let L_i be a list of workstations for capacity Q_i defined by Algorithm 8.2:
   For Q_1, $L_1 = (1,2,\ldots,m)$
   For Q_2, $L_2 = (1,2,\ldots,m,k_1)$
   ...
   For Q_p, $L_p = (1,2,\ldots,m,k_1,\ldots,k_{p-1})$
2. Determine the flow volumes between workstations, and from parts staging or preparation stations to each individual workstation, for $Q_1, Q_2, \ldots, Q_p$.
3. Let $i = 1$.
4. If L_i is empty, skip to step 9.
   If L_i is not empty, let X_i be an empty set.
5. Copy an element from L_i into X_i according to the sequential order in L_i.
6. Let all workstations, included in X_i, be placed in the same module and served by a single DAH. Based on the flow information from step 2, compute the average material handling delay and its standard deviation by using Equations (8.14) and (8.15).
7. Go to step 5 (i.e., one additional workstation will be placed in the module), if both of the following conditions are satisfied:

   (7.1) The average material-handling delay is no more than 60 seconds

   (7.2) The average delay plus 1.5 standard deviation does not exceed the mean process times of all workstations included in X_i.

If one or both conditions are violated, delete the last element from X_i and go to the next step. (In this case, a tentative module is created for Q_i.)

8. Record X_i for future reference, unless X_i is empty. If X_i is empty, stop the algorithm. (This means that no feasible solution can be found, and the DAH does not have enough capacity for a module with even a single workstation. Thus, the AS/RS is not the right choice for this line.)
9. If $i < p$, increase i by 1 and go to step 4. (Continue workstation placement with another tentative module.) If $i = p$, all tentative modules have been found. For each demand, exactly one module has been created. The workstations in the module, that correspond to demand Q_i, are identified by the list X_i.
10. Determine the list Z such that $Z = X_k$ is the shortest list among all X's, i.e.$|X_k| \leq |X_i|$ all $i = 1, 2, ..., p$. Destroy all X's except X_k. (A common module for all Q's has been defined.) Record the content of Z.
11. Delete all elements from L_1, L_2, ..., L_p, if these elements also appear in X_k.
12. If all X-lists become empty, all modules have been defined. Stop the algorithm. Otherwise, go to step 3 and create additional modules for nonempty L-lists.

At the completion of the algorithm, the contents of all Z-lists constitute a flexible line configuration. Suppose that r Z-lists have been recorded, that is, r modules have been created. Let Z_1, Z_2, ..., Z_r be the sequence of the lists. To increase line capacity, workstations are either added to the last module or placed in a new module. Furthermore, any addition of workstations can be done at the end of DAH aisle. Thus line disruption is kept at the minimum level.

For practical problems, capacity adjustment may take a larger increment than the ones used by Algorithm 8.3. Therefore, each addition may involve more than one workstation. These workstations may be placed within the same module or distributed into different modules. In this case, one can still add one station at a time and test DAH performance against the two design criteria. Suppose that workstations 1, 2, and 3 should be added to increase the total capacity from Q_{i-1} to Q_i. Then L_i will have three more elements than does L_{i-1}. These three elements, {1,2,3}, can be placed at the end of L_i. But the order of these three elements in L_i may be important. Since DAH performance is a function of its travel distance, their order may be determined by comparing the traffic volumes between each of these three workstations and other stations in the existing module. The workstation associated with the heaviest traffic should be placed first.

Algorithm 8.3 is illustrated in the following example, from Chow (1986).

Example 8.5 A magnetic disk assembly line consists of 8 sequential operations. The estimated mean process times (in hours) are 0.05, 0.10, 0.15, 0.15, 0.15, 0.35, 0,15, and 0.10, respectively. Assume that

1. $a_i = 0.95$ and $v_i = 1$ for $i = 1, \ldots, 8$.
2. The effective production hours, $T = 20$.
3. The peak demand is 200 pieces per day.

Algorithm 8.2 leads to the result shown in Table 8.8. A total of 8 workstations are required, one for each operation, to produce 54 pieces per day. If the demand is increased but is below 108, one additional workstation should be added to operation 6. For a demand between 109 and 126, extra workstations of operations 3, 4, 5, 6 are needed. This adding process continues until the line capacity reaches 217.

In this application, no buffer is allowed at workstations. Instead, a common storage rack is placed between the DAH aisle and the workstations (see Figure 5.2). Therefore, the average material handling delay (below 60 seconds) is the only design criterion. When a work unit is completed at a workstation, two service requests will be submitted: (i) remove the completed unit, and (ii) replenish the workstation with a fresh unit. The replenished unit may be retrieved from the storage rack. The unit may also come from an upstream workstation. On the other hand, the removed unit may go directly to a downstream workstation or to the rack. Since the rack has a large storage space, the rack overflow condition almost never happens. This implies that a

Table 8.8 A Line Capacity Plan

Capacity	Workstation Types							
54	1	2	3	4	5	6	7	8
108	—	—	—	—	—	6	—	—
126	—	—	3	4	5	6	—	—
162	—	—	—	—	—	—	7	—
189	—	2	—	—	—	6	—	—
217	—	—	—	—	—	—	—	8

completed work unit will never occupy the workstation due to insufficient rack storage.

From a material handling point of view, the traffic volume is determined by the number of pick and place operations. When a work unit is handed from one workstation to another, the number of pick and place operations can be different, dependent on whether the rack is used for intermediate storage or not. If the unit goes directly to a downstream operation, only one pick and one place operations are needed. If the downstream workstation is occupied, then the work unit must be sent to the rack first and retrieved later. That is, two pairs of pick and place are required. The expected traffic volume due to job completion at operation i is given by

$$w_i = Q(1 + u_{i+1}) \tag{8.17}$$

where u_{i+1} is the probability that all workstations of operation $i + 1$ are occupied upon a job completion at operation i.

If events happen in a random fashion, the probability of occupancy, u_i, can be approximated by the utilization factor of operation i. For a production demand, D,

$$u_i = \frac{D v_i t_i}{n_i T} \tag{8.18}$$

At the peak demand period, D = 200. It follows that $(u_1, u_2, \ldots, u_8) = (0.54, 0.54, 0.81, 0.81, 0.81, 0.95, 0.81, 0.54)$.

The DAH is also responsible for parts delivery. Regardless of its function, each workstation consumes exactly one parts tray for one work unit. Both full trays and empty trays are handled by the DAH.

Workstations have a standard size of 10 feet wide. Each module has an input-output (I/O) port, located at one end of the DAH aisle. All traffic between the workstations within the module and the outside must go through the port. Modules are connected by a conveyor, which can move objects continuously and has a much larger capacity than the DAH. Each I/O port has an accumulator capable of handling all incoming traffic from the DAH promptly. Therefore, a DAH is never blocked due to I/O port unavailability. All DAH's are identical and have the following hardware features:

Top speed	6.6 feet/second
Acceleration/deceleration	3.3 feet/second/second
Arm rotation speed	60.0 degrees/second
Pick time	4.5 seconds
Place time	4.5 seconds
Vertical travel time	1.0 second/foot

The module contents and the DAH performance derived from Algorithm 8.3 are presented in Table 8.9. The line configuration is illustrated in Figure 8.6.

It can be seen from Table 8.9 that the average material-handling delays are consistently below 60 seconds. The line starts with two modules and is gradually expanded into three modules. The last expansion is completed by adding a workstation of operation 8 to the third module. The total line capacity is 217 pieces per day, which already exceeds the peak demand.

Table 8.9 Line Configuration and DAH Performance

Demand	Module	Module Content	Average Delay	Utilization
54	1	(1,2,3,4,5,6,7)	21 seconds	0.39
	2	(8)	13 seconds	0.03
108	1	(1,2,3,4,5,6,7)	40 seconds	0.74
	2	(8,6)	13 seconds	0.12
126	1	(1,2,3,4,5,6,7)	32 seconds	0.65
	2	(8,6,3,4,5,6)	20 seconds	0.38
162	1	(1,2,3,4,5,6,7)	36 seconds	0.71
	2	(8,6,3,4,5,6,7)	28 seconds	0.59
189	1	(1,2,3,4,5,6,7)	40 seconds	0.74
	2	(8,6,3,4,5,6,7,2)	47 seconds	0.78
	3	(6)	13 seconds	0.05
217	1	(1,2,3,4,5,6,7)	46 seconds	0.79
	2	(8,6,3,4,5,6,7,2)	40 seconds	0.71
	3	(6,8)	13 seconds	0.11

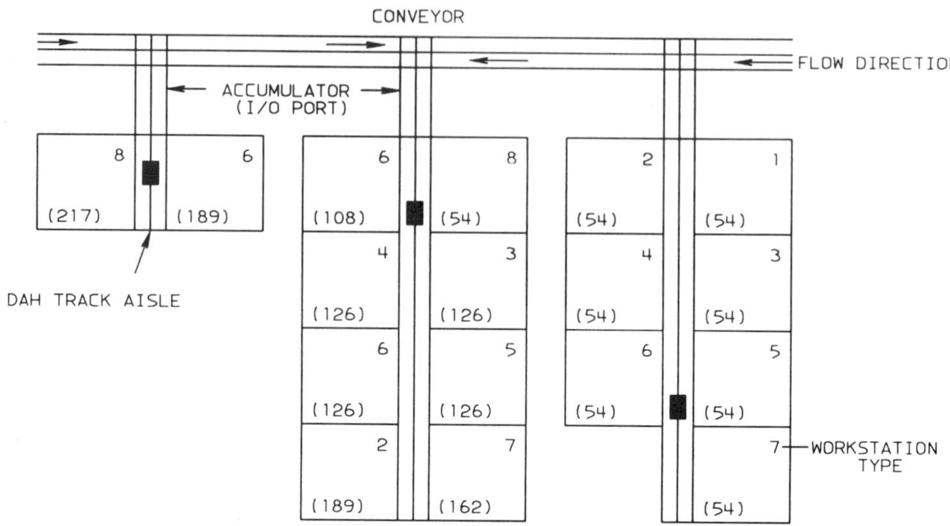

Figure 8.6 A flexible line.

When the product is about to be phased out, the contraction process is the opposite. First, a workstation of operation 8 is eliminated from the third module, and the capacity is reduced from 217 to 189. Next, module 3 is deleted and the last workstation in module 2 (i.e., the workstation of operation 2) is removed. The line capacity is further decreased to 162. This reduction process can be carried out until the line returns to 54 pieces per day. In this case, each operation has one and only one workstation. Any further elimination will be unacceptable unless no products are produced. □

The previous example illustrates a process for adjusting line capacity by gradually adding workstations to the existing line. If an addition involves multiple workstations, proper arrangement of workstations may improve material-handling performance. For example, since workstations of operations 3, 4, 5, and 6 are added to the second module at the same time as the line capacity is increased from 108 to 126, their locations may be changed for better DAH performance. This change will not affect the line expansion/contraction schedule. If the capacity increment is enlarged, then each expansion would involve a greater number of workstations, and a greater performance improvement can be expected. Since the design criterion is based on the material-handling delay, better DAH performance may lead to a smaller number of DAH's.

As mentioned before, another method for line expansion with minimal line disruption is to replicate the existing line. If tooling and space costs are high, the flexible line approach is usually more economical. This is illustrated in the next section.

8.5 Case Study—A Flexible Line

In Example 8.2, the assembly line consists of 13 operations. The assembly flow process is given in Figure 8.2. Tooling and space installation costs for different line sizes are given in Figure 8.3. For a given demand, the costs for different combinations of line sizes are shown in Figure 8.4. It can be seen that a large line tends to be more cost-effective. On the other hand, the cost saving diminishes as the line size increases. A large line may also be constrained by the building shape and the total available space. The design objective is to minimize the total manufacturing cost under a given space constraint.

Major cost items are:

1. Tooling
2. Space installation
3. Material-handling system
4. direct labor
5. Work-in-process
6. Facility operating expense

The tooling cost per operation is given by $d + n \times r$, where d is the design and debug cost, r the replication cost, and n the number of workstations. The space installation cost is directly proportional to the total space. To maximize space utilization, all workstations have the same depth of 10 feet, but their widths can vary. All workstations are 10 feet wide, except for operations 001, 007, and 009, where workstation widths are 2.5, 20, and 20 feet, respectively. All operations, except operation 001, must be performed in a clean room environment. The clean-room installation cost is $400 per square feet; for an ordinary factory environment, it is $140 per square feet.

The product size is about $2 \times 2 \times 2$ feet and it weighs about 80 pounds. From an ergonomic point of view, a manual material-handling system is not a feasible solution. Both tooling and space installation costs count for more than 50% of the total cost. Therefore, a flexible line concept is adopted. An AS/RS is selected for material handling. This will accomplish two goals: (i) dynamic capacity adjustment helps to achieve *just-in-time* investment, (ii) line modularization will ease the layout problem for a large line. Cost of a DAH is about $150,000, including the storage rack and DAH track.

Workstations will be placed in both sides of a DAH aisle. Hence, the total aisle length is about one half the sum of the workstation widths.

Labor cost is directly proportional to the number of operators. In this line, each manual operation needs exactly one operator. One operator, however, can handle multiple automated workstations. The cost rate of an operator is $14 per hour. The facility operating expense is a function of the size of the facility. For a cleanroom-environment, the expense is about $40 per square foot per year. For an ordinary factory environment, the cost rate is approximately $4 per square foot per year.

Since the cost of two 575 DGR lines is comparable to that of a single line of 1150 DGR at the peak demand (see Figure 8.3 and Table 8.4), the maximal line size is chosen to be 575 DGR. A line expansion schedule is established by looking at production demand pattern and line costs. Six different line sizes chosen for expansion are 48, 135, 231, 338, 484, and 575 DGR. That is, the initial size of the line is 48 DGR, and gradually expands to 135, 231, 338, 484, and 575 DGR. These values correspond to points 3, 7, 13, 18, 20, and 22 in Figure 8.3. It can be seen that these points are close to the envelope of the cost curve, which is indicated by the dashed line in Figure 8.3. Therefore, each expansion has a relatively small cost-to-capacity ratio. Table 8.10 shows the number of workstations per operation for different capacities.

Table 8.10 Number of Workstations per Operation for Different Capacities

Operation	Line capacity					
number	48	135	231	338	484	575
001	3	8	14	20	29	34
002	1	2	2	3	5	5
003	1	2	3	4	5	6
004	1	3	5	7	10	12
005	1	2	3	4	6	7
006	1	2	2	3	5	5
007	1	1	2	3	4	5
008	1	1	2	2	3	4
009	1	1	1	2	2	2
010	1	1	1	1	1	1
011	1	1	1	2	2	2
012	1	1	1	1	1	1
013	1	1	1	1	1	1

Each workstation is equipped with a rotary table which serves as the interface between the DAH and the workstation. Work units are placed on the table and can be transferred from the DAH to the workstation or from the workstation to the DAH. The table is also used as a local buffer between the DAH and the workstation. A new work unit may be placed at one end of the table and fed to the workstation by rotating the table. A completed unit is taken from the workstation in the same way. If a new unit and a completed unit are presented on both ends of the table, a table rotation causes a swap action. Whenever a new unit is taken by the workstation, it triggers a service request for a new work unit. As long as the time to replenish the buffer is less than the duration of the next operation cycle, the delay caused by the DAH is transparent to the workstation.

Figure 8.7 presents an expansion plan for a 575 DGR line, which consists of six modules served by six DAH's. Workstations are placed on both sides of DAH aisles. The operation number of each workstation is indicated by a single or double digit. The line is configured based on (i) DAH performance (i.e., material-handling delay), (ii) process flow, (iii) expansion schedule (based on a given demand pattern), and (iv) requirements of the working environment. Since operation 001 must be performed in a different manufacturing environment, all its workstations have been placed in an independent module. The initial line has only three modules, as shown by modules 1, 4 and 6 in Figure 8.7. The modules are expanded by moving the boundary line to the right. When the demand is over 135 DGR, modules 2 and 5 are created. Module 3 is introduced as the demand goes beyond 338 DGR. If the demand is higher than 575 DGR, a second line should be established. The average material-handling delay in each module is consistently below 60 seconds, and its 90th percentile is less than the minimal mean process time in the module. Modules are linked by a power roller conveyor, which is capable of serving 15 units per minute.

The manufacturing cost is estimated by the discounted cash flow method, using the following assumptions:

1. A product lifetime of five years
2. Six months to build the initial line
3. Build time of three months for each expansion
4. A discount rate of 10%
5. Company income tax bracket of 50%
6. Investment tax credit of 8%
7. Capital equipment depreciated over a five-year period with a schedule of 0.15, 0.22, 0.21, 0.21, 0.21
8. WIP cost treated as an interest loss with annual rate of 10%
9. Direct labor cost increases by 7% each year

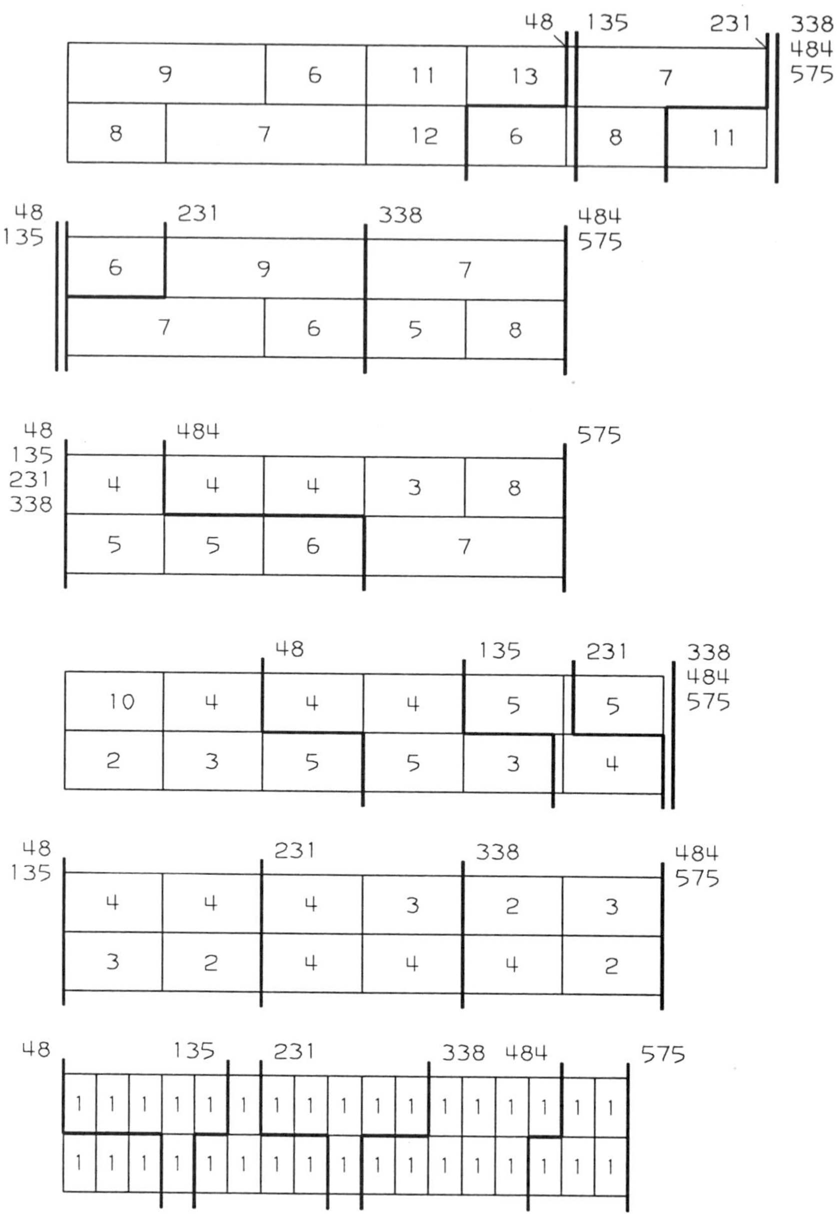

Figure 8.7 A line configuration based on the concept of line modularization.

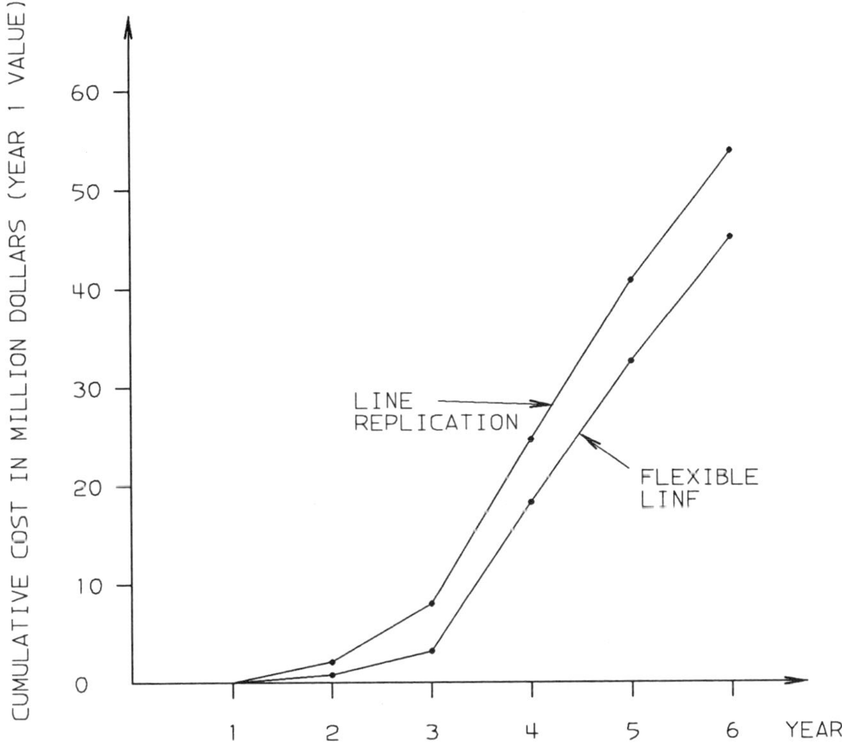

Figure 8.8 A cost comparison of two different line concepts.

The present values of accumulated cost are given in Figure 8.8, where two curves are shown. The lower one is the cost curve derived from two 575 DGR lines, based on the flexible line concept. The upper curve is the accumulated cost of five 230 DGR lines under the line replication approach, assuming the identical demand schedule and the same material handling system. The cost difference is about 15 to 20%. The saving from the flexible line approach is due to the fact that a small capacity increment delays the capital expenditure.

8.6 Remarks

This chapter has discussed three major subjects: line cost, line sizing, and line configuration. Two cost methods have been introduced. Although the discounted cash flow method is considered to be more precise than the total cost method, the former does not always provide accurate answers for practical problems. Major difficulties in the analysis of line cost come from (i) *uncertainty* such as demand, engineering changes, and inflation, (ii) *intangibility* such as yield, quality, and reliability. Sometimes, it is more important to determine the appropriate cost items than to choose a cost method.

The size of a line may affect facility requirements, line organization, production schedule, and line cost. The problem of line sizing is closely related to process and operation design. The method introduced in this chapter assumes that both the assembly process and the demand are fixed. Due to its fast run time, Algorithm 8.1 may be reinvoked for different demand assumptions and design concepts. The line designer, however, may have to select the best solution, based on both economic outlook and technical judgment.

A dynamic programming approach for the knapsack problem can be found in Bellman (1962). Algorithm 8.1, however, is based on the work of Gilmore (1966).

The flexible line concept allows workstation addition and deletion. This approach can be very useful for line adjustment involving workstation changes. Line flexibility is established by the appropriate selection of a material-handling system. The line designer must carefully analyze (i) feasibility of the selected system for both product and workstations, and (ii) system performance from a total-line point of view. Discussion in Section 8.4 is based on the work of Chow (1986).

An important topic that has not been discussed in this chapter is the information system for assembly lines. Works have been published in this area under the topics of *computer-integrated manufacturing* or *computer-aided manufacturing*; see Halevi (1980), Rembold (1984), and Rembold(1985), for example. To complete this chapter, design considerations are briefly discussed in the following paragraphs.

Information system design usually involves four major steps:

1. Defining functional requirements
2. Determining transaction rate and data volume
3. Proposing system architecture
4. Evaluating system performance

Negligence of performance evaluation is a common mistake. Many line managers may find system problems due to inadequate functional support, slow response time, and insufficient data storage. Problem corrections can take years and cost a fortune.

Queuing analysis (see Chapter 3) and simulation (see Chapter 9) are frequently used for performance evaluation. The reader may consult Kleinrock (1975), Kleinrock (1976), Kobayashi (1978), and Lavenberg (1983), for detailed information.

A manufacturing information system may consist of a standalone computer or a network of computers. In addition, a number of input/output peripheral devices are usually needed for data entry, information display, and documentation. In modern manufacturing systems, personal computers and local area networks have become increasingly popular. Due to their low cost and high computing power, personal computers have been used for providing manufacturing instructions to operators, testing function support, tooling or material-handling system control, data collection, and line monitoring. A personal computer may also serve as an interface between a workstation and the rest of the assembly line. Communication among these computers is through a local area network. Network performance behavior is a function of transaction rate, data volume, data transmission speed on the network, and protocol. Performance analysis of local area networks is given in Bux (1981).

Transaction rate and data volume are determined by data requirements and their distribution pattern. In an assembly line, data can be classified into at least five categories:

1. *Manufacturing instructions* Data files may be retrieved to provide all necessary information at workstation level to perform a complete operation. This file may be used for a new build product or for rework. The instruction set for new build products is fixed unless an engineering change occurs. The rework instructions, on the other hand, are dependent on the causes of product failure and, therefore, are normally generated by a testing or inspection operation, or from a failure analysis.

2. *Parts tracing* Often, in an assembly line, product failure may be due to defective parts. From a control point of view, the defective parts must be identified. Since parts from the same lot usually have uniform quality, a bad part implies that other parts in the same lot may have the same problem. If the number of defective parts exceeds a certain level, the entire lot should be rejected and defect information sent back to the parts supplier for corrections. If an assembly is composed of a large number of parts, parts tracing can be a complex problem.

3. *Quality control* In addition to defective parts, product failure may also result from a tooling problem, assembly procedure, operator's negligence, or handling damage. If product quality is monitored at the operation level, a trend of quality degradation may indicate that problems need to be corrected. Usually simple control charts may serve the purpose. Sophisti-

cated analysis may also be used for problem detection by relating product testing data to product design or assembly methods.

4. *Flow control* Flow control guides material movement from one place to another and regulates the flow rate. Usually this function has a direct communication link with the material-handling system. In many applications, flow control also deals with in-process inventory. Under a pull policy (see Section 6.3), for example, once a buffer overflow condition is detected, the flow control function may signal the material-handling system to block incoming traffic. Workstation failures may also be reported so that incoming work units can be directed to other stations. In a multiple-product environment, product verification at different operations may become necessary. Flow control must make sure that each operation receives, processes, and delivers the right product.

5. *Line measurement* Line monitoring and control is an important subject in operation management. For any sizable line, line performance behavior can be understood only through data collection and analysis. Line data may include operation cycle time, yield, operator performance, inter-failure time, repair time, downtime, material handling delay, workstation utilization, line throughput, the level of work-in-process, production lead time and line incidents (e.g., workstation failures, and safety violations). These data may be used as the input to the line-management functions, such as line scheduling, dispatching, and job assignment.

REFERENCES

Bellman, R. E., and S. E. Dreyfus (1962). *Applied Dynamic Programming*, Princeton University Press, Princeton, New Jersey.

Buffa, E. S., G. C. Armour, and T. E. Vollmann (1964). Allocating Facilities with CRAFT, *Harvard Business Review*, v. 42, pp. 136–159.

Bux, W. (1981). Local-Area Subnetwork: A Performance Comparison, *IEEE Transactions on Communication*, v. COM-29, pp. 1465–1473.

Chow, W. (1986). Design for Line Flexibility, *Institute of Industrial Engineers Transactions*, v. 18, pp. 95–103.

Gilmore, P. C., and R. E. Gomery (1966). The Theory of Computation of Knapsack Function., *Journal of Operations Research Society of America*, v. 14, pp. 1045–1074.

Halevi, G. (1980). *The Role of Computers in Manufacturing Process*, John Wiley, New York.

Klienrock, L. (1975), *Queuing Systems, Volume I: Theory*, John Wiley, New York.

Klienrock, L. (1976), *Queuing Systems, Volume II: Computer Applications*, John Wiley, New York.

Kobayashi, H. (1978). *Modeling and Analysis: An Introduction to System Performance Evaluation Methodology*, Addison-Wesley, Reading, Massachusetts.

Lavenberg, S. S., ed., (1983). *Computer Performance Modeling Handbook*, Academic Press, New York.
Rembold, U., and R. Dillmann, eds., (1984). *Methods and Tools for Computer Integrated Manufacturing*, Springer-Verlag, Berlin.
Rembold, U., C. Blume, and R. Dillmann (1985). *Computer-Integrated Manufacturing Technology and Systems*, Marcel Dekker, New York.

9

Line Simulation

An assembly line is composed of many individual components, such as operator, workstation, work-in-process, material handling system, information system, production scheduler, job dispatcher, WIP management policy, parts feeding scheme, human resource management policy, rework handling method, and so on. Line performance is the outcome of interactions between components. Performance evaluation involves two steps: (i) mathematical model, and (ii) model solution. Because of the large number of components, a model often has a complex structure and becomes mathematically intractable. For this reason, a simulation method is frequently considered, using a number of numerical experiments and summarizing the experiment outcomes. To achieve statistical stability, a great number of experiments is required. Therefore, a simulation procedure is usually carried out by a computer program. Although simulation has been proved to be an effective tool for performance analysis, it can be very time-consuming and costly. Line designers should have a good understanding of developing and using a simulation program. This chapter gives a brief introduction to this subject; both concepts and applications will be discussed.

Two different simulation approaches have been developed. One is the *trace-driven* (or *data-driven*) method, and the other the *distribution-driven* (or *self-driven*) method. For the trace-driven method, relatively detailed input files must be prepared, and gradually fed into a computer program which simulates the studied system numerically and gathers data for predetermined performance measures. The input files may be derived from observations of real systems, or created by using theoretical results. For each job, typical input includes arrival time, process time at each operation, flow pattern, and parts requirement. Other input may consist of machine failure and repair times, material-handling delay, parts-feeding time, etc. At the end of simulation, performance measures are summarized and reported.

The distribution-driven method is also called the *Monte Carlo* method. In this method, input data are generated from a set of predetermined statistical distributions, whenever needed. Distributions may be derived from observed data or theoretical results. When a simulation program is driven by statistical distributions, the input data can be presented in a very compact form. In particular, most theoretical distributions can be characterized by a few parameters. By changing these parameters, one can easily conduct a number of experiments under different conditions.

It is possible to construct a computer program that can run both methods. If input data is available from observing an existing line, the trace-driven method can be used for program verification. Then the verified program can be used for line performance-analysis by the Monte Carlo method. For a new production line, the trace-driven method is seldom used due to lack of detailed data. In this chapter, our discussion is concentrated on the Monte Carlo method.

9.1 Basic Concept of the Monte Carlo Method

The Monte Carlo method, at its original appearance, is a numerical approach for the solution of equations. Consider a simple integration problem:

$$a = \int_0^1 h(x)dx \tag{9.1}$$

This problem can be viewed as a problem of computing an expected value. Let X be a uniform random variable defined over $[0, 1]$, i.e.,its density function is given by

$$f(x) = 1 \qquad 0 < x < 1 \tag{9.2}$$

Hence,

$$E[h(X)] = \int_0^1 h(x)dx = a$$

If $X_1, X_2, \ldots$ are independent random variables with the same distribution, then by the law of large number it follows that

$$E[h(X)] = \lim_{n \to \infty} \frac{\sum_{i=1}^{n} h(X_i)}{n}$$

To evaluate $E[h(X)]$, numerical values of $\{X_i\}$ must be assigned. This can be done by observing the outcomes of a number of experiments that are subject to the same distribution given by (9.2). In computer simulation, a software program, called *random-number generator*, may be invoked to generate numerical values. Consequently, an approximate solution of Equation (9.1) can be obtained by simply computing the average value of $\{h(X_i) \mid i = 1,\ldots,n\}$. This solution is asymptotically equal to the true value of a. To see this, consider an estimate of a:

$$\hat{a} = \frac{1}{n} \sum_{i=1}^{n} h(X_i)$$

The variance of this estimate is given by

$$\begin{aligned} Var[\hat{a}] &= \frac{1}{n^2} \sum_{i=1}^{n} Var[h(X_i)] \\ &= \frac{1}{n} Var[h(X)] \end{aligned}$$

Since $Var[h(X)]$ is a constant, $Var[\hat{a}] \to 0$ as $n \to \infty$.
From the above discussion, a number of observations can be made:

1. The Monte Carlo method solves a problem by using a sampling technique.
2. Statistical distributions must be available for this method.
3. Random number generators for the given distributions must be available.
4. Computational effort is roughly proportional to sample size.
5. Accuracy of the solution is a function of sample size.

When dealing with a manufacturing line, a simulation procedure is much more complex and may involve the following major steps:

1. Define the problem structure, including input/output parameters and their relations.
2. Install random-number generators.
3. Establish a sampling plan—what should be recorded, when and how ?
4. Determine the run length of the simulation program.
5. Analyze output data and summarize the simulation results.

Since problem definition is application-dependent and model structure results from the modeler's thought process, a unique model does not exist. However, a good model should always have a fast run time and lead to an accurate solution. The following discussion will be limited to those areas where objective grounds can be established. The subject of random-number generation is considered first.

9.2 Random Number Generation

Strictly speaking, in a decimal system a random number is an element drawn from $S = \{1,2,3,4,5,6,7,8,9,0\}$ such that each element of S has an equal chance to be selected. If a random number between 0 and 1 is needed, then every point that lies in $[0,1]$ should be equally likely to be chosen. In this case, a random number may contain multiple digits. From a practical point of view, however, only a finite number of digits is required for simulation. When a computer is used, random numbers can be generated by a software program, based on some recursive formula. The generated numbers may not be "purely" random. Nevertheless, they appear to satisfy some major requirements of randomness, such as equal chance. For this reason, these numbers are called *pseudo-random numbers*. To simplify our terminology, our discussion will consider no difference between a random number and a pseudo-random number.

In computer simulation, random numbers are defined in a broad sense. A random number may be generated from a gamma distribution, a normal distribution, a Poisson distribution, etc. Accordingly, this number is called a gamma, a normal, or a Poisson random number. Sometimes, a uniform random number between 0 and 1 is simply called a random number. It can be seen later that many nonuniform random numbers can be derived from random numbers between 0 and 1.

9.2.1 Uniform Random Number

Several different methods have been proposed for numerical random number generation. One of the popular methods is called the *multiplicative congruential* method. Consider two positive integers a and b such that $b = na + c$, where n is a nonnegative integer and $0 \leq c < a$. Mathematically, this relation can be written as

$$c = b \bmod a$$

For example, if b = 10 and a = 3, then c = 1. Let x_i and x_{i+1} be two successive random numbers, a commonly used multiplicative congruential generator is given by

$$\begin{aligned} x_{i+1} &= 7^5 x_i \bmod (2^{31} - 1) \\ &= 16807 x_i \bmod 2147483647 \end{aligned} \tag{9.3}$$

A uniform random number between 0 and 1 is given by

$$r_i = x_i / (2^{31} - 1) \tag{9.4}$$

This method has been implemented as subroutines in FORTRAN, PL/1, SIMPL/1, and APL. In IBM systems, these subroutines are called GGL, GGL1, or GGL2.

9.2.2 Nonuniform Random Number

For a given (nonuniform) distribution, at least three different methods have been used for random-number generation: *inversion* method, *rejection* method, and *composite* method. Selection of a method for a given distribution is contingent on mathematical tractability and computational efficiency. For convenience, let X be a random variable with a distribution function F, f be its density (if continuous) or mass (if discrete) function, and R be a uniform random variable between 0 and 1. The random numbers generated from X and R are denoted by x and r, respectively.

Inversion Method

Let $Y = F(X)$. Since X is a random variable, so is Y. The distribution function of Y is given by

$$P[Y \leq t] = P[F(X) \leq t]$$

$$= P[X \leq F^{-1}(t)]$$
$$= F(F^{-1}(t))$$
$$= t \qquad 0 \leq t \leq 1$$

In the second equation, $F^{-1}(t)$ is the inverse function of F. Since F is a monotonously increasing function, $\{F(X) \leq t\} = \{X \leq F^{-1}(t)\}$. The last equation concludes that $Y = F(X)$ is a uniform random variable defined over $[0,1]$. This fact provides us a straightforward way to generate a random number with a invertible distribution function, F.

Algorithm 9.1

1. Generate a (uniform) random number, r.
2. Evaluate $x = F^{-1}(r)$. Then x is a random number from distribution F.

Rejection Method

Assume that X has a density function, f, defined on a real interval $[a,b]$ and its maximum value exists, i.e.,

$$M = \max_{a \leq x \leq b} f(x) \tag{9.5}$$

is finite. Then $f(x)/M \leq 1$ for all x in (a,b). Since $f(x)$ represents a relative frequency that the random variable takes its value at x and M is a simply a constant, $f(x)/M$ can be regarded as a "relative" probability that $\{X = x\}$ occurs. Let U be a uniform random variable between a and b. Then, by definition (see Section 2.1),

$$P[U \leq x \mid R \leq f(U)/M] = P[U \leq x, R \leq f(U)/M]/P[R \leq f(U)/M]$$

Since U is a uniform random variable over (a,b), $P[U \leq x, R \leq f(U)/M] = \int_a^x f(u)/M\, du$ and $P[R \leq f(U)/M] = \int_a^b f(u)/M\, du = 1/M$. Therefore,

$$P[U \leq x \mid R \leq f(U)/M] = \int_a^x f(u)\,du$$
$$= P[X \leq x]$$

Consequently, a random number of X can be generated by a pair of uniform random numbers: (R,U). U is accepted as the random number subject to distribution f, if and only if $R \leq f(U)/M$. This approach is summarized in the following algorithm.

Algorithm 9.2

1. Find $M = \max\{f(x) \mid a \leq x \leq b\}$.
2. Generate r. Let $u = a + r(b-a)$. (Thus u is a random number between a and b.)
3. Generate r.
4. If $r \leq f(u)/M$, then $x = u$ becomes a random number subject to distribution, $f(x)$. Otherwise, discard u, and go to step 2.

Two necessary conditions for using Algorithm 9.2 are (i) the existence of M and (ii) a finite interval, $[a,b]$. Although a theoretical distribution may not satisfy these conditions, there is no practical restriction. For instance, a normal distribution is defined over the entire real line. Let the mean and the standard deviation be m and σ, respectively. Then the interval $[a,b]$ can be chosen to be $m \pm 4\sigma$ and covers almost all the density. On the other hand, since $f(x)$ is a density function, $\int f(x)dx = 1$. This implies that the condition of $\{f(x) = \infty\}$ exists at most at a finite number of discrete points. Therefore, a large number M can be found such that $P[f(X) < M] \approx 1$.

Composite Method

If a random variable is a function of a number of random variables, then a composite method may be adopted. For instance, an Erlang random number is equal to the sum of exponential random numbers.

9.2.3 Examples of Nonuniform Random Number Generators

Some commonly used random-number generators will be discussed in the following paragraphs. Again, x represents the random number and r a uniform random number between 0 and 1.

Rectangular Distribution

$$f(x) = \frac{1}{b-a}, \qquad a \leq x \leq b$$

Then the random number $x = a + r(b - a)$.

Exponential Distribution

$$f(x) = \lambda e^{-\lambda x}, \qquad \lambda > 0, \quad x > 0$$

The distribution function is given by

$$F(x) = 1 - e^{-\lambda x}$$

By the inversion method, it follows that

$$x = -\frac{1}{\lambda} \ln r$$

Note that both r and $(1-r)$ are uniform random numbers between 0 and 1. For simplicity, $(1-r)$ in the above expression can be replaced by r.

Erlang Distribution

$$f(x) = \frac{1}{\Gamma(n)} (\lambda x)^{n-1} e^{-\lambda x} \lambda, \qquad \lambda > 0, \quad n = \text{positive integer}, \quad x > 0$$

Since an Erlang random variable is the sum of n independent, identical exponential variables, $x = y_1 + \cdots + y_n$, where y_i is the ith exponential random number.

If $n > 15$, an Erlang distribution can be approximated by a normal distribution.

Gamma Distribution

The density function of a gamma distribution is the same as an Erlang density, except that the parameter n may not be an integer, i.e., $n = k + v$, where k is the integer part and v is a value between 0 and 1. A gamma random variable with a parameter n is the sum of k independent exponential variables plus a gamma random variable with a parameter v. It then suffices to consider

$$f(x)=\frac{1}{\Gamma(v)}(\lambda x)^{v-1}e^{-\lambda x}\lambda, \qquad \lambda>0, 0<v<1, x>0$$

It is easy to show that $f(x)$ is a decreasing function in $(0,\infty)$, and that $\max\{f(x)\}=f(0)=\infty$. Hence, a rejection method can be established by letting a be sufficiently close to zero and (a,b) cover almost all density, e.g., $a=10^{-6}$, $b=v/\lambda+5\sqrt{v}/\lambda$ (mean plus 5 standard deviations), and $M=f(a)$.

Normal Distribution

$$f(x)=\frac{1}{\sigma\sqrt{2\pi}}e^{\frac{-1}{2}\left(\frac{x-\mu}{\sigma}\right)^2}, \qquad \sigma>0$$

The mean and the standard deviation are given by μ and σ, respectively. A normal random variable can be approximated by taking the sum of n uniform random variables. This is an application of the central limit theorem. Since $E[R]=1/2$ and $Var[R]=1/12$,

$$x=\left(\sum_{i=1}^{n}r_i-\frac{n}{2}\right)\sigma\sqrt{\frac{12}{n}}+\mu$$

Usually, $n=12$ is large enough for practical problems.

Geometric Distribution

$$p(x)=q(1-q)^x, \qquad 0<q<1, \quad x=0,1,\ldots$$

Since x is the number of failures before the first success, a geometric random number can be obtained by a composite method:

1. Let $x=0$.
2. Generate r, and let $x=x+1$.
3. If $r\le q$, then stop. x is the random number. Otherwise, go to step 2.

If q is small, however, this method may need a large number of iterations. In this case, an inversion method may be considered. First, note that

$$r = 1 - F(x\_) = (1-p)^x$$

Solving for x, we have $x = \ln r/\ln(1-q)$. At this stage, the value of x may not be an integer. The desired number is obtained by rounding-off. For a small q, x tends to be large. A round-off error is not a significant problem.

Poisson Distribution

$$p(x) = e^{-\theta}\frac{\theta^x}{x!}, \qquad \theta > 0, x = 0, 1, \ldots$$

A Poisson random number can be interpreted as the random arrival count during a time interval between 0 and 1 with an arrival rate θ. A composite method is defined by doing the following:

1. Let $x = 0$, and $z = 0$.
2. Generate an exponential random number with a mean θ. Denote this number by y. Increase the value of x by 1.
3. If $z + y > 1$ (the arrival occurs after time 1), stop. x is the Poisson random number. Otherwise, replace z by $z + y$ (the arrival occurs before time 1) and go to step 2.

Binomial Distribution

$$p(x) = \binom{n}{x} q^x (1-q)^{n-x} \qquad 0 < q < 1, x = 0, 1, \ldots, n$$

x is the number of successes observed in n independent trials with a probability of success q. Therefore, a composite method is devised:

1. Generate n uniform random numbers, $r_1, \ldots, r_n$. Let $x = 0$.
2. For $i = 1, \ldots, n$, if $r_i \leq q$, then increase x by 1.

For a large n, this procedure requires a considerable amount of computation. In this case, a binomial random variable can be approximated by a Poisson random variable if $q < 0.1$, or a normal random variable if $E[X] > 3SD[X]$.

Negative Binomial Distribution

$$p(x) = \binom{x+k-1}{x}(1-\theta)^x \theta^k, \qquad 0 < \theta < 1, k > 0, x = 0, 1, \ldots$$

Since a negative binomial random variable is the sum of k independent, identical geometric variables,

$$x = \ln\left(\prod_{i=1}^{k} r_i\right) / \ln(1-\theta)$$

Empirical Distribution

Sometimes, a theoretical distribution may not fit observed data. In this case, a straightforward method can be used for random-number generation. First, the observed data should be grouped into a number of classes. For discrete values, each class contains the number of observations with the same value. For continuous values, a pair of upper and lower bounds for each class should be defined. Let l_i and u_i be, respectively, the lower and the upper bounds of class i, and f_i be the frequency that an observed value lies in (l_i, u_i). For a discrete case, $l_i = u_i$. A random number can be generated in the following fashion:

1. Determine the number of classes, m, and the lower and the upper bounds: $\{l_i\}$ and $\{u_i\}$.
2. Count the number of observations that lie in (l_i, u_i) for i = 1, 2, ..., m. Denote this number by f_i.
3. Compute the total number of observations:

$$n = \sum_{i=1}^{m} f_i$$

4. Compute cumulated relative frequency. Let $b_0 = 0$ and

$$b_i = \sum_{j=1}^{i} f_j/n \qquad i = 1,2,\ldots,m \quad \text{(Note that } b_m = 1.)$$

5. Generate a uniform random number, r, between 0 and 1.
6. Determine the value of k such that $b_{k-1} < r \le b_k$, $k = 0, 1, \ldots, m$.
7. $x = l_k + r(u_k - l_k)$.

In this procedure, steps 1–4 are needed only for the first x. Once the class limits and the accumulated relative frequencies are determined, random numbers will be generated by executing steps 5–7. For a discrete number, $x = l_k = u_k$. Step 7 is not required.

For a given set of distribution parameters, random number generators can be invoked to create necessary data (such as operation process time, machine downtime, material flow, etc.) to drive a simulation program for line-performance evaluation. Simulation procedure is discussed in the next section.

9.3 Simulation Procedure

Simulation of a given manufacturing system usually involves a number of tasks:

1. Develop a simulation model. The model designer should determine the level of simulation detail, i.e., manufacturing components and their interrelations. Manufacturing components can be operators, parts, assemblies, workstations, material-handling systems, operation management policies, and so on. Their functional relations include process flow, parts flow, product flow, information flow, line layout, and operation logic.

2. Select simulation parameters. Input parameters, their distributions, and performance measures must be determined. Common input parameters can be operation process time, workstation interfailure time and repair time, number of workstations per operation, operation yield, time detractors such as lunch time and breaks, storage and buffer size, parts feeding scheme, working schedule, and operator behavior, etc. Output from a simulation program usually includes workstation utilization, operator utilization, line throughput, in-process inventory level, production lead time, sample distributions of process time, interfailure time, repair time, daily yield, or WIP. Sometimes overflow or underflow problems can also be detected by simulation. For instance, a simulation model can identify the

bottleneck or the underutilized areas. Blocking or starvation probability due to finite buffers can also be evaluated.

3. Choose a simulation time unit. A simulation program should maintain a clock with its time unit as a function of parameters. If operation process time is on the order of minutes, the time unit measured by seconds or 1/100ths of a minute should be sufficient. If process time is on the order of seconds, a time unit of a millisecond may be used.

4. Determine the simulation run length and termination condition. As discussed in Section 9.1, the accuracy of a simulation result is related to the sample size or the program run length. One of the fundamental problems in simulation is the optimal run length. A simulation run should be long enough to generate confident results. But, on the other hand, it becomes uneconomical if the run length is too long. This problem is usually resolved by using the concept of a confidence interval. Detailed discussion of it will be given in Section 9.4.

Simulation procedure may be illustrated by a simple example as follows.

Consider two successive operations. The first operation has a process time of S_1 minutes. Completed products are sent for inspection at operation 2. The inspection time is S_2 minutes per piece. Defective products are sent back for rework at operation 1. Rework time has the same distribution as S_1. The defect rate is assumed to be p. To reduce excessive work-in-process, an overflow condition is defined whenever the total number of work pieces in the system becomes K. New jobs arrive at the first operation with an interarrival time of T minutes. However, no new arrivals are allowed under an overflow condition.

The tooling and fixture used at operation 1 are subject to failure. The time between two successive failures is a random variable, denoted by Y. The assigned operator is also the repair person. Therefore, a repair action follows a failure immediately. The duration of a repair time is also a random variable, denoted by Z.

A material flow and the line configuration are illustrated in Figure 9.1, from which a simulation model can be directly derived. When a work unit arrives at operation 1, it may be processed immediately if the queue is empty, or it may have to wait until all preceding units in the queue have been processed. After having completed its work at operation 1, the work unit is moved to operation 2. Another waiting and processing cycle is repeated. Upon its completion, two possible flow directions exist. With a probability p, the work unit flows back to operation 1, i.e., defects are found. With a probability $1-p$, the unit leaves the system permanently. The flow direction can be determined by comparing the value of p with a uniform random number, r, over (0,1). If $r \leq p$, the unit is returned to operation 1. Otherwise, the unit leaves the system as a good product.

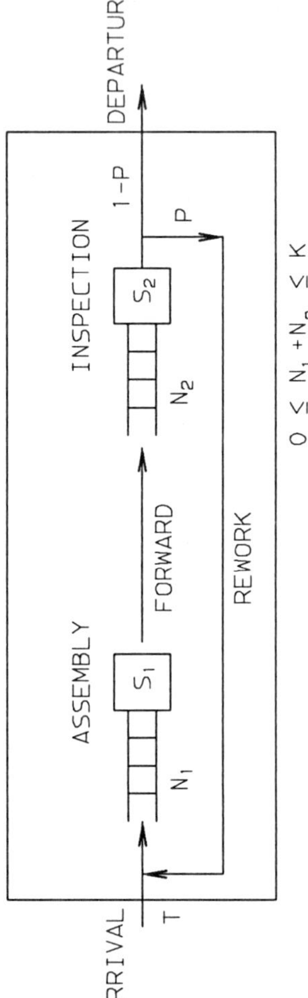

Figure 9.1 A simple assembly line model.

The upper bound of work-in-process is an important parameter. If the value of K is too small, workstations can be underutilized due to insufficient amount of work. On the other hand, a large K means a large amount of in-process inventory. Simulation can be used to determine the optimal value.

Since T, S_1, S_2 are measured by minutes, the time unit of the simulation clock may be counted by seconds, or 1/100 minute. A time scale less than this unit is insignificant, at least from a modeling viewpoint.

The input parameters of the simulation model include:

1. F: distribution of T (interarrival time)
2. G_1: distribution of S_1 (process time at operation 1)
3. G_2: distribution of S_2 (process time at operation 2)
4. G_3: distribution of Y, interfailure time at operation 1
5. G_4: distribution of Z, repair time at operation 1
6. p: product defect rate
7. K: upper bound of total units in the system
8. RL: simulation run length

The output parameters are determined by the problem-definition and system performance measures. Common measures are queue length at each operation, total work-in-process in the system, waiting time at each operation, production lead time, workstation utilization, and line throughput. These parameters are defined by:

1. N_1: queue length at operation 1
2. N_2: queue length at operation 2
3. N: total work-in-process in the system
4. W_1: waiting time at operation 1
5. W_2: waiting time at operation 2
6. W: production lead time
7. U_1: utilization of operation 1
8. U_2: utilization of operation 2
9. TP: line throughput

In addition to input/output parameters, a number of operating variables may be needed. For this specific problem, they are

1. R_0: time until the next arrival (from outside)
2. R_1: remaining process time at operation 1
3. R_2: remaining process time at operation 2
4. R_3: time until the next failure at operation 1
5. R_4: remaining repair time

6. CN_1: cumulative queue length at operation 1
7. CN_2: cumulative queue length at operation 2
8. CI_1: cumulative idle time of operation 1
9. CI_2: cumulative idle time of operation 2
10. CA: cumulative arrival count (from outside)
11. CF: cumulative job count from operation 1 to operation 2
12. CB: cumulative rework count from operation 2 to operation 1
13. CD: cumulative departure count (from the system)
14. t: reading from the simulation clock
15. M: a large number, e.g., 10^6

The time reading from the simulation clock is advanced each time an event occurs. An event can be an arrival, a departure, a job move from one operation to another. Thus, a time increment is equal to the time interval between two successive events. Consequently, the computational effort of a simulation program is a function of the number of events observed during RL. Major steps of such a simulation program are discussed in the following paragraphs.

1. *Initiation* A simulation run begins at $t = 0$ with an appropriately chosen initial condition. If the simulation run is used to study transient behavior, the initial condition should be carefully determined. For example, a change in process rate may cause a significant short-term fluctuation in both throughput and work-in-process. On the other hand, the initial condition is not important to long-term system behavior. In this case, an empty system is usually assumed for convenience. Hence,

$$N_1 = N_2 = CN_1 = CN_2 = 0$$

$$CA = CF = CB = CD = 0$$

$$CI_1 = CI_2 = 0$$

Since the system is empty, the first event must be an arrival. The arrival time is determined by generating a random number, T. If an operation is idle, its remaining process time is set equal to M. Consequently, no "departure" from an idle operation is possible until M time units later. Since M is, by definition, a large number, this practically means no departures. Thus, we have

$$R_0 = T \qquad R_1 = R_2 = R_3 = R_4 = M$$

2. *Determination of the next event* Compute $X = \min\{R_i \mid i = 0, 1, \ldots, 4\}$. (If two or more values can be the minimum, the first one is selected. For example, if $R_0 = R_2 = 5$ is the minimum value, then R_0 is considered to be the minimum one.)

3. *Statistics gathering* Advance the simulation clock by X units, i.e., $t \leftarrow t + X$. The cumulative queue length at operation 1 (or 2) is increased by $XN_1(XN_2)$. If $N_1 = 0$ (or $N_2 = 0$), then the cumulative idle time at operation 1 (or 2) is increased by X. The remaining arrival time and process times become $R_0 \leftarrow R_0 - X$, $R_1 \leftarrow R_1 - X$, and $R_2 \leftarrow R_2 - X$, respectively. If R_0 (or R_1, R_2) was previously set equal to M, a decrement of X practically has no effect. A more precise way to handle this situation is to perform a test:

$$\text{If } R_i \neq M, \qquad \text{then } R_i \longleftarrow R_i - X \qquad i = 0,1,2,3,4$$

Since R_3 is the time until the next failure, $R_3 \leftarrow R_3 - X$ if and only if operation 1 is nonempty. If $N_1 = 0$, R_3 remains unchanged. Likewise, R_4 is reduced by X only if a repair action is taken at operation 1.

Further statistics updating is contingent on the result from step 2. Several possibilities exist, and are considered in items 4–9 below:

4. *Arrival* In this case, R_0 is the minimum. Hence, $N_1 \leftarrow N_1 + 1$ and $CA \leftarrow CA + 1$. If $N_1 + N_2 = K$, an overflow condition is detected. The arrival stream should be temporarily shut off. This can be done by letting $R_0 = M$. Furthermore, if operation 1 was previously empty (i.e., $R_1 = M$), the arrival ought to be processed immediately. Therefore, a new process time needs be generated, i.e., $R_1 = s_1$ (a random number subject to distribution G_1), and $CI_1 \leftarrow CI_1 + X$. If $R_3 = M$, a new value should be assigned, i.e., $R_3 = y$ (a random number subject to distribution G_3).

5. *Forward* If R_1 is the minimum, a work unit moves forward to operation 2. Hence, $N_1 \leftarrow N_1 - 1$, $N_2 \leftarrow N_2 + 1$, and $CF \leftarrow CF + 1$. The total number of work units in the system remains unchanged.

If operation 2 was previously empty, a random process time should be generated, i.e., $R_2 = s_2$ (generated from G_2), and $CI_2 \leftarrow CI_2 + X$. Also, if the new value of N_1 becomes zero, $R_1 = M$. Otherwise, a new unit will be processed by operation 1, and hence $R_1 = s_1$.

The situation when R_2 is the minimum implies that an inspection cycle at operation 2 is completed. The completed work unit may either go back to operation 1 for rework or leave the system as a good product. Let r be a random number generated from a uniform distribution over (0, 1). If $r \leq p$ (the defect rate), the product needs rework. If $r > p$, the product is good. Contingent on the result of this comparison, line measurement is updated accordingly.

6. *Rework* This is opposite to case 5. First, $CB \leftarrow CB + 1$. Then N_1 is increased by 1 and N_2 is decreased by 1. R_1 and R_2 should be changed also. If operation 1 was previously empty, a new random number should be assigned to R_1 and R_3 resumes its aging process.

7. *Departure* In this case, the cumulative departure count is increased, i.e., $CD \leftarrow CD + 1$. But $N_2 \leftarrow N_2 - 1$. The value of R_2 is equal to M when $N_2 = 0$. $R_2 = s_2$ (inspect a new unit) when $N_2 > 0$. If $R_0 = M$, the value of R_0 should be replaced by a random number subject to the distribution of T.

8. *Failure* If $R_3 = X$, a workstation failure condition is detected. In this case, a repair time must be determined by generating a random number from the distribution of Z, say z. Then the remaining repair time, $R_4 = z$. The completion time of the work unit at operation 1 is prolonged by z time units. Hence, $R_1 \leftarrow R_1 + z$. Since no failure can possibly occur during a repair period, $R_3 = M$.

9. *Repair* If the next event is the completion of a repair action, i.e., $R_4 = X$, operation 1 is reactivated. The time until the next failure must be determined. Generate a random number according to distribution G_3, say y. Let $R_3 = y$ and $R_4 = M$.

10. *Output* If the simulation time exceeds the run length (i.e., $t > RL$), the program is terminated. Performance measures should be summarized as the program output:

a. Mean queue length: $\bar{N}_i = CN_i/t \; i = 1,2$
b. Average total work-in-process: $\bar{N} = \bar{N}_1 + \bar{N}_2$
c. Utilization: $U_i = 1 - CI_i/t \; i = 1,2$
d. Arrival rate: $\lambda = CA/t$
e. Line throughput: $TP = CD/t$
f. Throughput at operation 1: $TP_1 = CF/t$
g. Throughput at operation 2: $TP_2 = (CB + CD)/t$
h. Average waiting time: $W_i = \bar{N}_i/TP_i \; i = 1,2$
i. Average production lead time: $W = \bar{N}/TP$

The last two equations are implementations of Theorem 3.1. The average waiting times and the average lead time can also be directly obtained from simulation. For work unit i, arrival time (a_i) and departure time (d_i) are recorded. Then the lead time, $w_i = d_i - a_i$. The sample distribution of lead times is given by $\{w_i\}$. Usually a sample distribution is represented by frequency counts in disjoint classes. Hence, the number of data points is equal to the number of classes. If only the average lead time is needed, both computer storage and CPU time can be reduced by looking at the accumulated lead time. When unit i leaves the system, the accumulated lead time becomes $CW \leftarrow CW + (t - a_i)$, where t is the clock time at the departure epoch. The estimated average lead time is $W = CW/CD$.

Simulation results should always be verified. If measurements from the real system are available, one can compare the simulated results with measured data. Verification can also be done by examining the simulation output against theoretical results. For instance, if both the average queue length and the average waiting time are measured directly, their ratio should be close to the value of throughput. The product of the mean process time (obtained from the random-number generator) and the throughput should be close to the estimated utilization.

In addition to program errors, a common cause of a poor simulation result is that the program run length is too short. Determination of run length is the subject of the next section.

9.4 Run Length and Confidence Interval

A simulation result is a function of a number of random variables, such as interarrival time, process time, yield, interfailure time, repair time, etc. Numerical values of these variables are obtained from random number generators subject to predetermined statistical distributions. Each random number can be regarded as a sample from its corresponding distribution. Therefore, from a statistical viewpoint, a simulation process is merely a sampling process. To assure a good statistical estimation, two questions must be answered:

1. How is the run length determined, i.e., the sample size?
2. For a given run length, how good is the estimated value?

The answer is provided by looking at a confidence interval for the estimated parameter. Suppose that a set of statistics, $X_1, X_2, \ldots, X_n$ is a sequence of observations for a selected parameter. The true mean of the parameter is denoted by θ. If n is large enough, the statistic

$$\bar{X} = \sum_{i=1}^{n} X_i/n$$

has approximately a normal distribution. If k sets of independent observations are made, then their average values, $\bar{X}_1, \ldots, \bar{X}_k$ can be treated as k independent, identical normal random variables. Let m and S be the mean and the standard deviation of $\{\bar{X}_i\}$. Hence

$$m = \sum_{i=1}^{k} \bar{X}_i / k \tag{9.6}$$

$$S = \left[\sum_{i=1}^{k} (\bar{X}_i - m)^2 / (k-1) \right]^{1/2} \tag{9.7}$$

From Section 2.3, it is known that the statistic

$$Y = \frac{m - \theta}{S/\sqrt{k}} \tag{9.8}$$

has approximately a t-distribution with $(k-1)$ degrees of freedom. At the level of confidence of $(1-\alpha)$, a confidence interval for θ is $(m - t_{\alpha/2} S/\sqrt{k}, m + t_{\alpha/2} S/\sqrt{k})$, where $t_{\alpha/2}$ is a function of α and k. For example, if $\alpha = 0.1$ is chosen (i.e., a 90-percent confidence interval), the values of $t_{\alpha/2}$ are

k	2	3	4	5	6	7	8	9	10	11
$t_{\alpha/2}$	6.31	2.92	2.35	2.13	2.02	1.94	1.90	1.86	1.83	1.81
k	12	13	14	15	16	17	18	19	20	
$t_{\alpha/2}$	1.80	1.78	1.77	1.76	1.75	1.75	1.74	1.73	1.73	

To say that (a,b) is a 90-percent confidence interval for Y is equivalent to say that $P[a \le Y \le b] = 0.9$. The smaller the interval, the better the information it provides. Hence, the ratio of the width of a confidence interval to the mean can be used as an indicator. For example, a simulation run length can be determined by two conditions:

1. Confidence level, $1 - \alpha = 0.90$.
2. Width to mean ratio, $r = 0.10$.

Based on the accumulated data generated from the simulation program, one can periodically reconstruct the confidence interval and examine the r-value. If r is below 0.1, the program can be terminated.

Still two problems exist: First, the statistics collected during a simulation process are not always independent. For instance, the lead times of successive departures are highly correlated. A long lead time may mean a congested production line due to a large amount of work-in-process, machine failure, poor yield, or other resource-availability problems. Since products follow the same flow pattern, successive work units would experience a similar situation. For this reason, the observed statistics $\bar{X}_1, \ldots, \bar{X}_k$ may not be independent. The interval derived from a t-statistic may have a poor confidence level.

The second problem is related to the initial condition of a simulation process. For instance, the example of the two-stage production line in Section 9.3 assumes an empty line initially. Because of the existence of dependency in a simulation process, there is a transient period when the observed statistics are influenced by the initial condition. Unless investigation of transient behavior is the purpose of a simulation run, statistics collected during the transient period should be discarded for a steady-state analysis. Unfortunately, it is difficult to estimate the duration of a transient period.

Different methods have been proposed to deal with these problems. The following discussion introduces three simple methods of computing the confidence interval. For convenience, the average production lead time will be used to illustrate these methods.

9.4.1 Independent Replication Method

This approach merely uses different random number streams to repeat the same simulation process. Since repetitions are independent of each other, independent statistics can be collected from repetitions. The transient period is determined in a brute-force fashion. One may plot and inspect the observed data to determine the transient period. One may also assume that each repetition has collected sufficient data so that distortion due to transient behavior has little impact on the ultimate estimate.

Let $W_j(i)$ be the lead time of work unit i in the jth repetition during which N_j work units have been observed. The simulation run is repeated n times with different initial random numbers (see Section 9.2.1). The average lead times

$$\bar{W}_j = \frac{1}{N_j} \sum_{i=1}^{N_j} W_j(i) \qquad j = 1, \ldots, n$$

are independent statistics. By the central limit theorem, these averages are nearly normally distributed for large $\{N_j\}$. Using (9.6)-(9.8), a confidence interval can be derived from a t-distribution.

This method can also be used for other line measures. For example, each repetition can be further divided into a number of time slices. Utilization of each slice is computed. Then the "average" utilization for each repetition has asymptotically a normal distribution.

9.4.2 Batch-Mean Method

This method is similar to the independent replication approach, except that the simulation process involves a single run only (i.e., no repetition). The total run length is divided into n equal but disjoint time intervals. The number of observations during an interval may be called *batch size*. Statistics collected from different time intervals are considered to be independent. Since each interval has an equal length, the statistics have approximately the same distribution. The average lead time computed during each interval (a *batch mean*) is regarded as a normal random variable. Again, a t-distribution will give the confidence interval.

An advantage of this method is that only one transient period will be discarded. But the disadvantage is that the batch means may not be independent. Therefore, the actual confidence interval is usually larger than the estimated interval. In other words, the estimated confidence interval may have a lower confidence level than expected. To reduce correlation between batch means, the batch size must be sufficiently large. Determination of batch size, however, is still an open problem.

9.4.3 Regenerative Method

Sometimes, a specific state of the simulation process can be defined as a *regenerative point*. Once entering this state, the process starts all over again. Time between two successive regenerative points is a *regenerative cycle*. The processes during different regenerative cycles are stochastically independent and identical; that is, statistics collected from different cycles are independent and have the same distribution. For example, a regenerative cycle begins when a job arrives at a workstation and finds the workstation idle. The arrival-time epoch is a regenerative point. Then the work-in-process distributions during different regenerative cycles are independent and identical.

A regenerative cycle can be a continuous value such as a time interval, or a discrete value such as an event count. As before, let us take the average lead time as an example. Assume that N_j lead times have been measured during the jth regenerative cycle. The total production lead time is

$$A_j = \sum_{i=1}^{N_j} W_j(i) \tag{9.9}$$

The average lead time during cycle j is $\bar{W}_j = A_j/N_j$. Let

$$\bar{N} = \sum_{j=1}^{n} N_j/n$$

$$\bar{A} = \sum_{j=1}^{n} A_j/n$$

$$S_N^2 = \sum_{j=1}^{n} (N_j - \bar{N})^2/(n-1)$$

$$S_A^2 = \sum_{j=1}^{n} (A_j - \bar{A})^2/(n-1)$$

$$S_{NA} = \sum_{j=1}^{n} (N_j - \bar{N})(A_j - \bar{A})/(n-1) \tag{9.10}$$

These statistics represent two sample means, two sample variances, and a sample covariance, respectively.

Suppose that a simulation run experiences n regenerative cycles. The estimated average lead time becomes

$$\bar{W} = \frac{\sum_{j=1}^{n} A_j}{\sum_{j=1}^{n} N_j} = \frac{\sum_{j=1}^{n} A_j/n}{\sum_{j=1}^{n} N_j/n} = \frac{\bar{A}}{\bar{N}} \tag{9.11}$$

This statistic is an estimate of the mean lead time, $E[W]$. Since (N_1, A_1), ..., (N_n, A_n) are independent pairs, $(A_1 - E[W]N_1)$, ..., $(A_n - E[W]N_n)$ are independent, identical random variables. Consequently, for a large n,

$$\sum_{j=1}^{n} (A_j - E[W]N_j)/n = (\bar{A} - E[W]\bar{N})$$

is approximately a normal random variable with a mean zero and a variance

$$\begin{aligned} Var[A_1 - E[W]N_1]/n &\approx \sum_{j=1}^{n}(A_j - E[W]N_j)^2/n(n-1) \\ &\approx \sum_{j=1}^{n}(A_j - \bar{W}N_j)^2/n(n-1) \\ &= \sum_{j=1}^{n}[(A_j - \bar{A}) - \bar{W}(N_j - \bar{N})]^2/n(n-1) \\ &= (S_A^2 + \bar{W}^2 S_N^2 - 2S_{NA}\bar{W})/n \end{aligned}$$

Therefore, the statistic

$$Z = \frac{\bar{A} - E[W]\bar{N}}{\sqrt{(S_A^2 + \bar{W}^2 S_N^2 - 2S_{NA}\bar{W})/n}} \tag{9.12}$$

has asymptotically a normal distribution with a zero mean and a unit variance. From the normal table, a $(1-\alpha)$ confidence interval can be determined by the relation

$$1 - \alpha = P[-z_{\alpha/2} \leq Z \leq z_{\alpha/2}] \tag{9.13}$$

For example, when $\alpha = 0.1$, $z_{\alpha/2} = 1.96$. Thus, a 90-percent confidence interval for $E[W]$ can be derived from (9.12) and (9.13):

$$(\bar{A}/\bar{N}) \pm \frac{1.96}{\bar{N}}\sqrt{(S_A^2 + \bar{W}^2 S_N^2 - 2S_{NA}\bar{W})/n}$$

It can be seen from (9.10) and (9.11) that this interval is determined by $\{A_j\}$ and $\{N_j\}$.

The regenerative method is a single-run method. When this method is adopted, neither transient behavior nor the correlated statistics will present problems. On the other hand, for a complex manufacturing system this method may not be implemented easily. Selection of an effective regenerative point can be difficult. The simulation process must visit a regenerative point frequently so that a sufficiently large number of regenerative cycles can be observed. Even in a simple production line, this can be a serious problem.

9.3.4 Comparison of Different Methods

A numerical comparison of different methods may be of interest. An M/M/1 queuing system is considered so that the exact solution is known. Since the independent replication and the batch-mean methods are very similar, only the latter will be used for comparison.

From Section 3.2, it is known that a job's waiting time in an M/M/1 queue has an exponential distribution with a mean $1/(\mu-\lambda)$, where μ is the service rate (i.e., reciprocal of the mean service time) and λ is the arrival rate. The utilization factor is $\rho=\lambda/\mu$. The server can be in either of the two states: *busy* or *idle*. If a job arrives and finds an empty system (i.e., server is idle), then a *busy period* begins. On the other hand, an *idle period* starts with a departure leaving an empty system. A *busy cycle* consists of a busy period followed by an idle period, and can be treated as a regenerative cycle. A regenerative point is the arrival epoch that starts a busy period. Let

C = busy cycle
B = busy period
I = idle period
N = number of jobs served during C

In an M/M/1 queuing system, the interarrival time has an exponential distribution. Because of the memoryless property (see Section 2.2), an idle period has the same distribution. Thus, $E[I]=1/\lambda$. On the other hand, a busy period is the sum of N service times. $E[B]=E[N]/\mu$. The average busy cycle is given by

$$E[C]=1/\lambda+E[N]/\mu \tag{9.14}$$

Since N is also the number of arrivals during C, it follows that

$$E[N]=\lambda E[C] \tag{9.15}$$

Solving (9.14) and (9.15),

$$E[N] = \frac{\mu}{\mu - \lambda}$$
$$E[C] = \frac{\mu}{\lambda(\mu - \lambda)} \tag{9.16}$$

The average total waiting time during a busy cycle is

$$E[A] = E[W]E[N] = \frac{\mu}{(\mu - \lambda)^2} \tag{9.17}$$

For example, if $\lambda = 0.05$ and $\mu = 0.1$, then $E[W] = 20$, $E[N] = 2$, and $E[C] = E[A] = 40$.

Let

B = batch size
b = number of batches
c = number of regenerative cycles
$\bar{A}$ = estimated $E[A]$
$\bar{N}$ = estimated $E[N]$
D = number of job completions for a simulation run
CI_b = 90-percent confidence interval about $E[W]$ under a batch-mean method
CI_r = 90-percent confidence interval about $E[W]$ under a regenerative method

A comparison of regenerative and batch-mean methods is given in Table 9.1, where four different cases are shown. When $\rho = 0.5$, the simulation process reaches a steady-state condition very rapidly. The average busy cycle is 40 time units. During a cycle, two jobs will be served on the average. Consequently, 5,000 regenerative cycles take about 2×10^5 time units and generate about 10,000 departures. If a batch size of 100 is selected, a comparable simulation run under a batch-mean method should cover 100 batches. Consequently, our simulation program is terminated if any one of the following conditions is satisfied:

1. The number of regenerative cycles exceeds 5,000.
2. The number of departures exceeds 10,000.

Table 9.1 A Comparison of Regenerative and Batch-Mean Methods

	Case 1	Case 2	Case 3	Case 4
λ	0.05	0.09	0.09	0.09
μ	0.10	0.10	0.10	0.10
ρ	0.50	0.90	0.90	0.90
$E[W]$	20	100	100	100
D	9587	10000	96730	996223
B	100	100	200	1000
b	95	100	483	996
CI_b	19.15 ± 1.335	84.31 ± 13.724	98.39 ± 6.734	99.51 ± 3.348
c	5000	1184	10000	100000
$E[A]$	40	1000	1000	1000
$E[N]$	2	10	10	10
$\bar{A}$	36.79	692.33	951.68	991.40
$\bar{N}$	1.92	8.32	9.67	9.96
CI_r	19.19 ± 1.323	83.23 ± 19.386	98.39 ± 9.714	99.52 ± 3.827

3. The number of batches exceeds 100.
4. The simulated time exceeds 2×10^5.

It can be seen from Table 9.1 that as far as case 1 is concerned, the two methods have very little difference. Both of their confidence intervals cover the true value of $E[W]$. When a simulation method is used to solve a practical problem, the true value of $E[W]$ is unknown. Therefore, the estimated confidence interval should have a small width-to-mean ratio. This ratio is a function of simulation run length. Suppose that the batch size is large enough so that the computed average lead time of each batch is nearly independent and has asymptotically a normal distribution. The mean and the variance of batch mean lead times are respectively

$$\bar{W} = \frac{1}{b} \sum_{j=1}^{b} \bar{W}_j \tag{9.18}$$

$$S_W{}^2 = \frac{1}{b-1} \sum_{j=1}^{b} (\bar{W}_j - \bar{W})^2 \tag{9.19}$$

If a width-to-mean ratio is chosen to be δ, by t-distribution it follows that

$$\delta = \frac{2t_{\alpha/2}S_W}{\bar{W}\sqrt{b}}$$

where $t_{\alpha/2}$ is derived from a t-distribution with $(b-1)$ degrees of freedom and a confidence level of $(1-\alpha)$.

For example, if $b \geq 120$ (in this case a t-distribution is close to a normal distribution) and $\alpha = 0.1$ (i.e., a 90-percent confidence interval), $t_{\alpha/2} \approx 1.96$. For $\delta = 10\%$, the number of batches should be

$$b \approx \left[\frac{2 * 1.96 S_W}{0.1\bar{W}}\right]^2 = \left[\frac{39.2 S_W}{\bar{W}}\right]^2 \tag{9.20}$$

This approach can also be used to estimate the number of regenerative cycles needed for a given δ-value.

When $\lambda = 0.09$ and $\mu = 0.1$, the server becomes busy. If the number of departures is still 10,000, neither the estimated mean lead times nor their confidence intervals give accurate answers. The confidence interval obtained by the batch-mean method does not even cover the true value. Although this problem does not exist when the regenerative method is adopted, the confidence interval is too large. The width-to-mean ratio is $\delta = 2 \times 19.386/83.23 = 46.6\%$. In fact only 1184 regenerative cycles have been created. This value is very low as compared to $c = 5000$ when $\rho = 0.5$. When c is increased to 10,000 cycles, 96,730 completions are observed and δ becomes 19.7%. The estimated mean lead time is 98.39, very close to the true value $E[W] = 100$. When $c = 100{,}000$, $\delta = 7.7\%$ and $\bar{W} = 99.52$.

The results of Table 9.1 demonstrate that a long simulation run tends to have an accurate estimate. But this is done at the cost of computer time. The four cases in Table 9.1 consumed respectively 18, 18, 160, and 1638 CPU seconds on an IBM 4341 machine.

The estimated mean lead times under the batch-mean method and the regenerative method are very close to each other. Since the mean lead times from different batches are not truly independent, sometimes a 90-percent confidence interval may only have a coverage at a 70–80% level, particularly when the batch size or the number of batches is small. On the other hand, the mean lead times of different regenerative cycles are always independent, and a confidence interval should have a better coverage. Correlation between measures from different batches or regenerative cycles can be illustrated by looking at the *autocorrelation coefficient*. Consider a sequence of observations $\{X_j \mid j = 1,\ldots,m\}$. Their autocorrelation coefficient with a lag k is defined by

$$r_k = \frac{1}{S^2} \sum_{j=1}^{m-k} \frac{(X_j - \bar{X})(X_{j+k} - \bar{X})}{m-k}$$

where $\bar{X}$ and S^2 are the mean and variance of $\{X_j\}$.

Let $r_W(k)$, $r_B(k)$, and $r_R(k)$ be the autocorrelation coefficients of individual production lead times, average lead times in batches, and during regenerative cycles, respectively. Table 9.2 gives the computed r-values from simulation.

Autocorrelation of successive lead times is an increasing function of utilization, ρ. When $\rho = 0.5$, half of the time the server is idle. Suppose that at one time the system has a long queue. Since λ is considerably less than μ, the queue diminishes very quickly. Consequently, the autocorrelation coefficient has a small value for a time lag greater than 4. On the other hand, when $\rho = 0.9$, a long queue may exist for a long time. This phenomenon is clearly illustrated in Table 9.2. Due to the correlation between successive observations, a small batch size fails to provide a set of independent statistics. Furthermore, because of positive correlation, the sample variance of

Table 9.2 Autocorrelation Coefficients of Simulated Lead Times

	Case 1			Case 4		
λ	0.05			0.09		
μ	0.10			0.10		
B	100			1000		
b	95			996		
c	5000			100000		
k	$r_W(k)$	$r_B(k)$	$r_R(k)$	$r_W(k)$	$r_B(k)$	$r_R(k)$
1	0.762	0.036	−0.025	0.990	0.192	−0.003
2	0.605	−0.173	0.003	0.980	0.018	0.001
3	0.496	−0.238	0.035	0.971	0.007	0.003
4	0.406	−0.229	0.011	0.962	−0.043	−0.001
5	0.344	−0.184	−0.016	0.954	0.019	0.011
6	0.294	−0.229	0.031	0.945	0.035	−0.000

$\{\bar{W}_j\}$ is underestimated, and therefore the confidence interval is smaller than it should be.

The value of an autocorrelation coefficient is always between 1 and -1. If the magnitude of a coefficient is less than 0.2, correlation can be considered negligible. The results in Table 9.2 show that the batch sizes for both cases are already large enough. When the regenerative method is adopted, statistics from different cycles are independent. Therefore, $r_R(k) \approx 0$.

9.5 Applications

Development of a line simulation program may take from a few days to a few months, dependent on the level of modeling details, the chosen programming language, and the programmer's skill. A line model should cover enough details so that the model can adequately evaluate line-performance behavior. On the other hand, a model should be simple enough so that a simulation program can be quickly and accurately developed. Major components of a line simulation model may include:

1. Assembly workstations
2. Parameters for workstation availability (e.g., interfailure time, repair time, preventive maintenance, etc.)
3. Operators and their availabilities
4. Operation process times
5. Improvement processes
6. Assembly flow process
7. Assembly yield
8. Parts-flow pattern
9. Parts quality
10. Information flow
11. Information system (e.g., computer and network structure, etc.)
12. Material-handling system
13. Parameters for material-handling system availability
14. Line layout or configuration
15. WIP management policy
16. Parts-feeding policy
17. Rework management policy
18. Scheduling policy
19. Dispatching policy
20. Interface/interaction between components
21. Line-stop criteria
22. Line-performance measures
23. Simulation program termination condition

Not every line-simulation model considers all the components listed above. Selection of components is application-dependent. Model-input requirements can be derived from the selected components. For example, if a pull system is chosen for WIP management (see Chapter 6), buffer size between operations must be defined for the model. A model can be regarded as a functional relation between input and output, which can be numerically evaluated.

Although simulation has been widely used for manufacturing analysis, results may not be always satisfactory. Common problems include:

1. Opinion-based solutions
2. Lack of understanding
3. Wrong assumptions
4. Invalid input data
5. Misinterpretation of results
6. Unsuitable modeling techniques
7. Improper computer language

People sometimes prefer to make a decision based on their own opinions. If a simulation result does not coincide with their opinions, the former is often discarded. On the other hand, one can mistakenly expect that a simulation model would distinguish a small discrepancy. For instance, if the average repair time of a machine is one hour and the mean interfailure time is on the order of a hundred hours, then the machine reliability is 99% or better. In this case, improvement in machine throughput by doubling the mean interfailure time may not be detectable by simulation.

Another common problem is to use invalid input or to adopt unverified assumptions for a simulation model. Data collection is usually a very time-consuming job. When dealing with a nonexisting line, this can be a very frustrating situation for the line designer. Due to lack of information, invalid assumptions and input can only provide a wrong solution, and consequently lead to inappropriate business decisions. When using an existing simulation model, it is always advisable to understand the model first, including the input and output parameters, model assumptions and limitations.

Misinterpretation of output may also result from invalid input or assumptions, and lack of understanding of simulation technique. Some typical examples are:

1. Solution based on an unstable simulation process due to short program run time.

2. Assessment of a WIP management policy based on invalid operation process time distributions.
3. Experiments of resource allocation strategies based on "average" resource requirements.

The problems of simulation efficiency are often associated with unsuitable modeling techniques and improper computer language. If too many details have been included, a simulation model can be very complex. As a consequence, program development becomes a tedious task. A complex program is also susceptible to software errors. For these reasons, an elaborate model does not necessarily give an accurate answer.

Since an assembly line involves repetitive operations, a large number of codes in a simulation program will be executed over and over. Programming languages without compilation capability (e.g., APL and GPSS) will have slow program run time. On the other hand, a special program language designed for simulation (e.g., GPSS and SLAM) can reduce program development effort. Some languages have built-in functions for data collection and provide output analysis results. Selection of a programming language is dependent on the application and the programmer's experience. No single language is accepted by all simulation users. However, a good choice should consider at least the following four factors:

1. Program development effort
2. Program run time
3. Data-handling capability
4. Output analysis capability

From a line-design point of view, the value of line simulation is to allow the designer to experiment on a nonexistent line. This is illustrated in Example 9.1.

Example 9.1 A company is planning to produce a new product. Based on past experience, the variations of process time, product yield and machine reliability are identified as the three major manufacturing problems. Improvements in each of these areas are possible, but cost time and resources. Therefore, the line designer decides to prioritize these problems. An existing line that produces a similar product has been measured. The ranges of the measures at different operations are given in the following:

1. Coefficient of variation of process times (CV): 0.3–0.7
2. Mean time between failures (MTBF): 30–95 hours

3. Mean time to repair (MTTR): 1.5–2.6 hours
4. Production yield at test operations (YLD): 0.878–0.957

The following targets for improvement have been established by management, with input from the engineering department:

1. CV: 0.1–0.3
2. MTBF: 500 hours
3. MTTR: 2 hours
4. YLD: 0.925–0.970

Then a number of experiments are conducted by a line simulator. For each problem area, three cases (high, medium, low) are considered for experiments, including the existing case, the target case, and an intermediate case.

Process-time variation

1. V_h: $CV = 0.3–0.7$ (high)
2. V_m: $CV = 0.2–0.6$ (medium)
3. V_l: $CV = 0.1–0.3$ (low)

Reliability

1. R_h: MTBF = 500, and MTTR = 2
2. R_m: MTBF = 200, and MTTR = 2
3. R_l: MTBF = 30–95, and MTTR = 1.5–2.6

Yield

1. Y_h: YLD = 0.925–0.970
2. Y_m: YLD = 0.900–0.961
3. Y_l: YLD = 0.878–0.957

There are $3^3 = 27$ combinations. If each simulation run takes one-half hour of CPU time, a completed set of experiments will need 13.5 hours. An economical way is to experiment all three cases of a specific measure against different combinations of the other two measures. Consequently, only $3 \times 3 = 9$ cases need to be considered.

This arrangement can be illustrated in the following matrix.

$$\begin{array}{c} \\ V_l \\ V_m \\ V_h \end{array} \begin{array}{c} \begin{array}{ccc} R_l & R_m & R_h \end{array} \\ \begin{bmatrix} Y_m & Y_l & Y_h \\ Y_l & Y_h & Y_m \\ Y_h & Y_m & Y_l \end{bmatrix} \end{array} \tag{9.21}$$

The rows of the matrix correspond to the three levels of process-time variation, and the columns to the three levels of reliability, respectively. The permutation of different yield levels is such that each level will be considered exactly once in every row and in every column. Hence, (9.21) is one of the 12 (= 3! × 2!) possible patterns.

Each element of the matrix represents one experiment. For instance, the first case (at row 1 and column 1) is defined by (V_l, R_l, Y_m). The line throughputs estimated by the simulation model are given by

$$\begin{bmatrix} 178 & 177 & 189 \\ 164 & 176 & 173 \\ 164 & 163 & 157 \end{bmatrix}$$

The first row has an average value of $(178 + 177 + 189)/3 = 181.33$. This is the average throughput in the low variation case. Similarly, the average throughputs for the other two cases are 171 and 161.33, respectively. Let $v_i(r_i,y_i)$ be the average throughput for different levels of variation (reliability, yield), where $i = 1$, 2, and 3 for low, medium and high levels, respectively. Then

$$(v_1,v_2,v_3) = (181.33, 171.00, 161.33)$$
$$(r_1,r_2,r_3) = (168.67, 172.00, 173.00)$$
$$(y_1,y_2,y_3) = (166.00, 171.33, 176.33)$$

The average throughput of all nine cases is $\bar{t} = 171.22$. For each measure, the sum of squared deviations is defined by

$$SSV = \sum_{i=1}^{3}(v_i - \bar{t})^2 = 200.06$$

$$SSR = \sum_{i=1}^{3}(r_i - \bar{t})^2 = 10.28$$

$$SSY = \sum_{i=1}^{3}(y_i - \bar{t})^2 = 53.37$$

A large value of the sum of squared deviations means that a great improvement in line throughput can be expected. Since $SSV > SSY > SSR$, a reduction in process time variation is most important. On the other hand, the line may not benefit too much from a reliability improvement. □

9.6 Remarks

Compared to many other analysis tools, simulation requires less analytical skill and can be done in a relatively straightforward manner. If a manufacturing line has a simple structure (e.g., a small number of sequential operations and a single product type), the line and its operation logic can be well represented by a simulation model. For a complex line, some simplification is usually needed. However, a scientific procedure for modeling manufacturing lines essentially does not exist.

From a practical point of view, neither model construction nor program development can be a major obstacle. It is model implementation that is the problem. A good model must be accompanied with valid input data and output analysis capability. Although line information is often not totally available, simulation can still be a powerful tool for line study and decision-making. The line designer may take data from existing lines or create input data based on theoretical results. For example, the average process time of a given operation can be estimated by time and motion study. It is known that the coefficient of variation may vary between 0.1 and 0.65. If a manual operation involves large or heavy parts and a high degree of precision is expected, then the coefficient of variation is close to its upper bound. On the other hand, if operation cycles can be completed in a consistent manner, the coefficient will have a small value. Once the mean and the coefficient of variation are known, the operation process time distribution can be approximated by assuming a gamma, a log-normal or a Weibull distribution.

Many simulation languages or software packages have built-in functions for random-number generators subject to different distributions. These numbers can be derived from uniform random numbers. Uniform random numbers can be generated by the multiplicative congruential method introduced in Section 9.2.1. The reader may consult Bratley (1983) for other random-number generators and related discussion.

For nonuniform variables, their random-number generation methods are usually based on their individual characteristics. When dealing with a real-life problem, the observed data do not always have a physical meaning. In this case, the empirical distribution can be used. Sometimes, a *generalized lambda* distribution may be considered. X is said to be a generalized lambda variable, if

$$X = \lambda_1 + [R^{\lambda_3} - (1-R)^{\lambda_4}]/\lambda_2 \tag{9.22}$$

where R is a uniform variable between 0 and 1.

In the above expression, λ_1, λ_2, λ_3, and λ_4, can be determined by the mean, variance, skewness, and kurtosis, respectively. Tabulated values of these parameters are given in Ramberg (1979). Once appropriate values are chosen, a generalized lambda distribution can be used to approximate many theoretical distributions such as exponential, normal, gamma, Weibull, log-normal, and beta distribution. For instance, X has approximately a standard normal distribution when $\lambda_1 = 0$, $\lambda_2 = 0.1975$, and $\lambda_3 = \lambda_4 = 0.1349$. Since only one uniform random number is required, (9.22) can be evaluated very quickly.

Three different methods for output analysis have been discussed. The concept of regenerative method is well known in the theory of *stochastic processes*. In Ross (1970), for example, it is called the *Renewal Reward Theorem*. Applying this concept to the analysis of simulation output is discussed in Crane (1977) and Iglehart (1978). Other methods such as *spectral* method and *autoregressive* method can be found in Bratley (1983).

The design of the experiments in Example 9.1 is known as the *Latin Square* design. A procedure for a formal statistical analysis may be found in books on statistics or experimental design.

For detailed discussions about simulation methodology, the reader may consult Bratley (1983) and Fishman (1978).

REFERENCES

Bratley, P., B. L. Fox, and L. E. Schrage (1983). *A Guide to Simulation*, Springer-Verlag, New York.

Crane, M. A., and A. J. Lemoine (1977). *An Introduction to the Regenerative Method for Simulation Analysis* (Lecture Notes in Control and Information Sciences, Vol. 4), Springer-Verlag, Berlin.

Fishman, G. S. (1978). *Principles of Discrete Event Simulation*, Wiley, New York.

Iglehart, D. L. (1978). The Regenerative Method for Simulation Analysis, *Current Trends in Programming Methodology, Vol. III*, edited by K. M. Chandy and R. T. Yeh, Prentice-Hall, Englewood Cliffs, N. J.

Ramberg, J. S., E. J. Dudewicz, P. R. Tadikamalla, and E. F. Mykytka (1979). A Probability Distribution and Its Uses in Fitting Data, *Technometric*, v. 21, pp. 201–214.

Ross, S. M. (1970). *Applied Probability Models with Optimization Applications*, Holden-Day, San Francisco.

10

An Analysis of an Assembly Line Design

This chapter presents a study of an assembly line design, using the results discussed in the previous chapters. The values of line parameters are obtained from a combination of real-life assembly lines. The design problem and the line concepts are followed by a sequence of subjects such as process design, material-handling system, WIP management policy, and resource planning. Results from line simulation and other line measures are also discussed.

10.1 Problem and Approach

A high-technology product consists of over 100 parts. Both the size and the weight of the product are suitable for manual handling systems. The total assembly time, including testing and inspection, takes about 2 hours. Market research indicates a very competitive market and the first product shipment should be made within two years. The expected peak demand is from 250 to 500 pieces per day.

After having reviewed product characteristics, market situation, and manufacturing environment, a structure of the line design project is established. The following tasks are involved:

1. Design philosophy suitable for the given conditions
2. Line design concept
3. Assembly process and cycle time
4. Material handling system
5. WIP management policy

6. Parts logistics and feeding method
7. Human resource requirement
8. Shift schedule
9. Line layout

Because of complexity of the assembly work (a large number of parts) and limited time for tool design and debugging (less than 18 months), all workstations must have simple tools. On the other hand, the assembly line must be flexible, to accommodate demand change. The entire line design concept should be based on flexibility and simplicity.

10.2 Process and operation design

The assembly process involves an expensive testing operation which uses an automated tool that costs about $800,000. The test tool may test two pieces of product simultaneously. The total test time is 18 minutes, or the capacity is 9 minutes per piece. To facilitate design flexibility, it is decided that all assembly operations should be done manually, at least during the early production phase. There are other two testing operations, but their tooling costs are not very high. A preliminary estimate shows that the cost of a completed set of assembly tools is in the neighborhood of one million dollars.

To select an appropriate cycle time, the following factors have been taken into consideration:

1. Flexibility of manual operation
2. Utilization of high-cost tools
3. Job enrichment
4. Flow control strategy
5. Material handling effort
6. Operator performance behavior
7. Number of operators per line

A human operator is much more flexible than an automated tool. A manual job can be easily restructured. Since a high-technology product may experience a number of engineering changes, job restructuring is almost inevitable. If assembly tools are simple enough, line rebalancing after a major change may not be too costly. (See Section 7.2.4.)

Since the test equipment counts for about 50% of the total tooling cost, it should be fully utilized. On the other hand, the job content of a manual operation should be appropriately determined for both job interest and ergonomic reasons (see Sections 4.3 and 4.8). Past experience shows that a

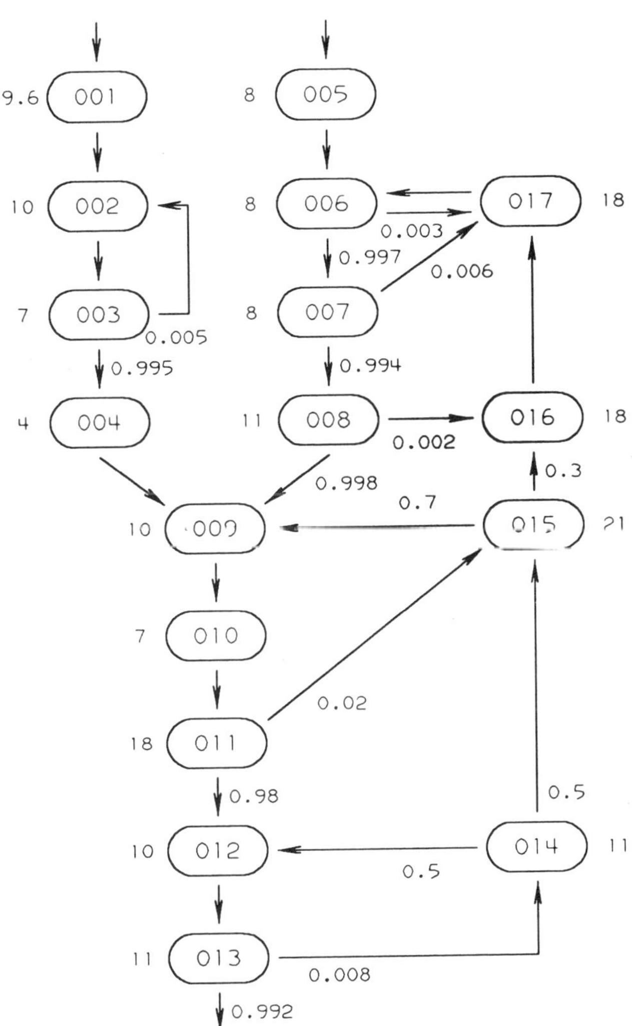

Figure 10.1 An assembly flow process.

cycle time of up to 10 minutes may still be appropriate. This suggests that a cycle time of 9 minutes may be considered.

It is possible to have one workstation per operation per line. Product can be passed from one workstation to its immediate neighbor during the assembly process. This will make the line flow control very simple. Furthermore, since the product can be manually handled (due to its weight and size), the assembly operators may perform the material-handling function.

From the discussion in Section 7.4, it is known that different operators perform differently. It is unnecessary to impose a fixed cycle time on tool designers. Instead, the mean process time for each operation is targeted between 7 and 10 minutes. In this way, tool designers would have a greater degree of freedom for their jobs. The 30% (= 10 – 7/10) difference can be reduced by using an intelligent job assignment policy, e.g.letting a high performer handle the bottleneck operation (see Section 7.4).

An assembly flow process, based on mean process times between 7 and 10 minutes, is shown in Figure 10.1, where the numbers beside the operation numbers (in the circles) are the estimated mean process times, and the numbers associated with links are the branching probabilities derived from estimated yields. It can be seen that some operation process times exceed 10 minutes.

The mean process times of rework operations (014–017) are greater than 10 minutes, but they can never be the line bottlenecks because of small start factors. The mean process times at operations 008 and 013 are 10% higher than the line cycle time. This may be corrected, if necessary, by having two high performers assigned to the workstations, or by improving the work methods of these operations. Operation 011 is the 18-minute testing operation. The two other testing operations are 003 and 013, respectively. Operations 003 and 004 both have short process times, and therefore may be served by the same operator. A line layout should allow an operator to access multiple operations. Not only can the line be better balanced, but line capacity can be adjusted by changing the number of operators.

The traffic flow from operations 005 through 013 can be regarded as the mainstream. Subassemblies built by operations 001–004 merge with the mainstream at operation 009, where the subassemblies are added to products coming from operation 008.

The total tooling cost is about 2.6 millions dollars for a completed tool set, including rework tools.

10.3 Material handling and WIP management

Most of the parts are small and distributed directly to workstations. However, a few large parts will be kitted in trays. Two types of kits will be used: type-1 kit has 4 parts used by operations 001 to 004, and type-2 has 5 parts for all other operations. A type-1 kit has a size of about 7×8 square inches, and a type-2 about 12×18 square inches. The use of kits serves at least two purposes:

1. Kits serve as the carries for moving the assemblies from station to station.
2. Space requirements for large parts at each workstation are reduced.

At each operation, the operator removes the work unit from the kit to the working bench for assembly, and then places the finished unit back into the same kit. The kits are handed over from operation to operation. Since assembly operators are responsible for WIP transfer, no dedicated material-handling system is needed.

For each assembly, two empty kits will be generated at operations 004 and 013, respectively. To return the empty kits, a conveyor driven by gravity is installed under the workbenches (see the next section) so that all empty kits are returned to the starting operations (i.e., operations 001 and 005) to be collected and refilled.

Parts quality level is measured by the number of defective parts per million (PPM). Since kits will be rejected for bad parts, the quality of kitted parts is particularly important. Let n be the number of parts in a kit and d be the defect rate. The probability that a kit will be rejected due to a parts quality problem is given by

$$p(n) = 1 - (1 - d)^n$$

Suppose that the peak demand is 500 pieces per day and no more than one kit should be rejected due to a parts quality problem. Since the two types of kits have 4 and 5 parts, respectively, the parts quality level is defined by

$$500p(4) + 500p(5) = 500[2 - (1 - d)^4 - (1 - d)^5] \leq 1$$

The greatest value of d that satisfies the above inequality is approximately 0.0002. Hence, the parts quality level should be 200 PPM or better.

Parts are directly fed to the line; that is, there is no warehouse operation and no receiving inspection. The parts staging area uses flow racks for the storage of 1.5 days' worth of parts.

A pull policy is adopted for WIP management (see Sections 6.3 and 6.4). Between two adjacent workstations, a finite buffer will be installed for WIP storage. The buffer sizes may vary, depending on the process time distributions. Based on the complexity of the manual job at each operation, the coefficient of variation of process time is estimated by the production engineers.

Table 10.1 Process Time and Buffer Size

Operation number	Mean process time (min)	Coefficient of variation	Buffer size
001	9.6	0.30	2
002	10.0	0.30	2
003	7.0	0.40	2
004	4.0	0.30	2
005	8.0	0.10	1
006	8.0	0.40	1
007	8.0	0.40	2
008	11.0	0.60	4
009	10.0	0.60	2
010	7.0	0.20	2
011	18.0	0.15	2
012	10.0	0.40	4
013	11.0	0.15	—

The buffer sizes are estimated by invoking Algorithm 6.6, under the conditions that (i) throughput reduction due to the finite buffer is no more than 5%, and (ii) the upper limit of buffer size is four work units. Table 10.1 presents the computational results, along with the operation number, the mean process time, and the coefficient of variation. The numbers shown in the last column are the estimated buffer size between operations and their immediate successors. For instance, a buffer size of 4 is needed between operations 008 and 009. The pull policy, however, does not apply on the rework areas. At any given workstation, defective products are removed from the working bench to a cart, and sent to the corresponding rework station. Theoretically, no storage limitation is imposed for rework.

10.4 Workstation design and facility planning

A standard working bench of 3 × 5 feet is chosen. The bench height can be adjusted to meet ergonomic requirements.

Design requirements of workstations are defined for engineers. First, for each operation, tools and fixtures should be confined to a given space so that each bench can hold up to 4 kits (i.e., a buffer for 4 work units). Second, tools and fixtures should be simple enough that both the function and the layout of a working bench can be easily changed.

Line layout should support the following principles:

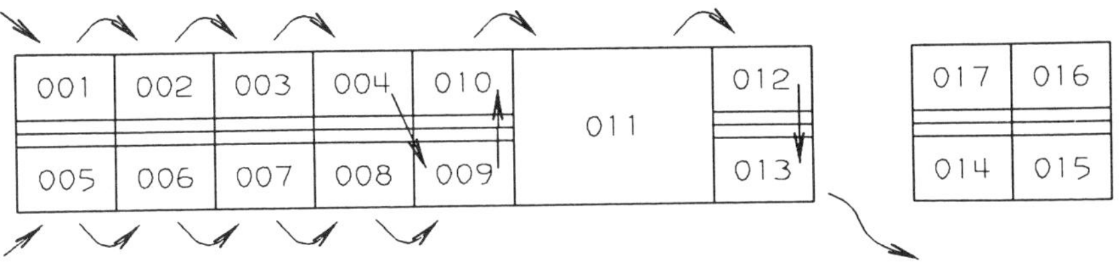

Figure 10.2 A layout.

1. Space economy
2. Minimal material-handling effort
3. Consistency with the WIP management policy
4. Flexibility for engineering changes
5. Operator accessibility to multiple operations
6. Good visibility for management

A line layout is illustrated in Figure 10.2, where operation numbers are assigned to working benches. Operation 011 requires a relatively large space for the testing equipment. Other working benches are all identical. Between the two rows of working benches is a common utility chassis. All utility lines, such as electricity, vacuum or air compression hose, liquid nitrogen, etc., go through the chassis. Utility outlets are placed below the bench surface so that wires and hoses will not tangle with tools and fixtures. Furthermore, because a common chassis is used, only one utility drop area is needed. This implies that facility construction can start without having all the details from line layout.

The purposes of using the standard working bench and the common utility chassis are (i) to save design cost, (ii) to shorten the time needed to build the production facility, and (iii) to increase line flexibility for process or engineering changes.

If a line contains exactly one workstation per operation, the space requirement will be 1500 square feet per line. This number includes space for operator, maintenance, and pedestrian aisles. The space installation cost is about $750,000 dollars per line.

10.5 Human Resource

Each operation has exactly one operator, except that (i) one operator handles all rework operations, (ii) one operator covers both operations 003 and 004 because of their short process times. Three additional operators may be required for part preparation (i.e., unpacking the parts coming from the vendors and preparing kits), parts feeding (i.e., distributing parts to workstations), and packaging finished goods. When the line is running at its full speed, a total of 16 operators should be assigned for assembly, testing, rework, parts handling, and packaging. It is anticipated that the absenteeism rate is about 10%, including vacation, sick leave, personal business, etc. Therefore, two floaters (16×0.1) should be added. With one additional technician, the total number of direct labor workers per line will be 20. This is considered to be an appropriate size for one department. If there are multiple lines, each can be placed under one line manager. Since all lines are identical, line performance can be compared easily.

The labor cost is approximately $36,000 per person per year, or $720,000 ($= 36{,}000 \times 20$) per shift per year.

10.6 Capacity and resource planning

The start factors are computed on the basis of the flow process and yield factors. Let v_i be the start factor and t_i be the mean process time (in minutes) of operation i. The yield capacity is equal to $60/(v_i t_i)$ per hour. The line capacity is simply the minimal yield capacity among all operations. If the line is operating at its full capacity, workstation utilization can be computed by dividing the line capacity by individual yield capacity. Average process times, start factors, yield capacity, and workstation utilization are given in Table 10.2.

The average workstation utilization is 0.79 without counting the rework operations, and 0.66 with rework operations included.

Line capacity is 5.4 pieces per hour. Two bottlenecks are found at operations 008 and 013, respectively. If both operations have their mean process time reduced to less than 10 minutes, the line capacity becomes 5.9 pieces per hour, with a bottleneck at operation 009. Furthermore, since all rework stations are lightly utilized, only one rework operator is needed per line.

Because of time detractors (e.g., lunch, break, maintenance, machine failure, calibration, setup, engineering changes, and personal delay), the total productive times per day for 2-shift or 3-shift operations are respectively 14.3 or 18.7 hours (see Section 7.2.3). Hence, the line capacity per shift is 38.6 pieces under the 2-shift case, and 33.7 pieces under the 3-shift case. A comparison of capital expense and labor cost under these two different strategies is given in Table 10.3.

It can be seen that when the demand is between 250 and 500, the 2-shift strategy will have more capital expense but less labor cost. To compute the payback period for the 2-shift strategy, let us normalize the amount of spending with respect to the same capacity. For instance, the normalized

Table 10.2 Process Time, Start Factor, Yield Capacity and Workstation Utilization

Operation	Process time	Start factor	Yield capacity	Utilization
001	9.6	1.0000	6.7	0.87
002	10.0	1.0050	6.0	0.91
003	7.0	1.0050	8.5	0.63
004	4.0	1.0172	14.8	0.37
005	8.0	1.0000	7.5	0.72
006	8.0	1.0185	7.4	0.73
007	8.0	1.0155	7.4	0.73
008	11.0	1.0094	5.4	1.00
009	10.0	1.0245	5.9	0.92
010	7.0	1.0245	8.4	0.65
011	18.0	1.0245	6.5	0.83
012	10.0	1.0081	6.0	0.91
013	11.0	1.0081	5.4	1.00
014	11.0	0.0081	673.4	0.01
015	21.0	0.0245	116.6	0.05
016	18.0	0.0094	354.6	0.02
017	18.0	0.0185	180.2	0.03

capital expense for 2-shift case is $7.8 \times 250/232 = 8.4$ million dollars when the capacity is 250 pieces per day. The normalized figures are shown in Table 10.4.

The payback period can be estimated by taking the ratio of the additional capital expense to the annual labor saving. If the peak demand is 250, 375, and 500 pieces per day, the payback period will be 0.14, 2.10 and 4.00 years,

Table 10.3 Cost Comparison between Two Different Shift Strategies

Cases	Two-shift			Three-shift		
Number of lines	3	5	7	3	4	5
Number of shifts	6	10	13	7	11	15
Total capacity	232	386	502	235	370	505
Capital expense	$7.8 M	$13.0 M	$18.2 M	$7.8 M	$10.4 M	$13.0 M
Annual labor cost	$4.2 M	$7.0 M	$9.1 M	$4.9M	$7.7 M	$10.5 M

Table 10.4 Normalized Line Cost

Cases	Two-shift			Three-shift		
Total capacity	250	375	500	250	375	500
Capital expense	\$8.4 M	\$12.6 M	\$18.1 M	\$8.3M	\$10.5 M	\$12.9 M
Annual labor cost	\$4.5 M	\$6.8 M	\$9.1 M	\$5.2 M	\$7.8 M	\$10.4 M

respectively. Since capital equipment is depreciable, additional tax saving from depreciation is also expected.

The shift strategy should be volume-dependent. If the peak demand is close to 250 per day, the 2-shift case is more economical. When the peak demand is near 500 per day, the line management should consider allowing 3 shifts, instead of adding new lines. Fortunately, altering the shift strategy does not involve physical line changes. Only the working schedule needs to be adjusted. For example, under the 2-shift strategy, the first shift may start at 6:00 a.m. and end at 2:30 p.m., and the second shift from 4:00 p.m. to 12:30 a.m. For a 3-shift operation, the schedule might be 7:00 a.m.–3:30 p.m., 3:00 p.m.–11:30 p.m., and 11:00 p.m.–7:30 a.m. Since 30 minutes lunch time is included, each shift lasts 8.5 hours.

It is planned that the production demand will reach about 250 pieces per day in five quarters or 15 months. The demand pattern is given in Table 10.5.

As the demand is increasing, the total capacity may be adjusted by adding shifts. However, capacity expansion must follow two rules:

1. No expansion is allowed unless the demand exceeds the capacity by more than 10%.
2. In the process of expansion, at most two shifts (or one line) can be added in each quarter.

Because of demand uncertainty, adding shifts may cause an overinvestment problem. The 10% allowance would delay the action of investment and may reduce the risk of overinvestment. On the other hand, to increase

Table 10.5 Production Demand Schedule

Quarter	1	2	3	4	5
Demand	25	83	165	235	250

Table 10.6 Line Capacity over Time

Quarter	1	2	3	4	5
Capacity	25	91	166	215	232
Shift	1	3	5	6	6

line production by 10% is not an unmanageable problem. This can be done by improving performance at the line bottleneck, bettering line discipline and supervision, and even using overtime. The second rule is based mainly on facility constraints. It takes longer than three months to add an additional line.

A two-month training period is required. Hence, operators must be recruited two to three months earlier than the time a shift becomes needed for production.

The planned capacity and the number of shifts are given in Table 10.6.

10.7 Line Performance Analysis

Queuing analysis and simulation can be used to characterize line behavior. The kitting operation and parts distribution are simple enough for analytic works. Since these two operations involve simple motions, neither learning period nor yield improvement are significant. Only ultimate performance will be analyzed in the following sections.

Kitting

Two kits are needed for each assembly. From Table 10.6, at the peak production stage, every two shifts or 14.3 hours, 232 pieces of assembly will be produced on the average. The average total number of kits to be prepared is $2 \times 232 + 1 = 465$ per day, including two kits per assembly and one kit for possible reject and replacement due to parts quality. The average time for preparing a kit is 1.5 minutes. Suppose that one kit station is installed. Since the kit demand is generated almost independently from 6 workstations, for operations 001 and 005, the service requests, as seen by the kitting station, may be approximated by a Poisson process. (This assumption is used for convenience. For most practical problems, the number of workstations should be at least 8 to justify the assumption.) Hence, the kitting operation may be analyzed by using an M/G/1 queuing model (see Section 3.3). The station is equivalent to the server with a mean service time of 1.5 minutes. The arrival rate is $465/(14.3 \times 60) = 0.54$

per minute. Therefore, utilization is $1.5 \times 0.54 = 0.81$. The service time has a coefficient of variation of 0.28. From Equations (3.10) and (3.11), it follows that the mean queue length is 3.25 and the average waiting time is 6.02 minutes. The queue length distribution can be approximated by using Equation (3.12). The 99th-percentile of the queue length distribution is approximately 10. This implies that between the lines and the kitting station a buffer of a size equal to 10 kits should be allowed. When a kit is drawn from the buffer, a new kit should be produced by the kit station. The probability of an empty buffer is roughly $1 - 0.99 = 0.01$. Since the buffer size is not large and the kitting station is 81% utilized, it is feasible to have one station supporting all 3 lines.

On the other hand, since some 150 kits must be delivered in each shift, it will not be economical to manually deliver one kit at a time. It is then decided that carts will be used. A cart can hold 6 kits, 3 of each kind. The average preparation time at the kitting station becomes 9 minutes per cart, with an arrival rate of 0.09 per minute. Although the value of utilization remains unchanged, queuing behavior may be different. Since the service time per cart is the sum of service times of 6 kits, the coefficient of variation tends to be smaller—roughly reduced by a factor of $\sqrt{6}$. The interarrival time of empty carts, as seen by the kitting station, also becomes more regular. Since the line cycle time is 10 minutes and each cart has 3 assembly sets, it is expected that a cart will be sent back for refill every 25 to 35 minutes.

The above results suggest a kitting policy that can be summarized as follows:

1. One kitting station can support 3 lines for 232 pieces per day (2 shifts).
2. Each line holds 2 carts of kits. If one cart becomes empty, the line operators can continue their assembly with the second cart. The empty one is sent back to the kitting station for refill.
3. The kitting station constantly provides 3 carts of kits. Upon the arrival of an empty cart, a cart of kits is released and the kitting station will refill the empty one.

Parts Distribution

In the previous discussion, it was mentioned that all non-kitted parts will be manually delivered to the workstations. Performance behavior of parts distributors can be analyzed in a similar fashion. A parts distributor is regarded as the server in an M/G/1 queuing system. Job arrivals are generated by parts requests from the workstations. A completed delivery trip constitutes one service cycle. Parts are delivered in trays 14×15 inches in size. For 232 assemblies, on the average a total of 127.3 trays will be delivered. This number is derived from a parts packaging study, based on the assumption that no workstation should have more than two tray types (see Section 6.2). The average round trip distance is 116 feet. The average travel speed of a parts distributor is assumed to be 2 feet per second. This number is derived from the average walking speed of a person who is pushing a cart plus any personal fatigue and delay. Hence, the average travel time for a round trip is $2 \times 58/2 = 58$ seconds.

If two trays are delivered at a time, the average total number of trips in two shifts is 63.6. A delivery cycle involves (i) one round trip, starting at the parts preparation area, (ii) loading and unloading trays, and (iii) refilling empty trays for the next delivery. A load or an unload operation takes about 10 seconds. Parts refill takes about 1.5 minutes on the average. The total amount of time for parts distribution is $63.6 \times 58/60 + 127.3 \times (4 \times 10/60) + 127.3 \times 1.5 = 336$ minutes. Therefore, the average round trip time is $336/63.6 = 5.3$ minutes.

Another type of delivery is for the kitted parts. A delivery cycle includes (i) placing all empty kits collected at operation 001 or 005 on a cart, (ii) moving the cart to the kitting station, and (iii) taking a cart with full kits to operation 001 or 005. The estimated mean cycle time is 1.3 minutes. Since each cart has 6 kits and a total of 464 kits is needed for 232 assemblies, the average number of trips per day is 77.3.

Consider both types of deliveries. The mean cycle time is $(77.3 \times 1.3 + 63.6 \times 5.3)/(77.3 + 63.6) = 3.1$ minutes. The delivery frequency is $(63.6 + 77.3)/(14.3 \times 60) = 0.164$ per minute. If all parts delivery is handled by one person, then utilization is equal to $0.164 \times 3.1 = 0.51$. The estimated coefficient of variation of a round-trip time is about 0.38. Using Equations (3.10)–(3.12), the following results are obtained:

1. The mean queue length is 0.81.
2. The average waiting time is 4.9 minutes.
3. The 99th-percentile of the queue length distribution is about 4.

The 99th-percentile can be regarded as the maximum number of uncompleted parts requests from workstations observed at any random time. If a workstation has just used up a tray, the maximum waiting time for replenishment is approximately equal to $(1 + L_m)E[S] = 5 \times 3.1 = 15.5$ minutes, where L_m is the 99th-percentile of the queue length and $E[S]$ is the average trip time.

It should be pointed out that the above approach for the estimation of the maximum waiting time is not an exact method. Therefore, this approach can only provide a design guidance. If the distribution of the trip time is available, it is possible to compute the waiting time distribution under the Poisson arrival assumption. (See Section 3.3.)

The above discussion leads to the following operation policy:

1. One operator can handle parts distribution for a production demand of 232 pieces per day.
2. The parts distributor is responsible for delivering parts trays, and for collecting and refilling empty trays.
3. For each tray type used by a workstation, two trays are needed. If one tray becomes empty, the assembly operator can take parts from the other tray while the empty one is being refilled.
4. The number of parts in each tray should last at least 15 minutes. This means that before the second tray becomes empty, the first tray should have been already refilled and delivered.

The last statement defines a parts packing requirement that minimizes the parts shortage problem. Although the requirement is satisfied in our case, it is worth some discussion. If this requirement is violated, some adjustment is needed. Several possible approaches exist: (i) increase the number of parts in trays by enlarging the tray size or packing density, (ii) alter the number of trays per delivery, (iii) hire more parts distributors, or (iv) change the material-handling method. Whichever approach is taken, it will result in a new value of the maximum waiting time. Therefore, success is not always guaranteed. For this reason, an iterative process may become necessary. Moreover, these adjustments may affect other line components. For example, a larger tray requires more workstation space, and consequently workstation design or even line layout may need to be modified.

From a resource-utilization point of view, it may not be very economical to have the parts distributor busy for only 50% of the time. However, low utilization means quick response to needs for parts deliveries, and leads to low parts inventory and high workstation utilization. The saving from line performance improvement can easily outweigh the loss resulting from low utilization of the parts distributor.

Line Operator Utilization

When the production rate is at 232 pieces per day, a total of 3 lines, 2 shifts per line, is required. If operations 003 and 004 use the same operators, the total number of operators needed to cover all operations is 90 (72 people for operations 001–013, 6 for rework operations 014–017, and 12 floaters. Parts handling is not included.) The mean total process time can be derived from the results shown in Table 10.2, and is given by $(v_1t_1 + \cdots + v_{17}t_{17}) = 124.3$ minutes, where v_i is the start factor at operation i and t_i is the mean process time. Thus, by Equation (1.2), the operator utilization is $124.3 \times 232/(90 \times 8 \times 60) = 0.67$. This value is lower than the average workstation utilization (0.67 vs. 0.79), due to the floaters who count for about 13% of the total labor force.

Line Simulation

To verify the validity of the design work, a computer program is developed to simulate line behavior. A software package called the Research Queuing Package (RESQ) is used. This PL/I-based package allows users to define a queuing or a simulation model in a structural manner and provides the model solution automatically (Sauer 1982, MacNair 1985, Chow 1985). The program development time is about three days. It takes about 10 minutes of CPU time on an IBM 3082 machine to achieve a confidence interval to a mean ratio less than 10%.

Three simulation runs have been conducted:

Case 1: *Base case* The process flow and the mean process times are given in Figure 10.1. Buffer sizes between operation are shown in Table 10.1.

Case 2: *Large buffer* Buffer sizes are increased to 20. The probability of blocking or starving is negligible.

Case 3: *Short process time* The mean process time of operation 008 is reduced from 11 to 8 minutes. Since the mean process time is changed, the buffer size between operations 008 and 009 is reduced from 4 to 2.

Line throughput, the average production lead time, and the average amount of work-in-process are given in Table 10.7, where the numbers

Table 10.7 Throughput, Lead Time, and Work-in-Process

Measures	Case 1	Case 2	Case 3
Line Throughput	211 (201,221)	229 (227,231)	230 (223,237)
Average Lead Time	3.52 hrs (3.38,3.66)	13.60 hrs (13.40,13.80)	3.84 hrs (3.76,3.92)
Average WIP	16.1 (15.8,16.4)	78.8 (77.7,79.9)	20.7 (20.2,21.5)

in parentheses define the 90% confidence intervals. A production lead time is the total elapsed time when an assembly has moved from operation 001 to operation 013 as an acceptable finished product. Only assemblies in the mainstream are counted as the work-in-process. Subassemblies at operations 001–004 are not included.

The base case can only produce 211 pieces per day. If the buffer size is increased, line throughput can be increased to 229 pieces per day. The previous estimate, based on Table 10.2, is 232 pieces per day under the assumption of unlimited buffers. The bottleneck is found at operation 008. When the mean process time at operation 008 is reduced to 8 minutes, line throughput becomes 230 pieces per day. Although both Cases 2 and 3 have about the same throughput, the former has much higher lead time and WIP level.

The difference between Cases 1 and 2 is due to buffer size. The throughput difference is expected to be 5%, according to Section 10.3. The simulation result, however, shows that the loss is about 7.7%. This small discrepancy is not surprising as both Algorithm 6.6 and the simulation run provide only approximate solutions.

Workstation utilization is given in Table 10.8. It can be seen that the simulation results for Cases 2 and 3 are comparable, except at operation 008. Both cases have higher utilizations than does Case 1. The average

Table 10.8 Workstation Utilization

Operation	Case 1	Case 2	Case 3
001	0.82	0.90	0.89
002	0.85	0.93	0.93
003	0.59	0.64	0.65
004	0.34	0.36	0.37
005	0.67	0.73	0.74
006	0.68	0.72	0.74
007	0.67	0.73	0.73
008	0.94	1.00	0.65
009	0.83	0.90	0.91
010	0.58	0.63	0.64
011	0.78	0.82	0.83
012	0.81	0.88	0.89
013	0.90	0.96	0.98
Average	0.728	0.785	0.765

workstation utilization is given in the bottom row of the table. Note that the average utilization in Case 2 is very close to 0.79, as obtained from Table 10.2.

Other Line Measures

Comparing the results from resource planning to the demand curve, additional measures are obtained by using Equations (1.3), (1.4), and (1.5), respectively:

1. *Line utilization*

$$u = \frac{1}{5}\left(\frac{25}{25} + \frac{83}{91} + \frac{165}{166} + \frac{215}{215} + \frac{232}{232}\right) = 0.98$$

2. *Line flexibility*

$$f = \frac{1}{5}\left(\frac{25}{25} + \frac{83}{91} + \frac{165}{166} + \frac{215}{235} + \frac{232}{250}\right) = 0.95$$

The impacts of line size may be illustrated by comparing two simple cases. In the first case, the line size is doubled, and in the second the line size is cut by half. The two basic planning rules, stated in Section 10.6, are still valid, except that the maximum number of shifts that can be added for each quarter is 1 and 4 for Case 1 and 2, respectively. The number of shifts, line capacities over time, and measures for flexibility are given in Table 10.9, where line flexibility is clearly a decreasing function of line size.

Table 10.9 Measures for Line Flexibility under Different Line Sizes

		Case 1		Case 2	
Quarter	Demand	Shift	Capacity	Shift	Capacity
1	25	1	49	2	25
2	83	2	122	6	90
3	165	3	199	10	166
4	235	3	215	13	233
5	250	4	232	13	252
Flexibility		0.77		0.98	

3. *Scale-up index*

$$d = \frac{1}{5}\left(\frac{25 + 91 + 166 + 215 + 232}{232}\right) = 0.63$$

To test the capability of scale-up, let us ignore the demand requirement so that the line capacity can be increased as quickly as possible. However, the rule of a 2-shift increment per quarter still applies. The schedule for line capacity, shown in Table 10.6, implies that the "relative" yield factors over time are 0.6466, 0.7845, 0.8586, 0.9267, and 1.0000, respectively. These are normalized values, assuming that the last quarter has a yield of 100%. For example, the relative yield in the first quarter is (25/1)/(232/6) = 0.6466. Then the maximal line capacities become as follows:

Quarter	1	2	3	4	5
Capacity	50	121	199	215	232
Shift	2	4	6	6	6

The new scale-up index will be 0.70, or about 11% higher.

REFERENCES

Chow, W., E. A. MacNair, and C. H. Sauer (1985). Analysis of Manufacturing Systems by the Research Queuing Package, *IBM Journal of Research and Development*, v. 19, pp. 330–342.

MacNair, E. A. and C. H. Sauer (1985). *Elements of Practical Performance Modeling*, Prentice-Hall, Englewood Cliffs, New Jersey.

Sauer, C. H., E. A. MacNair, and J. F. Kurose (1982). *The Research Queuing Package, Version 2: CMS Users' Guide*, IBM Research Report RA 139, IBM Research Division, Yorktown Heights, New York.

Author Index

Subject Index